Handbook of Particle Technology

Handbook of Particle Technology

Edited by
Daniel Santiago

Willford Press
www.willfordpress.com

Published by Willford Press,
118-35 Queens Blvd., Suite 400,
Forest Hills, NY 11375, USA

ISBN: 978-1-64728-357-5

Cataloging-in-Publication Data

Handbook of particle technology / edited by Daniel Santiago.
p. cm.
Includes bibliographical references and index.
ISBN 978-1-64728-357-5
1. Particles. 2. Particles (Nuclear physics). 3. Size reduction of materials.
4. Physics. I. Santiago, Daniel.
QC793.2 .H36 2022
539.72--dc23

For information on all Willford Press publications
visit our website at www.willfordpress.com

Willford Press

Contents

Preface

This book aims to highlight the current researches and provides a platform to further the scope of innovations in this area. This book is a product of the combined efforts of many researchers and scientists, after going through thorough studies and analysis from different parts of the world. The objective of this book is to provide the readers with the latest information of the field.

The branch of science which deals with the handling and processing of particles and powders is termed as particle technology. It deals with the production, modification, handling and usage of a broad range of particulate materials. These particles can be wet or dry, as well as vary in size from nanometers to centimeters. Some of the major areas of study associated with particle technology are the behavior of solids in bulk, separation of particles through the processes such as tabling, magnetic separation and sieving, and particle size analysis. It is also closely related to the field of mineral processing and petrochemical industry. This book unfolds the innovative aspects of particle technology which will be crucial for the progress of this field in the future. Also included herein is a detailed explanation of the various concepts and applications of this field. This book will also provide interesting topics for research which interested readers can take up.

I would like to express my sincere thanks to the authors for their dedicated efforts in the completion of this book. I acknowledge the efforts of the publisher for providing constant support. Lastly, I would like to thank my family for their support in all academic endeavors.

Editor

1

Ash and sulphur removal from bitumen using column flotation technique: Experimental and response surface methodology modeling

Yasser Vasseghian[1*], Mojtaba Ahmadi[1], Mohammad Joshaghani[2]

[1]*Chemical Engineering Department, Faculty of Engineering, Razi University, Kermanshah, Iran*

[2]*Faculty of Chemistry, Razi University, Kermanshah 67149, Iran*

HIGHLIGHTS

- Column flotation technique is employed to remove ash and sulphur from bitumen.
- Pine oil was found as frother in this process.
- RSM is capable of optimizing the process.

GRAPHICAL ABSTRACT

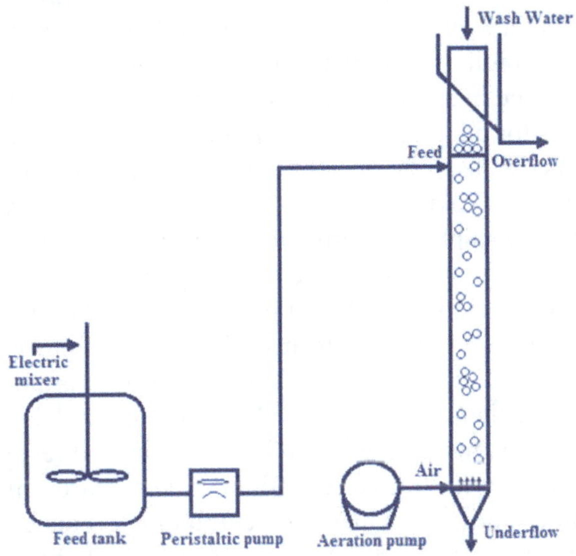

ARTICLE INFO

Keywords:
Ash removal
Pyrite sulphur removal
Bitumen
Column flotation process
Response surface methodology

ABSTRACT

This study investigates removing ash and pyrite sulphur from bitumen by column flotation process. Central composite design (CCD) of response surface methodology (RSM) was applied for modeling and optimization of the percentage of ash and pyrite sulphur removal from bitumen. The effects of five parameters namely the amounts of collector and frother agents, particle size, wash water rate and feed rate on percentage of ash and pyrite sulphur removal from bitumen were investigated. The used bitumen sample has 26.4% ash and sulphur content of 9.6% (6.81% in the pyrite sulphur form). All the tests were carried out under aeration rate of 4L/min and pulp containing 5% of solid using pine oil and kerosene as frother and collector agents, respectively. The coefficient of determination, R^2, showed that the RSM model can specify the variations with the accuracy of 0.971 and 0.975 for ash and pyrite sulphur removal from bitumen, respectively, thus ensuring a satisfactory adjustment of the model with the experimental data. The RSM was used to optimize the process conditions, which showed that initial amount of collector of 2.00 $kg/t_{bitumen}$, amount of frother of 0.2 ppm, particle size of 101.29 mesh, wash water rate of 0.5 L/min and feed rate 1.26 L/min were the best conditions. Under the optimized conditions, the maximum percentage of ash and pyrite sulphur removal from bitumen was 88.74% and 90.89%, respectively.

* Corresponding author.
E-mail address: y_vasseghian@yahoo.com.

1. Introduction

Bitumen is a glossy black substance consisting of different hydrocarbons with high molecular weight. Bitumen is produced by the oxidation of petroleum. Bitumen is a heterogeneous mixture of chemical compounds including about 90% of hydrocarbons, between 1% and 6% sulfur, less than 1.5% of oxygen and nitrogen molecules and a few ppm of metal components such as vanadium, nickel and iron [1]. This substance can be categorized based on various colors, hardness, density, volatile materials.

There are fertile mines of bitumen in Canada, Venezuela, Russia, Australia, and Iran. Most of Iran's bitumen can be used in solid and in the same mine form. The major mines of bitumen in Iran are located in Kermanshah, Ilam, Khuzestan provinces and other west and southwest regions. Bitumen extracted in the regions is mainly used in insulation industries, manufacture of coatings for oil and gas pipes, coking, fuel, preparation of special asphalt of road and etcetera. But, Iran bitumen has weaknesses which most importantly, is having high sulphur and mineral impurities contents. Among all the elements in the bitumen, sulphur has the most effect to limit the bitumen utilization as a clean fuel [2]. Different forms of sulphur are organic, pyrite and sulfate, which depending on the type of sulphur, there are many methods to remove them. Generally, more than half of the sulphur in bitumen is in the pyrite format [3]. Bitumen also contains inorganic minerals which are commonly called ash. The main minerals in coal are: silicates or shales (kaolinite type), quartz and sandstone, pyrite and siderite carbonates and anchorite. International standards extent the ash and sulphur value in the coal to be less than 7% and 6% respectively [4].

There are many methods for reduction of the ash and sulphur content of bitumen. Among them, the flotation techniques are widely used [5- 7]. Flotation is a separation process that depends on the difference in the surface properties of substances. Column floatation is a process utilized to selectively separate hydrophobic minerals suspended in a solution by attaching them to air bubbles and transferring them into froth layers. This is attained by using surfactants and wetting agents. It is considered the cheapest and most widely used method for separation of valuable minerals [8].

A pilot-scale flotation column was applied by Barraza and Piñeres to procreate vitrinite-rich fractions from some coal samples from Guachinte and Yolanda (south western Colombia). Maximum ash removal was obtained 71.7% and 76.5% for Guachinte and Yolanda coals respectively. Furthermore, maximum sulphur removal was acquired 63.2% and 75.4% for Guachinte and Yolanda coals respectively. In their study, Yolanda coal was afforded the highest concentration of vitrinite. It was in the order of 99.8% at neutral pH and when using the maximum frother concentration [9]. In another study, using column flotation, Piñeres and Barraza concluded that increasing of froth and aeration rate reduce coal combustible recovery [10]. Tao et al. researched flotation of a hard-to-float fine coal with a high content of ash by grinding-recleaning to roughing cleaning coal and agglomeration-floatation processes. Experimental results demonstrated that grinding-recleaning to roughing cleaning coal improved the cumulative yield from 50.87% up to 55.53% and alleviated the product ash content from 11.76% down to 10.74%. Whereas the agglomeration-floatation, the least ash of clean coal is 10.69%, with 58.72% yield, 7.85% better in yield and 1.07% lower in ash content [11]. Ashiwani et al. found that blended frothing molecules of short chain alcohol and polyglycol ether have a dramatic impact on the surface activity and flotation performance in term of ash reduction and improvement in coal yield [12]. The effect of solids pulp percentage in coal column flotation studied by Angadi et al. The impact of different variables on solids and water flow to the flotation froths are considered. They found out that increasing the concentration of frother reduces the size of the air bubbles. Also, increasing the surface area of the bubbles improve the flotation efficiency [13]. Dey et al. surveyed flotation behavior of weathered coal with a low content of ash in mechanical and column flotation cell. They found out that cleaning of the rougher concentrate is necessary to reduce its ash content, whereas the single-stage column flotation is found to be better which yields 49.6% concentrate at 12% ash [14]. Vasseghian et. al. employed flotation and leaching methods to remove ash and sulphur from bitumen by sulphuric acid. Using combination of above two methods, they succeeded to remove 47% of total sulphur and 61% of ash under optimum conditions [15].

In this study, ash and pyrite sulphur removal from bitumen was investigated in column flotation and correlated with modeling and optimization studies we used the central composite design (CCD) of response surface methodology (RSM) for analysis of operational conditions namely amount of collector, amount of frother, particle size, wash water rate and feed rate. It should be noted that using RSM to optimize and evaluation of interactive effects between variables for ash and sulphur removal from bitumen is a novel method.

2. Materials and methods

2.1. Materials and analytical tests

The bitumen samples were supplied by mines, in Kermanshah/ (Iran). Pine oil as frother was supplied by Boyakhsaz company/ (Iran) in liquid form. Kerosene as collector was purchased from the National Iranian oil products Distribution Company/ (Iran). Nitric acids with purity of 65% and hydrochloric acid (HCl) with 37% volumetric purity were provided from the Merck Company / (Germany). Desiccators (WG Dry model Box-503, Merck), Atomic absorption (GBC-932, GBC Australia) were used to prepare and analyze the sample and final product.

2.2. Experimental apparatus

The novel apparatus was designed and built to study ash and pyrite sulphur removal from bitumen. As shown schematically in Figure 1, a 10 cm- diameter flotation column of 2.5 m height made from Plexiglas material was employed in this study. A steel framework was used to keep the stability of the column. A polyethylene vessel with volume of 20 L equipped with an electric mixer was designed and used to provide initial load. Pulp feeder point was placed at 65 cm below the top of column. Pulp supplied was pumped into the column through a peristaltic pump (IP 55, WATSON-MARLOW, UK) with maximum power of 2.5 L/min. Wash water was inserted through a shower for washing foams in the column and separation of undesirable material from bubble-particle which was applied. Air was supplied through an internal sparger with 20 cm in diameter and 25 cm in height which was located at the bottom of the column. Tailings tube of the column was passed from the bottom of sparger.

2.3. Experimental procedures

The particle sizes of less than 250 μm were prepared by crushing the bitumen samples using filter. The pulp 5% was prepared using 50 g of a prepared bitumen sample (with specified characteristics). Then was washed using 500 mL water at 30 °C and decanted into a vessel. The paste mixture was

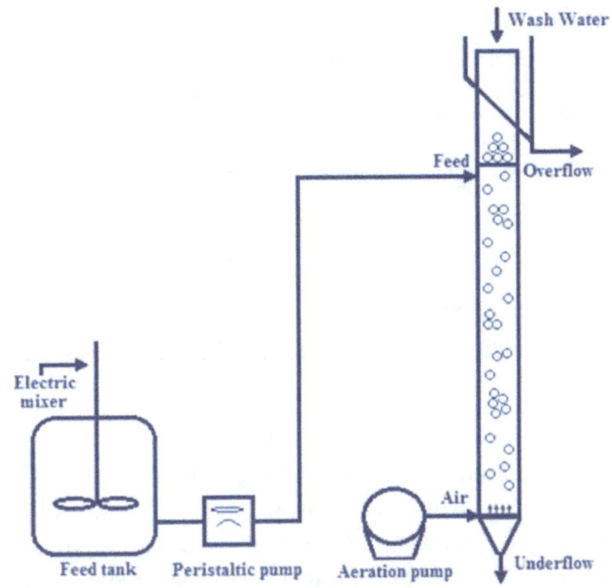

Fig. 1. A schematic illustration of the column flotation apparatus.

rested for 1 h. The volume was reached to 1 L by adding distilled water to the vessel. This mixture was stirred for 3 min. The collector was added to the vessel and the mixture was further stirred for 3 min. The foaming agent was added and mixed for 3 min. The mixture was transported to the column through a peristaltic pump and the process was started. The flotation time was fixed at 10 min when the foam was collected on the top of the column (froth zone). The concentrate was washed with wash water with a flow rate of 0.3 L/min for 5 min to wash the hydrophilic impurities along with air bubbles. Consequently, the concentrate was dried in the oven for 1 h at 110 °C and pyrite sulphur, total sulphur and ash contents of the dried samples were determined by the method published in our previous work [15].

2.4. Response Surface Methodology (RSM)

RSM is a collection of statistical and mathematical techniques beneficial for developing, improving, and optimizing processes [16]. The most comprehensive applications of RSM are in the special situations where several input variables potentially effect some performance measure or quality characteristic of the process. So, performance measure or quality characteristic is called the response. The field of RSM includes of the experimental strategy for probing the space of the process or independent variables, empirical statistical modeling to expand an appropriate approximating relationship between the performance and the process variables, and optimization methods for finding the values of the process variables that

Table 1.

Independent variables and their levels for the central composite design used in the present study.

Independent variables	Unit	Symbols	Level of factors				
			- α (-2)	-1	0	1	α (2)
Amount of collector	$Kg/t_{bitumen}$	X_1	0.5	1	1.5	2	2.5
Amount of frother	ppm	X_2	0.1	0.2	0.3	0.4	0.5
Particle size	mesh	X_3	50	100	150	200	250
Wash water rate	L/min	X_4	0.2	0.3	0.4	0.5	0.6
Feed rate	L/min	X_5	0.5	1	1.5	2	2.5

produce favorable values of the response.

In this study, a central composite design (CCD) was employed in order to optimize the ash and pyrite sulphur removal. Five factors were considered: amount of collector, amount of frother, particle size, wash water rate and feed rate. Table 1 summarizes the levels for each factor involved in the design strategy. Table 2 shows the standard array for five factors and 46 experiments. It also shows the run order and the observed responses. The obtained model was evaluated for each response function and the experimental data (percentage of ash and pyrite sulphur removal) were analyzed statistically applying analysis of variance (ANOVA).

$$y = \beta_0 + \sum_{j=1}^{k} \beta_j X_j + \sum_{j=1}^{k} \beta_{jj} X_j^2 + \sum_{j<j}^{k} \sum_{j=2}^{k} \beta_{ij} X_i X_j \qquad (1)$$

Where y is the predicted dependent variable, β0, βj, βjj, and βij are the regression coefficients for intercept, linear, quadratic and interaction terms, respectively, and Xi, and Xj are the independent variables. The Design expert statistical software (Design Expert 7.0.0.1, Statease, USA) was used for design of experiment, regression and graphical analyses of the obtained data, analysis of the measured responses and determining the mathematical models with best fits.

3. Results and discussion

3.1. The result of RSM model

The adequacy of the model is tested using the sequential f-test, lack-of-fit test and the analysis-of-variance (ANOVA) technique using the Design expert statistical software (Design Expert 7.0.0.1, Statease, USA) to obtain the best-fit model. In this present study, to investigate about the competency of models among various models (linear, 2FI, quadratic and cubic) to present ash and pyrite sulphur removal from bitumen, two different tests including of the sequential model sum of squares and model summary statistics were performed on the experimental data and the results are shown in Tables 3 and 4, respectively. The associated p-value of less than 0.05 for the model (i.e., α=0.05, or 95% confidence level) indicates that the model terms are statistically significant [17]. The lack-of-fit value of the model indicates non-significance, as this is desirable.

The ANOVA table for the reduced quadratic model is shown in Tables 3 and 4 for percentage of ash and pyrite sulphur removal from bitumen respectively. The reduced model results indicate that the model is significant (p-value less than 0.05). The other adequacy measures, i.e., R^2, adjusted R^2 and predicted R^2, are in reasonable agreement and are close to 1, which indicate the adequacy of the model [18]. The adequate precision compares the signal-to-noise ratio; a ratio greater than 4 is desirable [17]. The value of adequate precision ratio of 20.196 and 21.430 for ash and pyrite sulphur removal from bitumen, respectively, indicate adequate model discrimination. The lack-of-fit f-value of 1437.87 and 31.34 for ash and pyrite sulphur removal from bitumen, respectively, imply that the lack-of-fit is not significant relative to the pure error.

The final mathematical models for ash and pyrite sulphur removal from bitumen, which can be used for prediction within same design space in terms of coded factors, are given as follows:

(2)

$$\text{Ash removal (\%)} = +48.99 +6.21X_1 -10.72X_2 -1.85X_3 +0.43X_4 -0.33X_5 +1.15X_1X_2 +0.95X_1X_3 -0.026X_1X_4 -0.31X_1X_5 +0.43X_2X_3 +0.41X_2X_4 +0.80X_2X_5 -0.61X_3X_4 -0.36X_3X_5 -0.58X_4X_5 +6.28X_1^2 +5.62X_2^2 +4.88X_3^2 +4.45X_4^2 +1.17X_5^2$$

Table 2.
CCD with experimental and predicted values.

Run	Type	Independent variables					Experimental (%)		RSM (%)	
		X_1	X_2	X_3	X_4	X_5	Ash removal	Pyrite sulphur removal	Ash removal	Pyrite sulphur removal
1	Factorial	2	0.2	100	0.3	1	89.81	91.73	88.41	90.54
2	Center	1.5	0.3	150	0.4	1.5	50.63	56.24	48.99	54.30
3	Factorial	2	0.2	200	0.3	1	86.91	89.37	87.66	91.10
4	Factorial	1	0.2	100	0.3	1	80.85	84.69	79.51	82.44
5	Factorial	1	0.2	200	0.3	1	76.71	80.54	74.97	79.10
6	Axial	2.5	0.3	150	0.4	1.5	82.12	86.87	86.52	89.52
7	Axial	1.5	0.1	150	0.4	1.5	89.19	90.80	92.91	93.78
8	Factorial	1	0.2	100	0.5	2	81.22	85.61	79.90	85.47
9	Factorial	2	0.2	100	0.5	2	90.40	92.52	87.46	89.32
10	Factorial	1	0.4	100	0.5	2	58.87	67.33	57.72	64.29
11	Factorial	2	0.4	100	0.5	1	69.66	72.58	69.99	73.72
12	Axial	1.5	0.3	150	0.4	0.5	51.39	55.73	54.33	57.91
13	Axial	1.5	0.3	250	0.4	1.5	61.10	68.50	64.82	70.61
14	Factorial	2	0.4	200	0.3	2	68.53	72.69	67.73	72.18
15	Factorial	2	0.2	100	0.5	1	90.74	92.80	90.77	92.68
16	Axial	0.5	0.3	150	0.4	1.5	59.22	64.60	61.68	67.43
17	Factorial	2	0.2	100	0.3	2	87.80	89.61	87.42	88.81
18	Center	1.5	0.3	150	0.4	1.5	50.60	56.18	48.99	54.30
19	Factorial	1	0.4	100	0.5	1	59.18	64.47	56.61	61.08
20	Factorial	2	0.4	100	0.3	1	70.52	73.29	65.98	69.58
21	Factorial	1	0.2	200	0.3	2	77.29	81.21	73.79	77.98
22	Factorial	1	0.2	200	0.5	1	78.22	82.09	75.02	79.72
23	Axial	1.5	0.3	150	0.6	1.5	65.01	70.49	67.65	73.39
24	Factorial	2	0.4	200	0.5	2	68.57	72.70	66.98	71.46
25	Factorial	1	0.4	100	0.3	2	58.62	62.27	55.94	60.08
26	Factorial	1	0.2	100	0.3	2	79.70	82.41	79.76	83.26
27	Center	1.5	0.3	150	0.4	1.5	50.77	55.12	48.99	54.30
28	Factorial	2	0.2	200	0.3	2	85.60	86.68	85.24	87.43
29	Factorial	2	0.2	200	0.5	1	87.80	90.02	87.60	90.00
30	Axial	1.5	0.3	50	0.4	1.5	69.10	73.51	72.24	76.88
31	Axial	1.5	0.3	150	0.2	1.5	61.71	67.69	65.93	70.27
32	Center	1.5	0.3	150	0.4	1.5	50.82	55.13	48.99	54.30
33	Factorial	2	0.4	100	0.5	2	68.47	72.41	69.86	74.39
34	Factorial	1	0.2	200	0.5	2	67.08	74.68	71.50	76.96
35	Factorial	1	0.4	100	0.3	1	51.02	53.49	52.49	55.23
36	Factorial	2	0.4	100	0.3	2	68.21	72.02	68.19	71.88
37	Factorial	1	0.4	200	0.5	1	51.33	55.71	51.37	56.19
38	Factorial	1	0.4	200	0.3	2	52.12	57.11	51.69	56.48
39	Factorial	2	0.2	200	0.5	2	86.45	87.50	82.85	84.70
40	Factorial	1	0.4	100	0.3	1	51.00	53.50	52.49	55.23
41	Axial	1.5	0.5	150	0.4	1.5	46.89	50.82	50.03	53.32
42	Factorial	2	0.4	200	0.3	1	68.80	73.19	66.96	71.82
43	Factorial	1	0.2	100	0.5	1	81.53	85.80	81.98	86.29
44	Factorial	1	0.4	200	0.5	2	52.82	57.51	51.05	57.47
45	Factorial	2	0.4	200	0.5	1	69.01	74.03	68.54	72.73
46	Axial	1.5	0.3	150	0.4	2.5	49.11	54.14	53.02	57.45

(3)

$$\text{Pyrite sulphur removal (\%)} = +54.30 +5.52X_1 -10.11X_2 -1.57X_3 +0.78X_4 -0.11X_5 +1.56X_1X_2 +0.97X_1X_3 -0.43X_1X_4 -0.64X_1X_5 +0.42X_2X_3 +0.5X_2X_4 +1.01X2X5 -0.81X_3X_4 -0.48X_3X_5 -0.41X_4X_5 +6.05X_1^2 +4.81X_2^2 +4.86X_3^2 +4.38X_4^2 +0.85X_5^2$$

From Tables 3 and 4, it is obvious that, linear terms (X_1, X_2, X_3), interactive term (X_1X_2) and quadratic terms (X_1^2, X_2^2, X_3^2, X_4^2) have the largest effects on ash and pyrite sulphur removal form bitumen due to its higher F values as well as low p-values.

Process variables (linear, interaction and quadratic) effects in ash and pyrite sulphur removal from

Table 3.

ANOVA analysis for the percentage of ash removal from bitumen.

Source	Std. Dev.	R^2	Adjusted R^2	Predicted R^2	p-Value	PRESS	Remarks
Adequacy of the model tested							
Linear	8.23	0.696	0.657	0.635	<0.0001	3248.65	
2FI	9.25	0.711	0.567	0.516	<0.0001	4302.08	
Quadratic	3.22	0.971	0.947	0.888	<0.0001	997.57	Suggested
Cubic	4.02	0.982	0.918	-1.312	<0.0001	20574.13	Aliased

Source	Coefficient Estimate	Sum of squares	Degree of freedom	Standard error	Mean square	F Value	p-Value
Model	48.99	8637.92	20	1.55	431.90	41.70	<0.0001
X_1	6.21	1513.75	1	0.51	1513.75	146.14	<0.0001
X_2	-10.72	4513.16	1	0.51	4513.16	435.70	<0.0001
X_3	-1.85	132.47	1	0.52	132.47	12.79	0.0015
X_4	0.43	7.32	1	0.51	7.32	0.71	0.4084
X_5	-0.33	4.20	1	0.51	4.20	0.41	0.5298
X_1X_2	1.15	41.17	1	0.58	41.17	3.97	0.0472
X_1X_3	0.95	27.38	1	0.58	27.38	2.64	0.1165
X_1X_4	-0.026	0.021	1	0.58	0.021	0.0020	0.9643
X_1X_5	-0.31	3.00	1	0.58	3.00	0.29	0.5952
X_2X_3	0.43	5.67	1	0.58	5.67	0.55	0.4662
X_2X_4	0.41	5.27	1	0.58	5.27	0.51	0.4824
X_2X_5	0.80	19.91	1	0.58	19.91	1.92	0.1778
X_3X_4	-0.61	11.21	1	0.58	11.21	1.08	0.3081
X_3X_5	-0.36	3.92	1	0.58	3.92	0.38	0.5438
X_4X_5	-0.58	10.63	1	0.58	10.63	1.03	0.3208
X_1^2	6.28	1091.59	1	0.61	1091.59	105.38	<0.0001
X_2^2	5.62	874.89	1	0.61	874.89	84.46	<0.0001
X_3^2	4.88	661.00	1	0.61	661.00	63.81	<0.0001
X_4^2	4.45	548.51	1	0.61	548.51	52.95	<0.0001
X_5^2	1.17	38.06	1	0.61	38.06	3.67	0.0667
Residual		258.96	25		10.36		
Std. Dev.	3.22						
Mean	68.53						
C.V.%*	4.70						
Adeq Precision	20.196						

*C.V.% is Coefficient of Variation.

Table 4.
ANOVA analysis for the percentage of pyrite sulphur removal from bitumen.

Source	Std. Dev.	R^2	Adjusted R^2	Predicted R^2	p-Value	PRESS	Remarks
Adequacy of the model tested							
Linear	7.89	0.683	0.644	0.619	<0.0001	2998.78	
2FI	8.71	0.711	0.566	0.523	<0.0001	3748.83	
Quadratic	2.79	0.975	0.955	0.906	0.0021	737.77	Suggested
Cubic	3.14	0.987	0.944	-0.625	0.0008	12776.00	Aliased
Source	**Coefficient Estimate**	**Sum of squares**	**Degree of freedom**	**Standard error**	**Mean square**	**F Value**	**p-Value**
Model	54.30	7666.85	20	1.34	383.34	49.10	<0.0001
X_1	5.52	1197.95	1	0.45	1197.95	153.44	<0.0001
X_2	-10.11	4017.78	1	0.45	4017.78	514.61	<0.0001
X_3	-1.57	94.68	1	0.45	94.68	12.13	0.0018
X_4	0.78	23.92	1	0.45	23.92	3.06	0.0923
X_5	-0.11	0.51	1	0.45	0.51	0.066	0.8001
X_1X_2	1.56	76.29	1	0.50	76.29	9.77	0.0045
X_1X_3	0.97	28.90	1	0.51	28.90	3.70	0.0658
X_1X_4	-0.43	5.70	1	0.50	5.70	0.73	0.4012
X_1X_5	-0.64	12.68	1	0.50	12.68	1.62	0.2142
X_2X_3	0.42	5.40	1	0.51	5.40	0.69	0.4134
X_2X_4	0.50	7.87	1	0.50	7.87	1.01	0.3249
X_2X_5	1.01	31.78	1	0.50	31.78	4.07	0.0545
X_3X_4	-0.81	19.94	1	0.51	19.94	2.55	0.1226
X_3X_5	-0.48	7.16	1	0.51	7.16	0.92	0.3473
X_4X_5	-0.41	5.24	1	0.50	5.24	0.67	0.4205
X_1^2	6.05	1012.54	1	0.53	1012.54	129.69	<0.0001
X_2^2	4.81	642.11	1	0.53	642.11	82.24	<0.0001
X_3^2	4.86	655.18	1	0.53	655.18	83.92	<0.0001
X_4^2	4.38	532.53	1	0.53	532.53	68.21	<0.0001
X_5^2	0.85	19.81	1	0.53	19.81	2.54	0.1238
Residual		195.18	25		7.81		
Std. Dev.	2.79						
Mean	72.55						
C.V.%*	3.85						
Adeq Precision	21.430						

*C.V.% is Coefficient of Variation.

bitumen with respect to sum of squares (SS) of each variable obtained from ANOVA were also investigated in this study and the results are shown in Figure 2 and 3, respectively.

It can be seen from the Figures 2 and 3, that linear effect of process variables have the greatest impact (61.71%) on process, followed by quadratic and interactive effects of process variables (32.14% and 1.28%, respectively). Similarly, in pyrite sulphur removal the effects of linear, quadratic and interactive terms were 53.36, 28.63 and 2.01%, respectively. Furthermore, The residual error measures amount of variation in the response left unexplained by the model and its effect was low (4.87% and 16% for ash and pyrite sulphur removal form bitumen, respectively) in column flotation process.

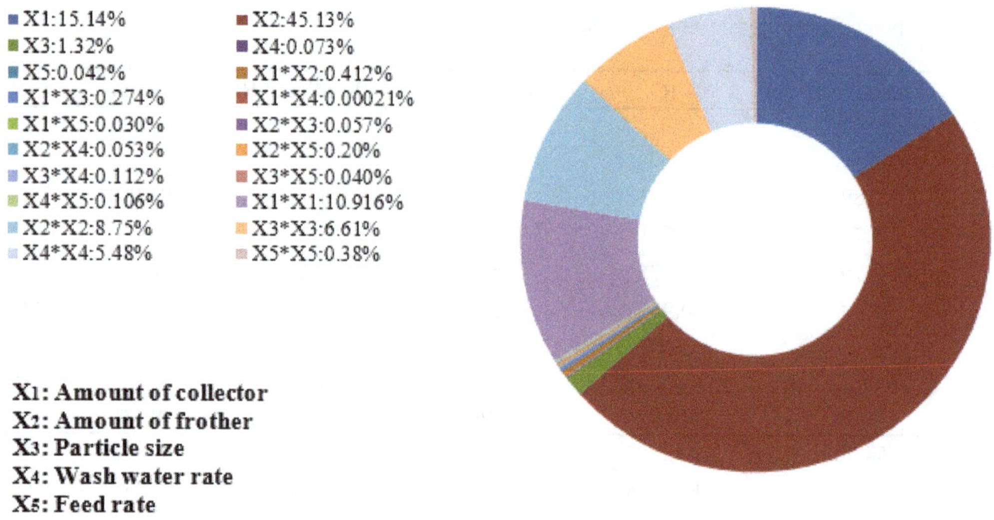

Fig. 2. Linear, interactive and quadratic effect of process variables on ash removal from bitumen.

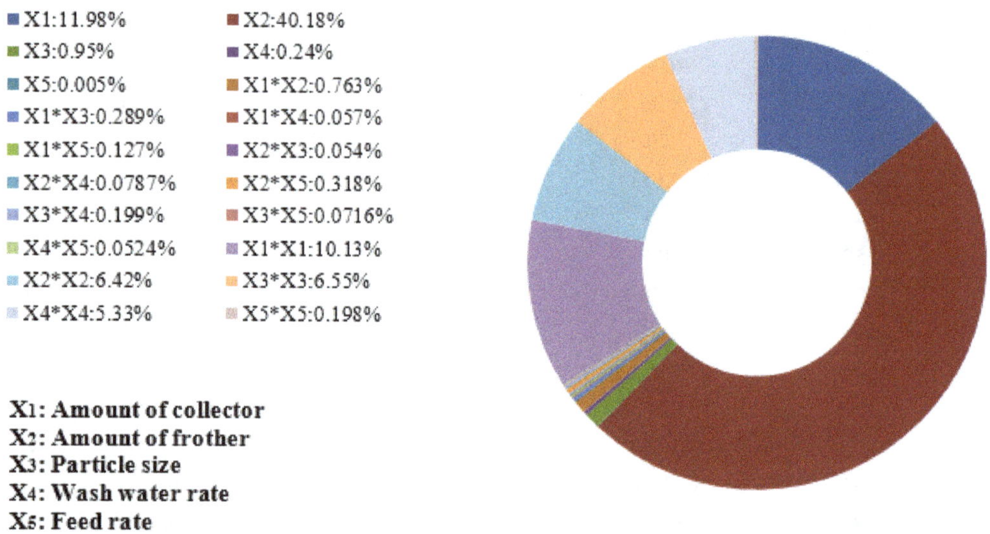

Fig. 3. Linear, interactive and quadratic effect of process variables on pyrite sulphur removal from bitumen.

3.2. Effects of process parameters on the responses

From the response surface analysis, it is clear that the variables had both positive and negative effects on ash and pyrite sulphur removal from bitumen. In order to distinguish the effects of the variables on the removal yields, 3D graphs were developed. Figures 4 and 5 illustrate the response surface plots of ash and pyrite sulphur removal from bitumen as a function of the combined positive or negative effects of significant terms in the RSM model.

3.2.1. Collector agent and ash-pyrite sulphur removal from bitumen

Figures 4-5(a-d) imply that the amount of collector has a positive effect on both the ash and pyrite sulphur removal from bitumen [19, 20]. This is because the absorption of collector on bitumen increases the hydrophobicity and consequently the floating property of particles. In the other words, the amount of pulp that goes to the froth zone increases because the contact angle of bitumen particles with water improves [13].

3.2.2. Frother agent and ash-pyrite sulphur removal from bitumen

The results show that frother agent had the most significant effect on both the ash and pyrite sulphur removal from bitumen (β2, -10.72 and -10.11 for ash removal and pyrite sulphur removal, respectively). Decreasing frother agent, the ash and pyrite sulphur removal from bitumen also increases [20] (Figure 4 -5(a, e-g)). Frother rise causes the increase ash and sulphur in the froth zone because mineral matter have very little time for hydrophilic and enter the tailings part resulting stick to the bubbles and come with them to the froth zone [21]. Moreover, Interaction of hydrophilic part of pine (frother) oil and the hydrated mineral matter can cause an increase in the hydrophobicity of ash particles. This later improves the recovery process. The pine oil addition reduces the surface tension at the liquid–air interface, resulting the production of a finer bubble size distribution which improves flotation rates and recovery values. The role of frothers in flotation is to generate smaller air bubbles. The increase in frother dosage steadily increases the solid flow rate.

3.2.3. Particle size and ash-pyrite sulphur removal from bitumen

The ash and pyrite sulphur removal from bitumen declined with the rise of particle size (See Figures 4-5(b, e, h and i)), most possibly due to the lowered particle contact with air bubbles and detracted particles ability to raise bubbles to the froth zone resulting in a decrease in removal yields. Ceylan and Zeki Küçük reported that the ash and pyrite sulphur removal from bitumen can be varied as a factor of particle size [22]. The efficacious area for heat and mass transfer increases when the particle size parameter is reduced. This last causes a growth in conversion factor of pyrite to sulfate which consequently improves the sulphur removal [23]. Small particles have some benefit compared to large particles. For example flotation and recovery of small particles are much better than large particles. Also, susceptibility of small and large particles is different for each reagent. For instance, the large particles need more collectors to have the same value of hydrophobicity as those of small particles; this increases the operating costs owing to collector and frother consumption increase. In general, the flotation of large particles is feasible only in the presence of oil collectors and higher aeration rates and longer time compared to smaller particles.

3.2.4. Wash water rate and ash-pyrite sulphur removal from bitumen

The wash water rate positively affects the ash and pyrite sulphur removal, although its effect is very low. (Figure 4-5(c, f, h and j)). This can be explained by the increase in water ratio to air ratio and the large bubbles explode in the froth zone at higher wash water rate.

3.2.5. Feed rate and ash-pyrite sulphur removal from bitumen

(Figure 4-5(d, g, i and j)) presents ash and pyrite sulphur removal percentage vs. feed rate. It is obvious that feed rate has a negligible impact on reducing of the ash and sulphur percentages (β2= -0.33 for ash removal and β2= -0.11 for pyrite sulphur removal), When feed rate is increased, removal yields reduces, which this is most possibly due to an increase in suspended solids in the froth zone and turbulency in the pulp.

3.2.6. Interaction effects

As is clear from Tables 3 and 4, the only interaction between process variables for ash and pyrite sulphur removal from bitumen is X1X2 (p-value < 0.05). There was appreciable interaction between amount of collector and amount of frother. At low collector values, amount of frother was high when the ash and pyrite sulphur removal from bitumen touched low amount. Though, at higher collector levels, amount of frother was low when the ash and pyrite sulphur removal from bitumen achieved higher quantity (Eqs. 2 and 3). The adsorbed kerosene on bitumen emulsifies the pine oil and reduces the quantity of pine oil available for frother action [24]. The attraction between kerosene and pine oil is possible due to hydrophobic interaction. According to Aston et al. [25], low concentration of pine oil will not change the hydrophobicity that is enough to affect flotability. But when particles of varying degree of hydrophobicity are present, some high carbon content particles are likely to be depressed due to decrease in hydrophobicity and a group of high mineral matter particles are floated due to increase in the hydrophobicity.

3.3. Optimization of ash and pyrite sulphur removal from bitumen using RSM

Numerical optimization method [26] was applied to optimize the process parameters on the maximum ash and pyrite sulphur removal from bitumen using Design expert software. According to the second order polynomial equation, the optimum conditions for ash and pyritic sulphur removal from bitumen included amount of collector of 2.00kg/tbitumen, amount of frother of 0.2ppm, particle size of 101.29mesh, wash water rate of 0.5L/min and feed rate 1.26L/min. Under these conditions, the predicted ash and pyrite sulphur removal from bitumen were 88.74% and 90.89%, respectively. The optimal condition determined by the RSM optimization approach was used to confirm the

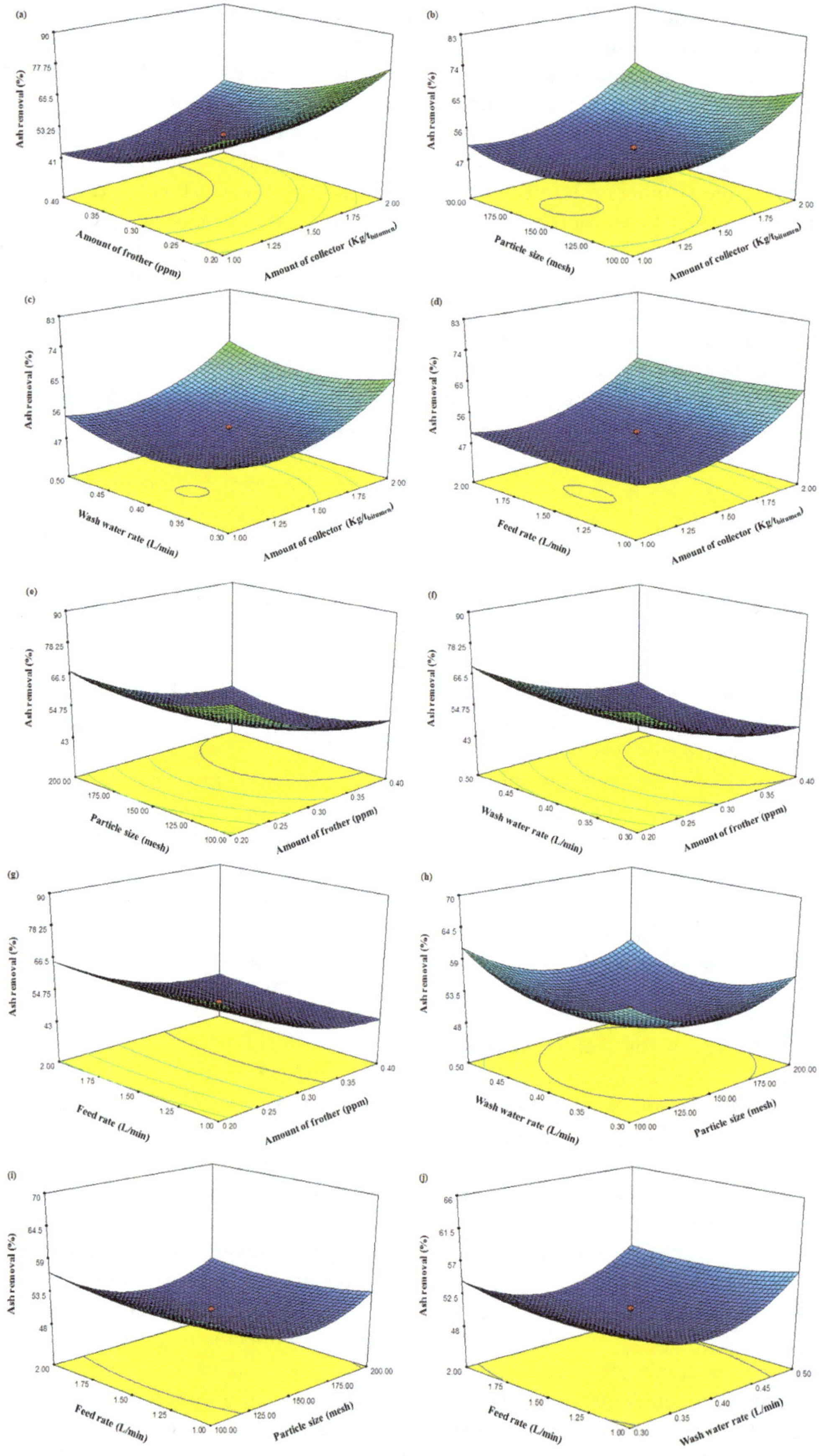

Fig. 4. Response surface plot for ash removal estimation.

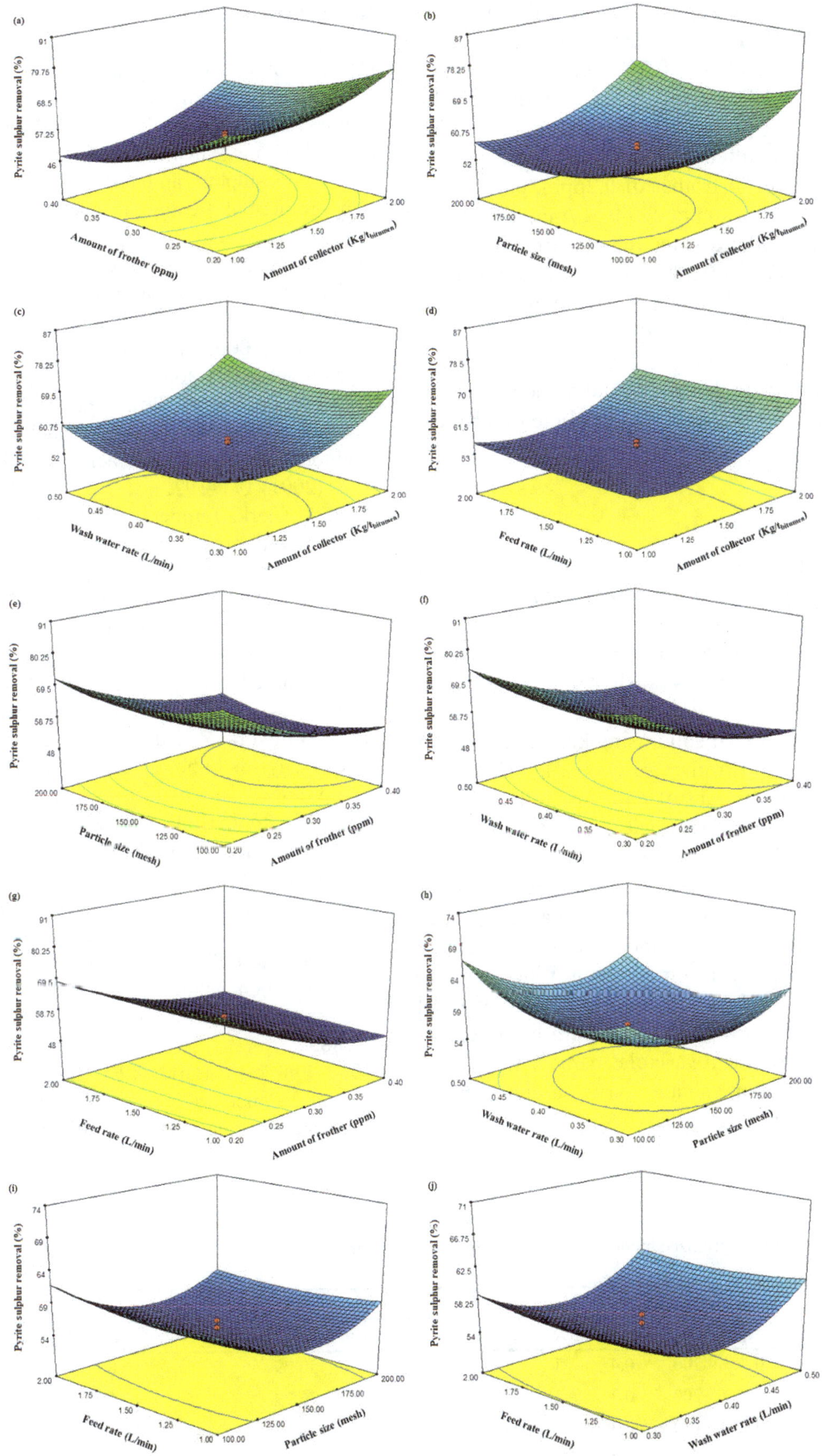

Fig. 5. Response surface plot for pyrite sulphur removal estimation.

triplicate experiments. The average ash and pyrite sulphur removal from bitumen using the triplicate experiments were 88.85± 0.21% and 90.93± 0.53%, respectively, which was very close to predicted value. For convenience purposes, the optimum conditions were slightly modified- amount of collector of 2kg/tbitumen, amount of frother of 0.2ppm, particle size of 100mesh, wash water rate of 0.5L/min, feed rate of 1L/min. The results revealed the experimental values equal to 90.74% and 92.80% for ash and pyrite sulphur removal from bitumen, respectively.

The predictability of the optimized models was investigated using five independent experimental runs. Table 5 summarized the results and demonstrated excellent confidence between the predicted and measured value.

4. Conclusions

Column flotation process was employed for ash and pyrite sulphur removal from natural bitumen extracted from Kermanshah mines. A bitumen sample was used with sulphur content of 9.6% (6.81% in the pyrite sulphur form) and 26.4% ash. All the experiments were performed with aeration rate of 4L/min and pulp containing 5% of solid using kerosene and pine oil as collector and frother, respectively. The results indicate that the amount of collector influenced the ash and pyrite sulphur removal from bitumen. Moreover, the second factor i.e. the ash and pyrite sulphur removal from bitumen decreases with the increase of amount of frother. It was observed that the particle size and feed rate has an opposite effect on ash and pyrite sulphur removal from bitumen. Whereas, wash water rate has positive effect on ash and pyrite sulphur removal from bitumen. It should be noted that the effect of two recent variables namely wash water rate and feed rate on ash and pyrite sulphur removal from bitumen are negligible.

References

[1] M. Erol, C. Colduroglu, Z. Aktas, The effect of reagents and reagent mixtures on froth flotation of coal fines, Int. J. Miner. Process. 71 (2003) 131–145.

[2] R.C. Timpe, M.D. Mann, J.H. Pavlish, P.K.K. Louie, Organic sulphur and hap removal from coal using hydrothermal treatment, Fuel Process. Technol. 73 (2001) 127–141.

[3] M. Abdollahy, A.Z. Moghaddam, K. Rami, Desulfurization of mezino coal using combination of flotation and leaching with potassium hydroxide/ methanol, Fuel 85 (2006) 1117–1124.

[4] M.S. Karen, B. John, A. Thomas, O. Donnell, G. David, Production of Ultra Clean Coal Part I Dissolution behavior of mineral matter in black coal toward hydrochloric and hydrofluoric acids, Fuel Process. Technol. 70 (2001) 171–192.

[5] K.A. Clark, Temperature effects in the conditioning and flotation of bitumen from oil sands, in terms of oil, Recovery and physical properties, Canadian Patent No. 289,058 (issued Apr. 23, 1929).

[6] L.L. Schramm, E.N. Stasiuk, D. Turner, The influence of interfacial tension in the recovery of bitumen by water-based conditioning and flotation of Athabasca oil sands, Fuel Process. Technol. 80 (2003) 101–118.

[7] V. Wallwork, Z. Xu, J. Masliyah, Bitumen recovery with oily air bubbles, Ca. J. Chem. Eng. 81 (2003) 993–997.

[8] A. Al-Otooma, M. Allawzi, A.M. Al-Harahsheh, M. Al-Harahsheh, R. Al-Ghbari, R. Al-Ghazo, H. Al-Saifi, A parametric study on the factors affecting the froth floatation of Jordanian tar sand utilizing a fluidized bed floatator, Energy 34 (2009) 1310–1314.

[9] J. Barraza, J. Piñeres, A pilot-scale flotation column to produce beneficiated coal fractions having high concentration of vitrinite maceral, Fuel 84 (2005) 1879–1883.

[10] J. Piñeres, J. Barraza, Effect of pH, air velocity and frother concentration on combustible recovery, ash and sulphur rejection using column flotation, Fuel Pro-

Table 5.
The predictability of the optimized models using five independent experimental runs.

Run	Independent variables					Experimental (%)		RSM Prediction (%)	
	X_1	X_2	X_3	X_4	X_5	Ash removal	Pyrite sulphur removal	Ash removal	Pyrite sulphur removal
1	1	0.4	200	0.4	1	46.01	50.29	46.05	50.47
2	1	0.2	200	0.4	1	72.78	76.79	70.55	75.02
3	1.5	0.2	150	0.3	1.5	71.82	74.49	69.76	73.33
4	1.5	0.3	150	0.3	1.5	55.20	59.27	53.01	57.90
5	2	0.4	200	0.4	1	63.98	69.21	63.30	67.88

cess. Technol. 97 (2012) 30–37.

[11] X. Tao, Y. Cao, J. Liu, K. Shi, J. Liu, M. Fan-maomming, Studies on characteristics and flotation of a hard-to-float high-ash fine coal, Procedia Earth Planet. Sci. 1 (2009) 799–806.

[12] K.G. Ashiwani, P.K. Banerjee, A. Mishra, Influence of chemical parameters on selectivity and recovery of fine coal through flotation, Int. J. Miner. Process. 92 (2009) 1–6.

[13] S.I. Angadi, J. Ho-Seok, S. Nikkam, Experimental analysis of solids and water flow to the coal flotation froths, Int. J. Miner. Process. 110–111 (2011) 62–70.

[14] S. Dey, G. Manjari Paul, S. Pani, Flotation behavior of weathered coal in mechanical and column flotation cell, Powder Technol. 246 (2013) 689–694.

[15] Y. Vasseghian, N. Heidari, M. Ahmadi, G.R. Zahedi, A.A. Mohsenipour, Simultaneous ash and sulphur removal from bitumen: Experiments and neural network modeling, Fuel Process. Technol. 125 (2014) 79-85.

[16] H.M. Raymond, D.C. Montgomery, Response Surface Methodology: Process and Product Optimization using Designed Experiment, second ed., A Wiley- Interscience Publication, Hoboken, 2002.

[17] D.C. Montgomery, Design and Analysis of Experiments, eighth ed., John Wiley and Sons, New York, 2012.

[18] J. Antony, Design of Experiments for Engineers and Scientists, second ed., Elsevier Science & Technology Books, ISBN: 9780080994178, 2014.

[19] N.K. Pradyumna P. Reddy, N. M. Vibhuti, Interpretation of interaction effects and optimization of reagent dosages for fine coal flotation, Int. J. Miner. Process. 75 (2005) 83-90.

[20] M. S. Jena, S.K. Biswal, M.V. Rudramuniyappa, Study on flotation characteristics of oxidised Indian high ash sub-bituminous coal, Int. J. Miner. Process. 87 (2008) 42-50.

[21] F. Boylu, S.J. Laskowski, Rate of water transfer to flotation froth in the flotation of low-rank coal that also requires the use of oily collector, Int. J. Miner. Process. 83 (2007) 125-131.

[22] K. Ceylan, M. Zeki Küçük, Effectiveness of the dense medium and the froth flotation methods in cleaning some Turkish lignites, Energ. Convers. Manage. 45 (2004) 1407-1418.

[23] S. Mukherjee, P.C. Borthakur, Chemical demineralization/desulphurization of high sulphur coal using sodium hydroxide and acid solutions, Fuel 80 (2001) 2037- 2040.

[24] T.C. Rao, K.J. Pillai, M. Vanangamudi, 1982. Statistical analysis of coal flotation- a prelude to process optimization. IX International Coal Preparation Congress, New Delhi, 1982.

[25] J.R. Aston, J.E. Lane, T.W. Healy, The solution and interfacial chemistry of nonionic surfactants used 229–256.

[26] J.P. Maran, B. Priya, Ultrasound-assisted extraction of pectin from sisal waste, Carbohydr. Polym. 115 (2015) 732-738.

2

Synthesis and statistical analysis of changing size of nano-structured PbO_2 during mechanical milling using Taguchi methodology

Maryam Omidvar [1*], Esmaeil Koohestanian[2], Omid Ramezani Azghandi[3]

[1] *Department of Chemical Engineering, Quchan Branch, Islamic Azad University, Quchan, Iran*

[2] *Department of Chemical Engineering, University of Sistan and Baluchestan, Zahedan, Iran*

[3] *Department of Mechanical Engineering, Ferdowsi University of Mashhad, Mashhad, Iran*

HIGHLIGHTS

- PbO_2 nano-structure was synthesized by ball milling method.
- Particle size for the nano-structured PbO_2 was about 50 nm.
- BPR parameter had the greatest impact on the size of particles.
- Minitab software revealed that the synthesis process be done in two hours.

GRAPHICAL ABSTRACT

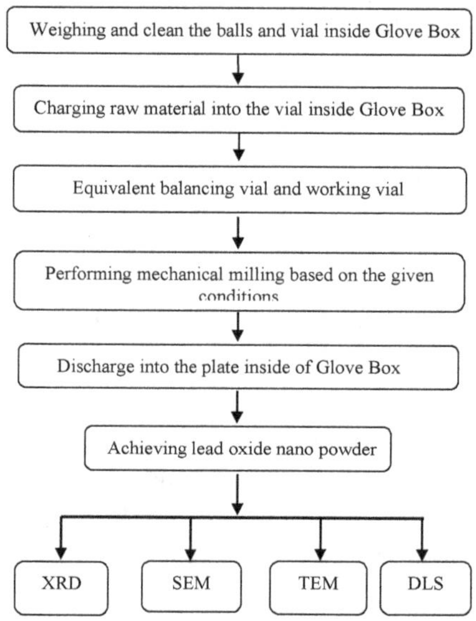

ARTICLE INFO

Keywords:
Synthesis
Mechanical milling
PbO_2
Nano-structure
Design of experiment

ABSTRACT

The research investigates synthesized nano-structured PbO_2 using ball milling. The structure and morphology of the samples were determined in the process of milling by means of XRD and SEM. The size of particles was estimated through DLS analysis. The TEM image of the synthesized powder verifies the achievement of nano dimensions. Design and analyses of the results using Taguchi methodology reveal that the size of synthesized nano-structured PbO_2 decreases as ball to powder ratio (BPR) increases while the average size of the particles increases as mechanical milling speed increases from 200 to 250 rpm. Considering the results of TEM, the particle size of the synthesized nano-structured PbO_2 by means of mechanical milling was estimated to be 50 nanometers. In addition, the even distribution and spherical morphology of the synthesized powder by this method is crystal clear in SEM images. Additionally, the result of the statistical analysis of particle size based on the effective parameters by means of Minitab software showed that BPR parameter had the greatest impact on the size of particles; BPR increase improved the objective parameter as compared with other parameters under study. According to the results obtained by Minitab software and considering the little influence of time on particle size decrease and in order to minimize the costs of synthesis, it is suggested the synthesis process be done in two hours and the BPR parameters be increased so as to decrease the size of particles.

*Corresponding author.

E-mail address: momidvar@iauq.ac.ir, maryam_omidwar@yahoo.com

1. Introduction

Mechanical milling is an easy and inexpensive method to synthesize non-crystalline metals and other non-equilibrium metals [1]. The major mechanism of mechanical milling is the recurrence of cold fusion and particle breakage which result in the production of nano-structure in the course of time [2]. Mechanical milling method makes the occurrence of chemical reactions and morphological changes in normal temperatures possible by means of synthetic acceleration [3]. As a result, many materials and structures can be produced in solid state [7-3]. Simplicity of equipment, needlessness of high temperatures and the implementation of production in just one step are the properties of mechanical milling which can economize the production of many materials and alloys [8].

Since the provision of optimal nano-synthesized materials is costly and time-consuming, the use of modern designing and lab-optimizing methods can reduce costs [9]. Accordingly, many methods have been devised which can be prioritized based upon process, cost and precision [10]. One of such methods commonly used in industrial scale is the Taguchi method which is generally applied as a practical one by various industries to improve the quantity and quality of products [11].

The purpose of this study is to synthesize the nano-structured PbO_2 by means of mechanical milling and to investigate the impacts of milling time, ball to powder ratio and the speed of milling process via Taguchi method. To investigate the micro-structure, particle distribution and surface morphology of this nano-structure, XRD, DLS and SEM were used. Also, to ascertain the achievement of nano-structure, TEM was applied. Eventually, the impact of effective parameters on the size of particles during the process of mechanical milling was administered via DLS analysis.

2. Experimental activities

2.1 Preparation of samples

8 samples including PbO_2 and commercial acetone purchased from Merck Company were applied. Mechanical milling was performed in a Fritsch Pulverisstte 5 planetary ball-mill. The vial and ceramic balls are made of alumina. To prevent unwanted reactions, all the steps of the experiment including the charging of raw materials to the vial, the placement of vial inside the miller and eventually the extraction of final product were administered under controlled argon atmosphere (Glove Box). In order to avoid an excessive temperature rise within the vial, 60 min ball milling was followed by a 10 min cooling interval. The design and analyses of tests was conducted by means of Taguchi method and Minitab software (version 17) considering three factors namely time, speed and BPR. The milling conditions are summarized in Table 1:

Table 1.
The conditions of the tests suggested by Taguchi method.

Sample No.	Time (h)	Speed (rpm)	BPR
1	2	200	20
2	2	200	30
3	5	250	20
4	5	250	30
5	2	250	20
6	2	250	30
7	5	200	20
8	5	200	30

The temperature of the laboratory was retained between 23 to 28°C. Raw materials were measured by means of a digital scale with a precision of 0.001 gram in a place which was absolutely hygienic. Also each experiment was repeated at least three times and the mean values were reported in the Table 2.

2.2 Micro-structure properties

The analysis of XRD was administered by means of XRD device D8 Advance model using Cu-Kα (λ=1.5418Å) latitude 2θ from 20 to 60 degrees. The width of X-Ray diffraction is affected by the size of crystallite and internal strain [12]. The analysis of particle size distribution was done prior to and following every stage by means of DLS in order to know the particle size and compare with the subsequent stage and to choose optimum and optimal synthesis conditions. In addition, the morphology of the nano-structured synthesized PbO_2 was investigated through SEM kyky EM3200 model. Also, to verify the achievement of nano-structure an image was taken of the synthesized particle by means of TEM.

3. Results and discussion

3.1 Results obtained by mechanical milling

The patterns of X-ray diffraction which belong to raw powder and milled powder in test 4 (after 5 hours of milling at 250 rpm and BRP 30) are shown in Figure 1. The obtained picks were compared with PbO_2 picks by means of X'pert HighScore. The results clearly show that by means of mechanical milling it is possible to obtain nano-structured PbO_2 from micron PbO_2 powder without a change in phase.

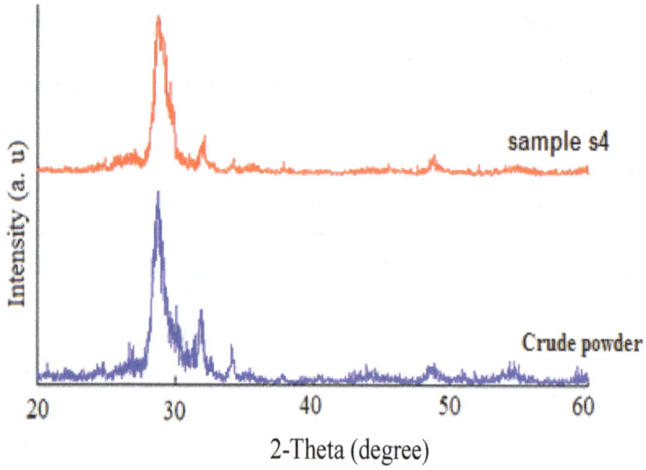

Fig. 1. The patterns of x ray diffraction in various states of mechanical milling.

Generally speaking, Figure 1 illustrates the spread, displacement and pick intensity decline as the time of mechanical milling rises. The spread of picks as milling time increases shows the decline of particle size [13]. To sum up, as time increases in mechanical milling, the picks become wider and their relative intensity declines. In order to further investigate size changes during the process of mechanical milling, DLS analysis was administered on all samples the results of which are summarized in Table 2.

By investigating Table 2, it can be concluded that the decline in particle size happened faster in the initial stages of mechanical milling, becoming slower in subsequent stages. The reason is that as the particle size declines and the borders widen, the ratio of surface to particle volumes becomes bigger and the number of atoms over the surface become larger; therefore, the number of free bonds in atoms increases and the average power which every atom receives from another declines and this causes the increase of network parameter in smaller particles. Furthermore, as BPR increases, on average, less powder mass is placed between balls in every hit, putting more strain on the powder from the balls and declining the size of particles.

Microscopic images of SEM samples prior to and following mechanical milling are shown in Figure 2. According to the images, following the milling, the size of nano-structure declines significantly with more evenness which is caused by an increase in the number of ball strikes and energy rise obtained from the balls in the process of mechanical milling. The morphology of the synthesized nano-structure is spherical and semi-spherical sized 100 nanometer on average which is one of the most ideal morphologies for applications such as battery industries due to having high surface to volume ratio in sphere shape.

Table 2.
The size of synthesized nano-structured particle by means of DLS analysis in various states of mechanical milling.

Sample No.	Time (h)	Speed (rpm)	BPR	Particle Size (nm)
0	0	0	0	1535
1	2	200	20	824.2
2	2	200	30	354.6
3	5	250	20	957.1
4	5	250	30	329.0
5	2	250	20	765.2
6	2	250	30	675.5
7	5	200	20	868.1
8	5	200	30	491.1

Figure 3 shows TEM image of PbO_2 over a course of five hours, with a BPR ratio of 30 and speed of 250 rpm. As can be seen in the picture, nano PbO_2 particles have spherical morphology. The particles are spread over an area of 20 to100 nanometer and the average nano-structure of the synthesized PbO_2 is 50 nanometer. The reason for the difference in TEM and SEM is that, during the investigation, preparation by TEM was done by means of Ultrasonic device and the particles were separated and consequently the irregular shape which is discernible through DLS is removed by Ultrasonic device and hence a better and clearer image. In addition, the reason for the difference in the size of particles by means of TEM and SEM can be attributed to the fluid ethanol film over the particles used as solvent in DLS analysis. Hence the reported amount is more accurate when done but the costly TEM analysis.

The results of test analysis by means of Taguchi method are shown in Figure 4. Based on the image, as BPR increases, the average nano-particle size declines. The reason is that the increase of BPR causes less powder mass to be placed among balls in every strike. As a result, more energy is placed upon powder on the part of balls, making the particles smaller. Also, according to this image, there is no significant change in particle size after two hours, and not only will that not decrease particle size, but the average size in all tests is very insignificant, showing the rise in milling time causes the phenomenon of cold fusion to defeat breakage.

Fig. 2. SEM images of PbO_2 in various mechanical milling states.

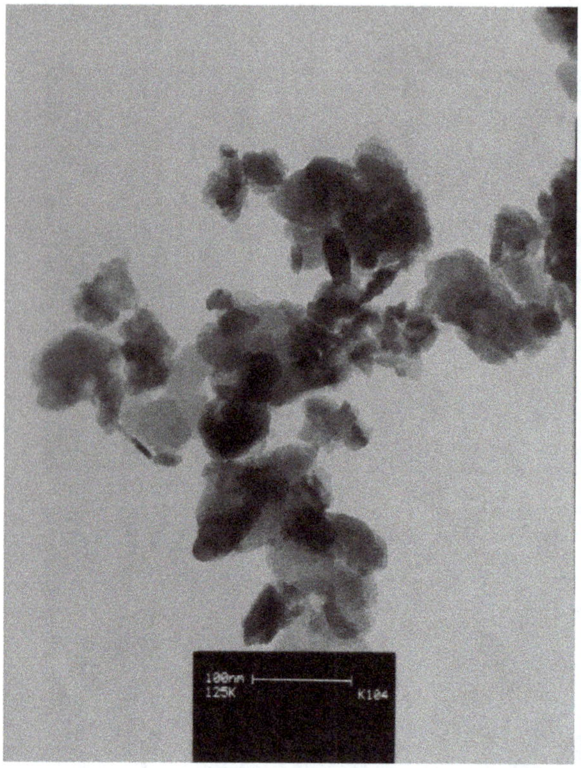

Fig. 3. TEM image of PbO_2 powder over 5 hours of mechanical milling, speed= 250 rpm and BPR=30.

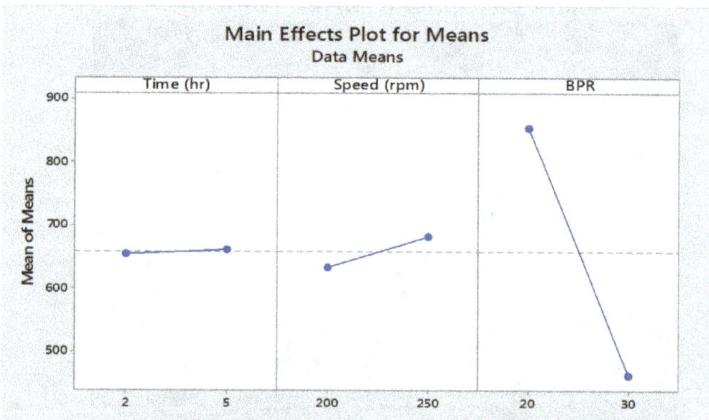

Fig. 4. The results of the average effect of the parameters under study on the average particle size.

Figure 5 illustrates the results of optimization analysis on the data in Table 2. According to this image, in which the parameters have been investigated in similar scales, the slope of every graph shows the effect of that parameter on the size of the particle. The parameter goal for the software should be defined as smaller is better. According to Figure 5, the BPR parameter has had the greatest effect in particle size change with BPR increase improving the goal parameter compared with other variables in question. The reason is that as BPR increases, on average, less powder mass is placed between balls in every strike; therefore, more energy is placed over the powder on the part of balls, making the particles smaller.

By means of a comparison between these results and raw powder analysis, it can be concluded that a decline in the size of particles occurs in the first two hours and an increase in the time of mechanical milling more than that causes the process of cold fusion to defeat breakage. According to the results in these images, it is suggested that the synthesis process be done in two hours in order to minimize synthesis costs and when needed the BPR parameter be increased to decline the size of particles.

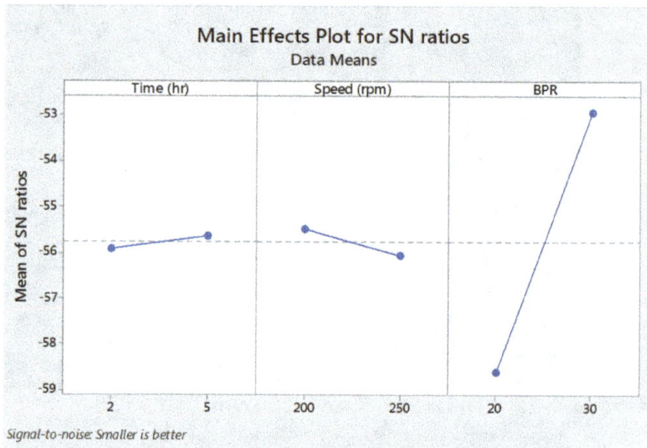

Fig. 5. The results of the average effect of the parameters under study on the average particle size.

4. Conclusions

The structure, size and distribution of particles and synthesized nano-structured morphology by means of XRD, DLS, SEM and TEM were investigated. The results of XRD clearly reveal that it is possible to obtain nano-structured PbO_2 from PbO_2 without a change in phase by means of mechanical milling. According to the results of XRD, it was revealed that when the time of mechanical milling increases, the size of particles decreases. TEM image displayed a united distribution of the synthesized particles. Based on the reported results, in this image, the particles were spread in the range of 20 to 100 nanometer and the average synthe sized nano-structured PbO_2 was 50 nanometer. In addition, the results of sample analysis, obtained through Minitab software, showed that BPR parameter had the greatest impact on the size of particles with BPR increase improving the goal parameter compared withother variables in question. According to the results obtained via Minitab software, it is suggested that the synthesis be done in two hours to reduce costs given the little effect of time parameter on the size of particles and if needed the PBR parameter be increased in order to decrease the size of particles. The optimal results obtained by means of this method are related to size, morphology, phase and the structure of the synthesized materials the crystallite size of which was reported to be 50 nanometer through TEM analysis. The morphology is spherical and semispherical which due to having high surface to volume ratio is the most ideal morphologies for various applications such as using in battery plates.

The findings in this part of the study certify the economic justification of this method for production and application in industrial scale. The process of mechanical milling can be done in one stage and contrary to chemical methods there is no need for complicated chemical and thermal procedures.

Acknowledgement

The authors consider it essential to appreciate Islamic Azad University of Quchan for sponsorship and allocation of laboratory equipment to conduct the experiment which was reported in this research.

References

[1] C. Xu, S. De, A. M. Balu, M. Ojeda, and R. Luque, Mechanochemical synthesis of advanced nanomaterials for catalytic applications, Chemical Communications 51 (2015) 6698-6713.

[2] A. Dushkin and M. Sopicka-Lizer, High-Energy Ball Milling: Mechanochemical Processing of Nanopowders, first eddition, Woodhead Publishing Limited, 2010.

[3] N. K. Hoa, H. Abd Rahman, and M. Rao Somalu, Preparation of Nickel Oxide-Samarium-Doped Ceria Carbonate Composite Anode Powders by Using High-Energy Ball Milling for Low-Temperature Solid Oxide Fuel Cells, Materials Science Forum (2016) 97-102.

[4] E. Ayoman and S. G. Hosseini, Synthesis of CuO nanopowders by high-energy ball-milling method and investigation of their catalytic activity on thermal decomposition of ammonium perchlorate particles, Thermal Analysis and Calorimetry 123 (2016) 1213-1224.

[5] C. E. Halbig, T. J. Nacken, J. Walter, C. Damm, S. Eigler, and W. Peukert, Quantitative investigation of the fragmentation process and defect density evolution of oxo-functionalized graphene due to ultrasonication and milling, Carbon 96 (2016) 897-903.

[6] Q. Han, R. Setchi, and S. L. Evans, Synthesis and characterisation of advanced ball-milled Al-Al_2O_3 nanocomposites for selective laser melting, Powder Technology 297 (2016) 183-192.

[7] T. Jayaramudu, G. M. Raghavendra, K. Varaprasad, G. V. S. Reddy, A. B. Reddy, K. Sudhakar, et al., Preparation and characterization of poly (ethylene glycol) stabilized nano silver particles by a mechanochemical assisted ball mill process, Applied Polymer Science (133) 2016.

[8] Koch, C. and Y. Cho, Nanocrystals by high energy ball milling. Nanostructured Materials 1 (1992) 207-212.

[9] S. M. Pourmortazavi, M. Rahimi-Nasrabadi, Y. Fazli, and M. Mohammad-Zadeh, Taguchi method assisted optimization of electrochemical synthesis and structural characterization of copper tungstate nanoparticles, Refractory Metals and Hard Materials 51 (2015) 29-34.

[10] D. C. Montgomery, Design and analysis of experiments: John Wiley & Sons, 2008.

[11] B. Nematollahi, M. Rezaei, and E. Nemati Lay, Synthesis of Nanocrystalline CeO_2 with High Surface Area by the Taguchi Method and its Application in Methanation, Chemical Engineering & Technology 38 (2015) 265-273.

[12] M. S. Asl, M. G. Kakroudi, and S. Noori, Hardness and toughness of hot pressed ZrB_2–SiC composites consolidated under relatively low pressure, Alloys and Compounds 619 (2015) 481-487.

[13] G. Zhuang, Z. Zhang, J. Guo, L. Liao, and J. Zhao, A new ball milling method to produce organo-montmorillonite from anionic and nonionic surfactants, Applied Clay Science 104 (2015) 18-26.

3

Effect of ion concentration on viscosity, electrical conductivity and deposit weight of doped nano alumina prepared by electrophoretic deposition

Mostafa Milani[1], Seyed Mohammad Mirkazemi[2], Seyed Mohammad Zahraee[1*]

[1] *Advanced Materials and Renewable Energies Department, Iranian Research Organization for Science and Technology, Tehran, Iran*
[2] *School of Metallurgy and Materials Engineering, Iran University of Science and Technology, Tehran, Iran*

HIGHLIGHTS

- Viscosity influenced by cation addition when I_2 concentration is in critical value.
- Viscosity is not a pH dependent parameter rather is the dopant concentration dependent.
- Smaller cations such as Mg^{+2} have higher influence on conductivity of suspension.

GRAPHICAL ABSTRACT

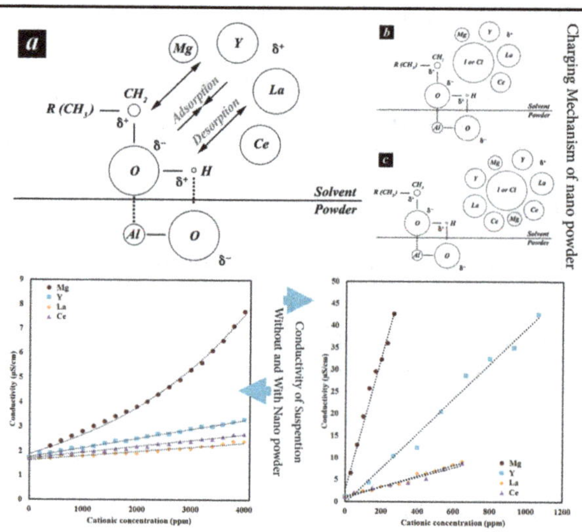

ARTICLE INFO

Keywords:
Electrophoretic deposition
ANOVA
dopant
viscosity
sintering aid

ABSTRACT

Viscosity, electrical conductivity and deposit weight were determined for Electrophoretic deposition (EPD) Mg^{2+}-, Y^{3+}-, La^{3+}- and Ce^{4+}-doped alumina's ethanolic suspensions prepared at dopant concentration between 350 to 1350 ppm. The concentration of XCly (X, y were: Mg, 2; Y, 3; Ce, 3 and La, 3, respectively) the charging salt, is also found to be a critical factor to control the viscosity. It is shown that the deposit weight is influenced by precursor concentration, but not by conductivity, viscosity or the pH of the suspension. All two way concentration interactions without Mg^{2+} and Ce4+ concentration simultaneous change are significantly in analysis of variance (ANOVA) model. The viscosity of suspension reached 2.5 mPa.s with Mg^{2+}-, Y^{3+}-, La^{3+}- and Ce^{4+}- decreased to 100, 100, 100 and 0 ppm in low iodine concentration (400 ppm), due to the most heavily cations that can adsorb to alumina surface with iodine adsorption but lighter Mg^{2+}- cations adsorb under the influence of OH groups excite on alumina surface. The interest in the present study is to achieve a model between viscosity and additive concentration.

*Corresponding author.
E-mail address: smzahraee@irost.org

1. Introduction

Electrophoretic deposition (EPD) is a colloidal shaping method for preparation of ceramic particle's deposits of overall thickness from nanometers to several millimeters [1]. EPD has the advantages of little training time, some restriction of the shape of the substrate, the ability to mass production and no requirement of binder burnout because of little or no organic matter. The EPD of ceramics was first studied by Hamaker in 1940 [2]. In all previous studies main task is the control of deposited mass onto electrode. Suspension characteristics, such as solid content, zeta potential, conductivity, viscosity, etc., are key parameters of the EPD mass controlling [3]. In Hamaker equation the deposition weight (w) during the EPD process can be described by:

$$w = fC\mu SEt \qquad (1)$$

where C, μ, S and t are the particle concentration, electrophoretic mobility of particles, deposition area and deposition time, respectively; E is the electrical field (E = U/d), U and d are the applied potential and distance between the electrodes, respectively. There is general agreement that further work needs to be done to develop a full, quantitative understanding of the fundamental mechanisms of EPD to optimize the working parameters for a broader use of EPD in materials processing [4].

Suspension viscosity has an important role in various shaping methods such as slip casting [5–7], spry freeze drying [8] and EPD [9–11]. Stuer and Bowen [12] investigated the effects of different concentrations of dopants (Mg^{2+}, La^{3+}, Y^{3+}) on the aqueous suspension rheological behaviour of alpha alumina suspensions to achieve high solid loadings and low viscosity suspensions. Their results show that dopant addition increases the viscosity, and also causes a yield stress to appear. Upon their results, in higher cation charge at constant cationic concentration, higher effect on the double layer thickness and hence on the measured shear stress was observed [12]. Biswas et al. [13] studied the viscosity of lanthanum doped alumina suspensions. La_2O_3 used as precursor for lanthanum. When La_2O_3 weight increased suspension viscosity increased.

In EPD process The Henry equation relates the electrophoretic mobility to the viscosity.

$$\mu = \frac{\zeta.\varepsilon_0.\varepsilon_{r,l}}{\eta} f(\frac{1}{\kappa}, a) \qquad (2)$$

where, ζ (mV), is the zeta potential, ε^0(8.8544×10^{-12} A^2s^2/Nm^2), is the vacuum dielectric constant, εr,l, is the solvent dielectric constant, η (Pa s) is the solvent viscosity and f(1/κ, a), is a function of the particle radius, a (nm), and the Debye length, 1/κ (nm). Depending on the viscosity of the solvent, suspensions with a similar dielectric constant and zeta potential shows large differences in the electrophoretic mobility of the particles. Given the Smoluchowski approach for thin double layer and large particles (a >> 1/κ) powders with a similar zeta potential move faster in solvents with lower viscosities,

$$\mu = \frac{\zeta\varepsilon_0\varepsilon_r}{\eta} \qquad (3)$$

In all cases, dopant level are constant and relationship between concentration and viscosity are not investigated. Also, since the electrophoretic mobility and viscosity have close relevance dopant level can determine electrophoretic mobility and final deposition yield. The present work investigated the viscosity of ethanolic suspension of doped alumina nano powder to understanding electro-deposition behavior of this suspension. In this way, relation between dopant concentration, viscosity and electrophoretic mobility investigated and developed an equation for viscosity of suspension in present of cationic dopants. As a result, the cationic concentration in the suspension was optimized by means of viscosity measurements. Mg^{2+}, Y^{3+}, La^{3+} and Ce^{4+} used as dopant cations and 1/2 fraction, 2-level factorial design with 4 center points used for experimental design [14].

2. Experimental

2.1. Suspension preparation

Ethanol (Ph Eur, CAS 100986, purity 99.6 vol.%) without further purification is used as the dispersing medium. The dispersant used is iodine (I_2, 99.8%) all were supplied by Merck Millipore, Germany. Specifications of the alumina nano powder was used presented in Table 1. Impurity analysis by the inductively coupled plasma atomic emission spectroscopy ICP–OES method was shown in Table 2.

Table 1.
Characterization of alumina powder

Supplier	Model	Purity (%)	BET surface area (m^2/g)	median particle size (nm)	Structure
US Nano, US	US3008	99+	19.95	80	corundum

Table 2.
Alpha alumina impurity determined by ICP-MS

Impurity	Mg	K	Ga	Na	Fe
Concentration (ppm)	411	20	40	10	10
	Y	Zn	La	Ce	Ca
	11	18	2	5	40

The amounts of magnesium oxide (MgO), yttrium (III) oxide (Y_2O_3), lanthanum (III) oxide (La_2O_3), and Cerium (IV) oxide (CeO_2) were varied as grain growth inhibitor to study their effects on viscosity and deposited weight of alumina nano powder. For all 20 different powders, the influence of the dopant concentration in constant potential and time were investigated according to an experimental matrix (Table 3). To dope the powder and prepare the suspensions for EPD, 50 g of alpha alumina nano powder was dispersed in 100 mL ultra-pure ethanol before addition of the desired amounts of a 400, 600 and 800 ppm iodine solution as a dispersant. After an ultrasonic bath (UB) treatment of 15 min, Mg^{2+}, Y^{3+}, La^{3+} and Ce^{4+} ethanolic solutions (purity > 99.99%, Sigma Aldrich chloride salts) were added. The amounts of doping agent (cationic ratio [doping element^{X+}]/[Al^{3+}]) introduced were 350 to 1350 ppm total cationic ratio. The final suspensions with an Operational pH (O.pH, pH in non-aqueous solvents) around 4 to 7 were stirred and UB treated for another 15 min before the electrophoretic deposition.

These different powders are referred to as I-Mg-Y-La-Ce as a XXXXX number, to indicate the cationic dopant as well as the dopant elements (Mg = magnesium, Y= yttrium, La = lanthanum and Ce = Cerium; all dopants were added in three concentration (Table 3) for all doping samples). After stirring and ultrasonication, the suspensions were loaded in a deposition cell.

Table 3.
Experimental matrix (according to sample code order)

StdOrder	RunOrder	Code	Iodine (I)	MgO(Mg)	Y_2O_3 (Y)	La_2O_3 (La)	CeO_2 (Ce)
9	14	41110	400	100	150	100	0
13	7	41133	400	100	150	300	300
10	17	41413	400	100	450	100	300
14	20	41430	400	100	450	300	0
11	8	43113	400	300	150	100	300
15	19	43130	400	300	150	300	0
12	16	43410	400	300	450	100	0
17	10	623211	600	200	300	200	150
18	18	623212	600	200	300	200	150
19	5	623213	600	200	300	200	150
20	6	623214	600	200	300	200	150
1	15	81113	800	100	150	100	300
5	3	81130	800	100	150	300	0
2	4	81410	800	100	450	100	0
6	2	81433	800	100	450	300	300
3	9	83110	800	300	150	100	0
7	7	83133	800	300	150	300	300
4	12	83413	800	300	450	100	300
8	13	83430	800	300	450	300	0
16	1	83433	800	300	450	300	300

2.2. Deposition process

After suspension preparation, samples were deposited on a stainless steel electrode with a Ti_6Al_4V counter electrode (50 mm×40 mm×2 mm). Electrodes were cleaned with dipping in 1 M HNO_3 acid, rinsed with deionized water and acetone. Electrophoretic deposition was carried out in a polytetrafluoroethylene (PTFE) cubical cell of about 50 × 40 × 44 mm³ and a capacity of 85 mL. The potential and distance between the electrodes is fixed at 50 V and 10 mm respectively. Electrodes were removed from the colloidal suspension, after 300 s of deposition processing, and were carefully removed and the deposits were left to dry at room temperature for 24 hours. These green bodies were easily removed from the electrode and were then weighed to determine the yield.

2.3. Characterization

Conductivity and pH of the suspension were measured at room temperature 10 min before and after the electrodes were immersed in the suspension. Conductivity was measured with a Cond 330i probe (WTW, Weilheim, Germany) and pH was measured with a pH meter (Mi180, Milwaukee, Szeged, Hungary) at room temperature (25.0 °C ± 0.5 °C). Three aqueous standards of pH 4, 7 and 10 were used for calibration due to the lack of standard buffer solutions for ethanol. Thus, the pH meter determines so-called "operational pH values" (O.pH) for nonaqueous suspensions. The theoretical background and method involving O.pH was described by Wang et al. [15]. The viscosity of suspensions was measured with a viscosity meter (type DV2T Viscometer, Brookfield, USA) that equipped by small sample adapter (SSA). During the rheology measurements, the setup and the suspensions were kept at 25 ± 0.1 °C by a thermostatic bath. The data acquisition cycle was: (1) ramp from 0 to 180 s^{-1} in 30 s, (2) hold at 180 s^{-1} for 60 s, and (3) decrease from 180 to 0 s^{-1} in 30 s. The data acquisition was performed three times on the same suspension with the same cycle with 1 min waiting time between the cycles.

2.4. Method

The viscosity and deposited weight results were analyzed by statistical analysis of variance (factorial ANOVA) [14]: by conducting different series of independent experiments, and combined into one matrix, the effect of the interacting parameters to be considered as well as giving the statistical significance of variations of properties as a function of the different experimental parameters. The effect of the parameters, cation concentration, iodine concentration, and possible interactions between them on the measured property (i.e. viscosity or deposited weight) are determined within a confidence interval of 95% ($p \leq$ alpha level) according to standard statistical analysis [14].

3. Results and discussion

3.1. Stability of suspension

Stability of suspension characterized by sedimentation time of half of powder in suspension. In all suspensions more than 180 min were required to sediment half of added powder. On the other hand in lower viscosity the sedimentation time increased and exceeded to 300 min.

3.2. Cation adsorption mechanism

Figure 1 presented conductivity measurement of alumina/iodine suspensions by adding cations from 0 to 4000 ppm separately. From all suspension parameters, pH and conductivity of the suspension could be measured as a macroscopic result to understand the cation adsorption mechanism. To understand the adsorption mechanism of cations in surface of particles in suspension, the conductivity of suspension with and without alumina addition were measured. Ethanol conductivity measurement by addition of cations indicated in Figure 2. As can be seen, for all cations the conductivity of suspension and ethanol increases with increasing of the doping concentration of cations. The increase in conductivity with cation addition at present of iodine can be explained by the following reactions:

$$CH_3 - CH_2 - CH_2 - OH + I_2 \rightarrow CH_3 - CH_2 - CHO + 2HI \quad (4)$$

$$MCl_x + xHI \xrightarrow{I_2} M^{+x} + xHCl + xI^- \quad (5)$$

The increment of the conductivity as a result of iodine addition and cations addition into the ethanol is due to the generation of ionic species (Cl^-, I^-, M^{+x} and H^+) in them. When alumina was adds to ethanol these ions released in the process can adsorb to alumina surface and formed electrostatic forces between particles created a stable suspension [16]. Decrement in electrical conductivity after addition of alumina in same cationic concentration (difference between Figure 1 and Figure 2) can be mainly attributed to the adsorption of ions on alumina surfaces which cannot contribute to electrical conductivity of suspension. When ethanol molecules were adsorbed on nano alumina surface powder OH and CH_2 groups creates a positive charge on this surface [17,18] and the Iodine and Chlorine anion were adsorbed firstly on the surface of the suspended particles making them negatively charged.

Then heavily cations electrostatically adsorbed to this anions. As a result this charged particles and some portion of non-adsorbed ions increased the conductivity of suspension by increasing of cation concentrations in same iodine concentration.
In competition, a drastic increase of conductivity is obtained by increasing Mg^{2+} concentration, while no significant difference of conductivity values is found when increasing Y^{3+}, La^{3+} and Ce^{4+}, concentrations from 0 to 4000 ppm, indicating that the increase of conductivity is mainly influenced by the addition of Mg^{2+}.

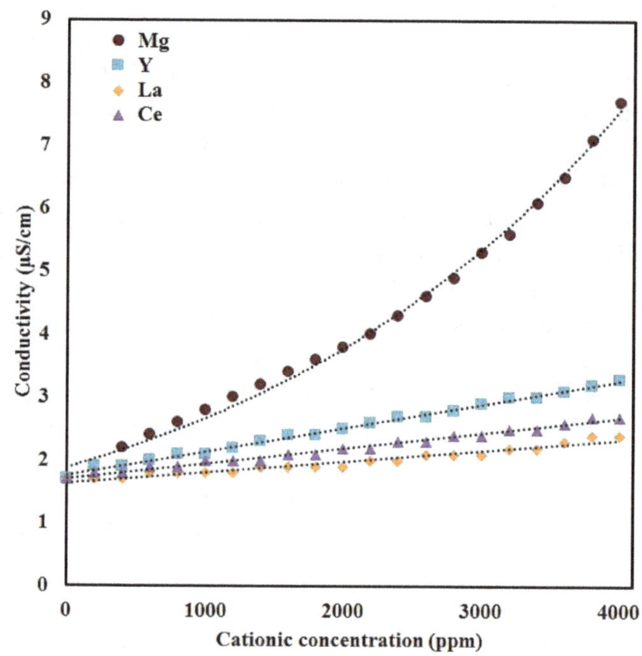

Fig. 1. Conductivity changes of suspension by separated cation concentration. Suspension made by 100mL ethanol, 400 ppm Iodine and 20 g alumina nano powder. Lanthanum concentration (♦), Yttrium concentration (■), Cerium concentration (▲) and Magnesium concentration (●).

According to this results if the concentration of I_2 ions is lower than the critical concentration there will be numerous free ions with high ability to move. It causes higher conductivity of suspension. When this concentration reaches to a critical value all of the alumina particles will be surrounded by all cations. The formed charged particles are bigger than free ions and could not participate in current transition, it results in low suspension electrical conductivity. The schematic of the formed species with different iodine concentration is illustrated in Figure 3. Iodine and Chlorine anions adsorbed to alumina surface. These large anions can adsorb cations. When iodine concentration equals to critical concentration the coating of cations have the most uniformity (Figure 3b). Figure 3a and c are showing the adsorption morphology in which iodine concentration is lower and higher than the critical value, respectively.

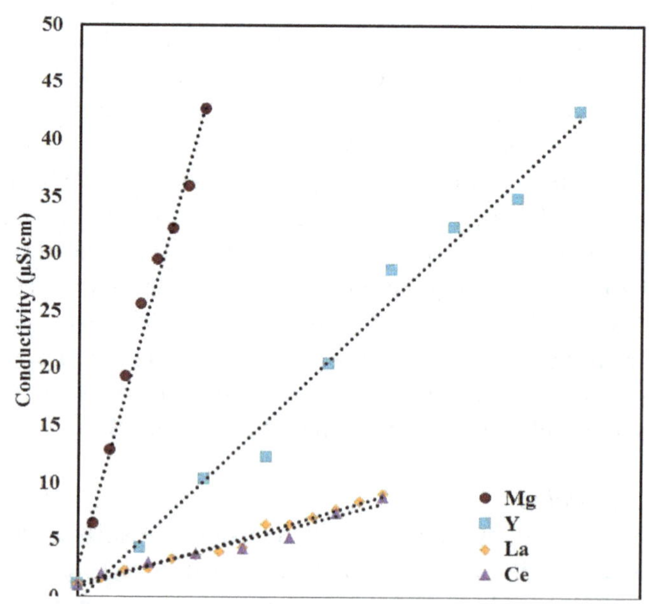

Fig. 2. Conductivity of ethanol by adding cations separately. Lanthanum concentration (♦), Yttrium concentration (■), Cerium concentration (▲) and Magnesium concentration (●).

This mechanism resulted the negative charged suspended alumina particles and promote the anodic deposition. On the other hand, the pH in all suspensions was approximately 5.8 ± 1.5, which is below the alumina IEP (isoelectric point) in ethanol [19] indicating that the particles were negatively charged in presence of iodine and proved anodic deposition. Another proof of the adsorption of cations on the surface of the suspended particles is discussed in below by chemical analysis results.

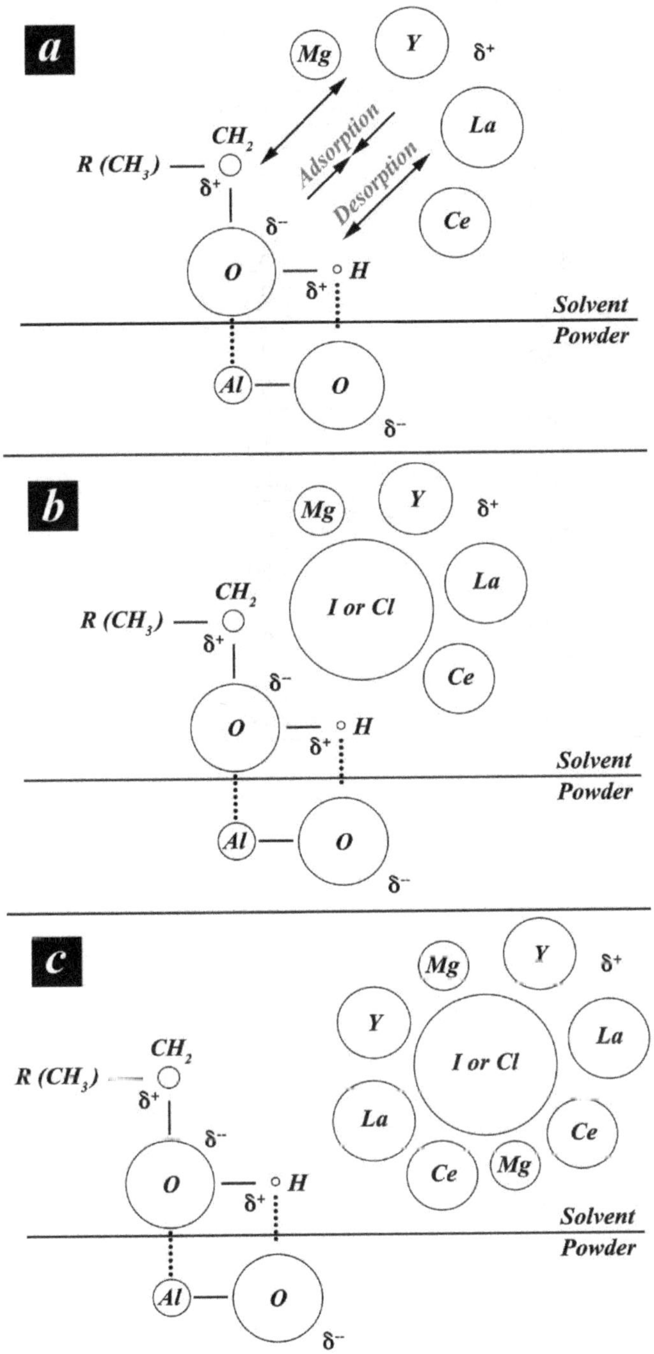

Fig. 3. A schematic representation of the alumina surface in ethanol in present of iodine anions and solved cation salts. Iodine concentration a) < critical concentration, b) = critical concentration and c) > critical concentration.

3.3.ANOVA Analysis

In the earlier work, we found that charged nano alumina particles by positively ions reached toward the electrode successfully [20]. In this investigation the EPD of nano alpha alumina using two level factorial design has been studied. The properties of the suspensions that were prepared with various conditions are summarized in Table 4. Design consisting of 4 center points and 16 axial points that rendered a total of 20 runs of experiment. In this study, the response variables measured were deposit weight (g) and viscosity (mPa.s), pH and conductivity (μS/cm) of suspension. All suspensions were stable in 3 h after preparation and no sedimentation occurred during this time. Low viscosity of the suspensions confirm this stability.Two level factorial design is considered as a useful method when the response is influenced by several variables. In addition to a useful mathematical and statistical technique for modeling and assessment of effects of factors at different levels and interactions between factors. This technique is applied in an experimental model to provide mathematical and statistical technique for modeling and analysis of engineering problems.

The viscosities of the suspensions were increased from 2.50 mPa.s to 6.10 mPa.s at shear rate of 120 s^{-1} (Table 4) and these increase were not associated with the increasing or decreasing of ionic concentrations. Figure 4 shows half normal probability plot of viscosity. Since the signs of the estimated effects are arbitrary, there are recommended the use of the half-normal probability plot of effects over the normal probability plot. In half normal plots the estimated effect of a factor is generally unimportant on or near a line close to zero, while the estimated effect of an important factor will generally move well out of line. According to Figure 4, all two interaction parameters have significant effect on viscosity without interaction between Mg^{2+} and Ce^{4+} concentrations. All primary factors except Mg^{2+} concentration categorized in not significant effects. Mg^{2+} has lower cationic radius and adsorb on surface of nano alumina powders and produce a high repulsive energy between them, but other cations with higher cationic radius have lower charge density and lower repulsive energy.

Figure 5 shows the Pareto plot for viscosity.

Table 4.
Experimental results (according to sample code order)

Code	pH	Conductivity (μS/cm)	Deposit weight (g)	Viscosity (mPa.s)
41110	5.70	3.9	4.40	2.50
41133	5.66	5.4	4.00	5.98
41413	7.32	4.6	4.11	2.90
41430	6.42	3.9	5.76	3.05
43113	6.35	14.9	4.15	3.92
43130	6.13	6.5	5.26	3.32
43410	5.11	12.5	4.60	3.12
623211	6.14	5.5	3.65	5.42
623212	6.10	5.5	3.87	6.04
623213	5.90	5.2	3.45	5.87
623214	6.18	5.1	3.49	6.10
81113	5.04	6.0	3.39	2.52
81130	5.69	7.5	3.72	3.70
81410	6.12	10.7	3.06	2.90
81433	5.70	7.5	3.26	3.50
83110	5.27	13.2	4.72	3.37
83133	5.77	15.2	4.32	2.87
83413	4.60	15.1	4.20	3.15
83430	5.30	19.0	4.15	3.25
83433	4.80	10.6	4.13	3.35

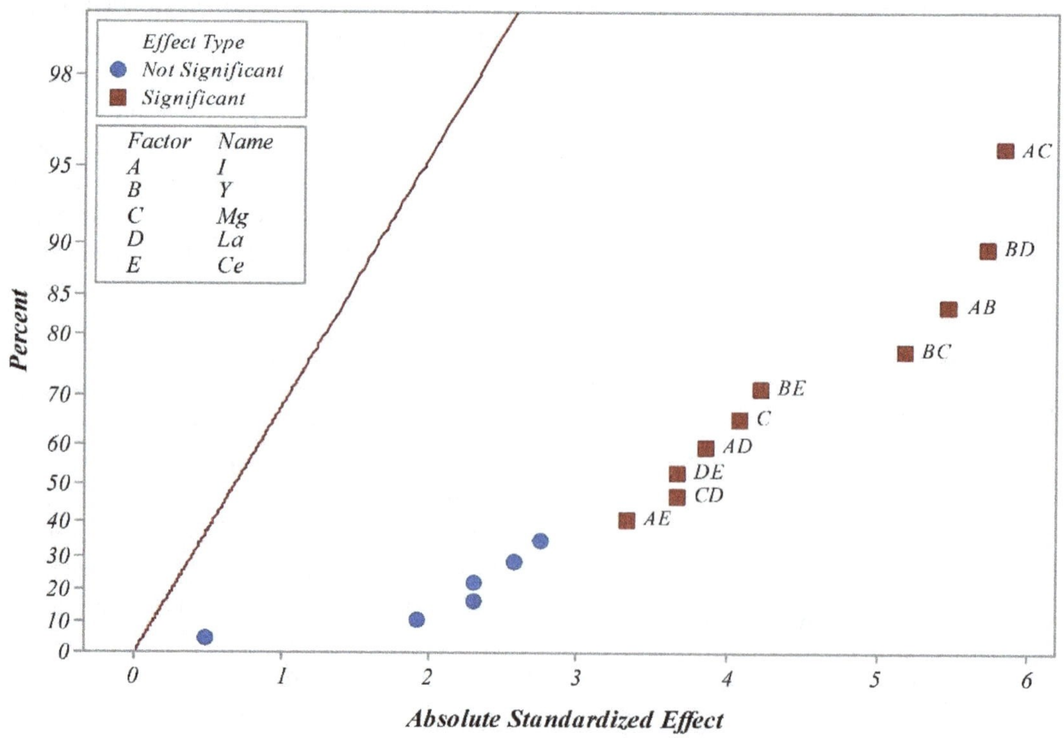

Fig. 4. Half normal plot of the standardized effects for viscosity, α = 0.05.

Pareto plot displays the absolute value of the effects and draws a reference line on the chart (3.182 in Figure 5). Any effect that goes beyond the reference line is potentially important. As before all two interaction factors except Mg × Ce and none of primary factors except Mg^{2+} concentration have significant effect. This result confirmed from the nonlinear 2D contour plots. Figure 6 shows a graph with graduated colors, hard blue for lower desirability and green for higher one. These 2D contour plots highlight the effect of additives concentrations on viscosity. It is clear that at higher iodine concentration, cation concentration played a major role in increasing the viscosity for Mg^{2+}, Y^{3+} and La^{3+} cations. In the case of Ce^{4+} at high iodine concentrations, viscosity cannot be influenced by Ce^{4+} concentration. In low iodine concentration viscosity decrease in low Mg^{2+} concentration and high Y^{3+}, La^{3+} and Ce^{4+} concentration. This is well in agreement with the mechanism will be discussed. When iodine concentration is higher than that of critical concentration viscosity increased by increasing in Y^{3+} and La^{3+} cations concentrations but decreasing in Mg^{2+} cation concentration. When Y^{3+} and La^{3+} cations that are larger cations add to suspension further iodine and these cations electrostatically adsorb on each other and increase viscosity. Cationic charge density can be explained this behaviour.

In iodine concentration lower than the critical concentration the surface of alumina nano powders cannot be completely covered by ions and enough repulsive forces cannot be produced to make stable suspension and reduce viscosity. Electrical conductivity of suspension confirm this mechanism.

Since the response function is linear, the linear model is employed. Equation 1 demonstrates the mathematical expression of the model.

$$Y = b_0 + \sum_{i=1}^{k} b_i X_i + \sum_{i=1}^{k}\sum_{j=1}^{k} b_{ij} X_i X_j + e \qquad (6)$$

where "i" and "j" determines the linear coefficients "b" is the regression coefficient, "k" is the number of experimental factors, and "e" is the random error.

The stability of the regression model is also estimated by an analysis of variance (ANOVA). The results shown in Table 5. In the first step, the probability of significance is determined. The effect of the independent variable is significant if the probability of significance (p) value is equal to or less than the selected alpha-level (here 0.05), and the insignificant variables are those with p values greater than the selected α-level [14].According to these, all two way interactions with the p value < 0.05 are significant except $Mg^{2+} \times Ce^{4+}$ cations concentrations with p value = 0.15. The model p value of 0.023 reveals that the model is still significant.

According to ANOVA model all interactions between iodine and cations i.e. $I \times M^{x+}$ are significant and showed that critical concentration of iodine is necessary to adsorb cations on alumina nano powder surface and reduced viscosity by repulsive electrostatic forces.

The final improved empirical models in terms of actual factors for viscosity is given in Eq. 7. Coefficients of individual factors and their interactions in the regression model shows how the response changes with respect to the interception.

$$\begin{aligned} \text{Viscosity} &= 3.155 + 0.1331\,I + 0.1419\,Y - 0.2106\,Mg - \\ &0.0244\,La - 0.1194\,Ce + 0.2831\,I{\times}Y - \\ &0.3019\,I{\times}Mg + 0.1994\,I{\times}La + 0.1719\,I{\times}Ce - \\ &0.2681\,Y{\times}Mg + 0.2956\,Y{\times}La + 0.2181\,Y{\times}Ce - 0.1894\,Mg{\times}La - \\ &0.0994\,Mg{\times}Ce + 0.1894\,La{\times}Ce - 0.267\,I{\times}Y{\times}Mg{\times}La{\times}Ce \end{aligned} \qquad (7)$$

The model summary statistics for viscosity (mPa.s) are given in Table 6. The value of R^2 is 98.76%, which is very close to 1. Likewise, the adjusted R^2 value are in close agreement.

Relation between viscosity and additive cations concentrations showed by Matrix plot in Figure 7. It is obvious that all sections have semi linear profile and indicates that by increasing the concentration span in this EPD process, the viscosity decreased, although the effect of the La^{3+} and Ce^{3+} cations concentrations are more significant than that of the Iodine, Mg^{2+} or Y^{3+} cations concentrations. Figure 7 also showed curvilinear relation between additive concentration and deposited weight. Maximum deposited weight with a simultaneous relatively minimum viscosity is considered as desirable conditions. The optimal setting suggests that these outcomes can be achieved by varying the all concentrations to zero level. Deposit weight had a lower, average and upper value of 3.06, 4.08 and 6.04 g respectively (Table 4). If the values of deposit weight are divided in to 3 range, half of samples had weight between 4.06 to 5.06 g. Analysis indicated that the 5-way interaction have a higher effect on the deposition weight but is not a significant parameter (Figure 8a). According to Hamaker model deposit weight had direct relationship with electrophoretic mobility that increased when viscosity decreased or zeta potential increased [2]. Although deposition weight increased with a decrease in viscosity, there are not a linear relation between them, considering in some cases zeta potential increased although viscosity increased.

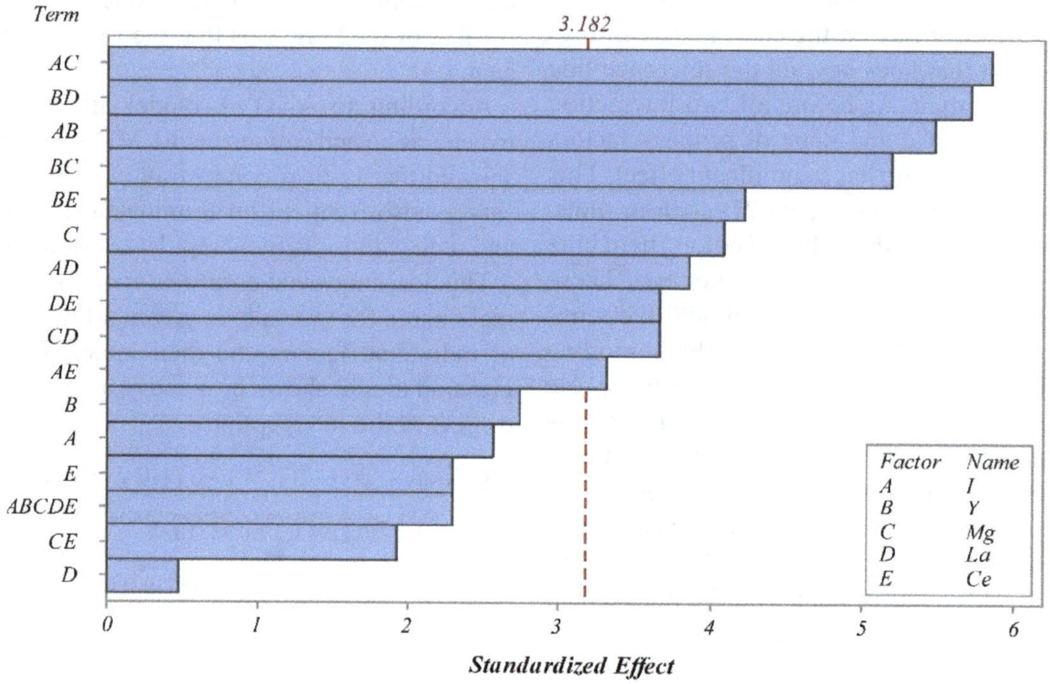

Fig. 5. Pareto plat of the standardized effects for Viscosity, α = 0.05.

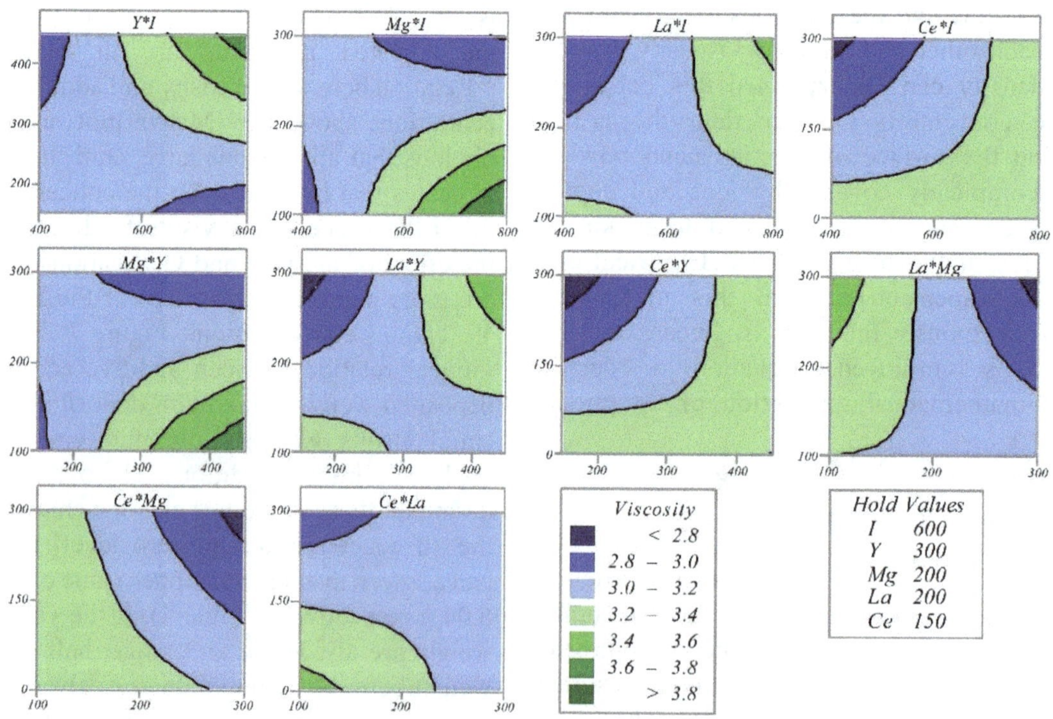

Fig. 6 . Contour plot of viscosity.

Table 5.
Analysis of Variance for viscosity model

Source	Degree of freedom	Adj SS	Adj MS	F-Value	P-Value
Model	16	10.2456	0.64035	14.97	0.023
Linear	5	1.5529	0.31059	7.26	0.067
I	1	0.2836	0.28356	6.63	0.082
Y	1	0.3221	0.32206	7.53	0.071
Mg	1	0.7098	0.70981	16.6	0.027
La	1	0.0095	0.00951	0.22	0.669
Ce	1	0.228	0.22801	5.33	0.104
2-Way Interactions	10	8.4647	0.84647	19.79	0.016
I×Y	1	1.2826	1.28256	29.99	0.012
I×Mg	1	1.4581	1.45806	34.09	0.01
I×La	1	0.636	0.63601	14.87	0.031
I×Ce	1	0.4727	0.47266	11.05	0.045
Y×Mg	1	1.1503	1.15026	26.9	0.014
Y×La	1	1.3983	1.39831	32.7	0.011
Y×Ce	1	0.7613	0.76126	17.8	0.024
Mg×La	1	0.5738	0.57381	13.42	0.035
Mg×Ce	1	0.158	0.15801	3.69	0.15
La×Ce	1	0.5738	0.57381	13.42	0.035
5-Way Interactions	1	0.2279	0.22791	5.33	0.104
I×Y×Mg×La×Ce	1	0.2279	0.22791	5.33	0.104
Error	3	0.1283	0.04277		
Total	19	10.3739			

Table 6.
Viscosity model summery

S	R-sq	R-sq(adj)	R-sq(pred)
0.206801	98.76%	92.17%	*

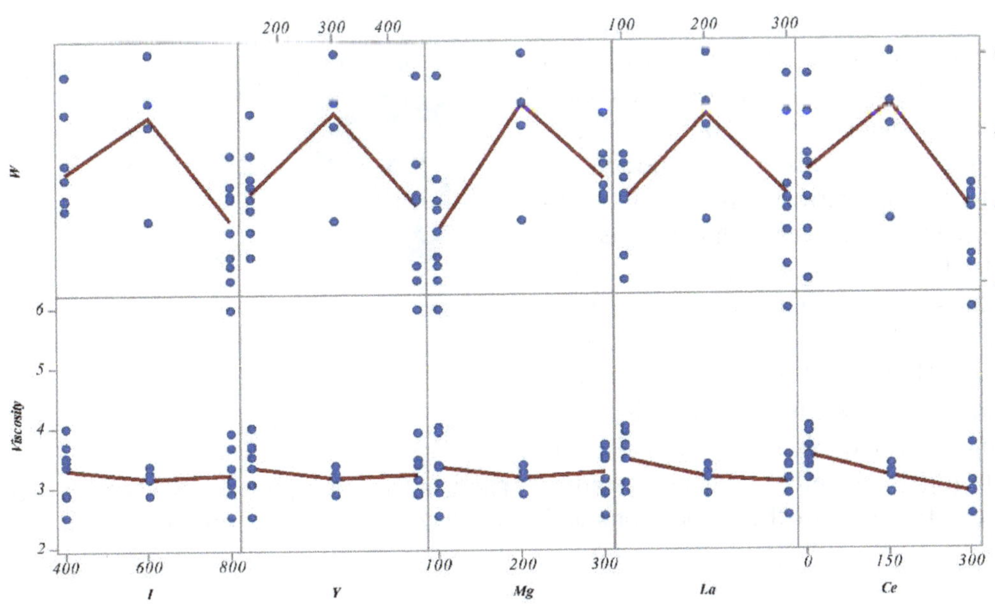

Fig. 7. Matrix plots of viscosity (mPa.s) and deposition weight (g) vs. additive concentrations (ppm).

According to observations of these research 2-way interaction between Mg^{2+} and Y^{3+} cations concentrations have significant effect on pH of suspension (Figure 8b). After that 5-way interaction have most effect (not-significant). pH of suspensions according to additive concentrations showed in Figure 9. pH changes have not a significant trend. On the other hand according to Figure 9 conductivity increased by increasing in I, Mg^{2+} and Y^{3+} cations concentrations. La^{3+} and Ce^{3+} cations concentrations have not significant effect. To summarize results between viscosity, deposited weight, pH and conductivity, Figure 10 exhibited Matrix plots of there. It seems deposit weight has been influenced by pH and conductivity although there are not a clearly relationship. Maximum deposited weight can be achieved in pH around 6 and electrical conductivity around 5 μS/cm. There is an important note that the yield obtained from the different suspensions with equal iodine/nano powder ratio can be attributed to what cations (Mg^{2+}, Y^{3+}, La^{3+} and Ce^{3+} cations concentrations in this work) change electrical conductivities and pH of the suspensions as a result of the different concentrations of added cation salts in the suspensions rather than to the different in amount of conductivities and pH values of the suspensions.

The pH in all suspensions was approximately 5.8 ± 1.5, which was below the all reported for alumina IEP in ethanol [15,21,19], indicating that the particles were positively charged [20]. The dissociation of iodine resulted the negative ions, which then would reverse the charge of the suspended alumina nano powder and promote the anodic deposition. The effect of additive concentration on electrical conductivity of suspension were also investigated from the developed mathematical model. Figure 8c show Pareto plots of these model. According to Figure 8c Mg^{2+} cation concentration has significant effect on the electrical conductivity. Consequently, the ionic load increased with the increasing concentration of additives. The higher concentration of additives in the suspension implied a higher fraction of ions and thus there expected the suspension possessed a lower pH (for Cl^-) and higher electrical conductivity. But it is not a linear relation between that. The experimental values obtained from ICP measurement were in good agreement with the calculated values in preparation step of the suspensions and No Fe, Cr or Ti and V (arising from the substrate or counter electrode) was detected in the deposited layer.

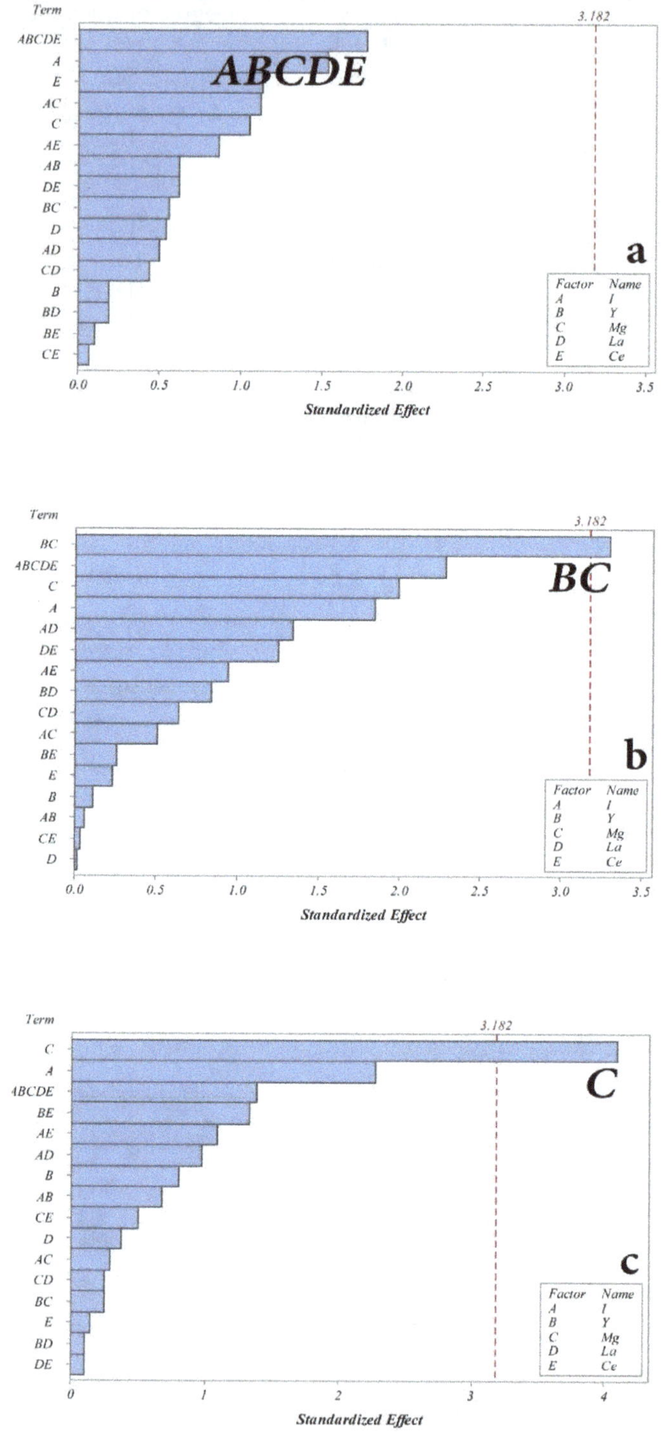

Fig. 8 . Pareto plat of the standardized effects for deposition weight (a), pH (b), conductivity (c). α = 0.05.

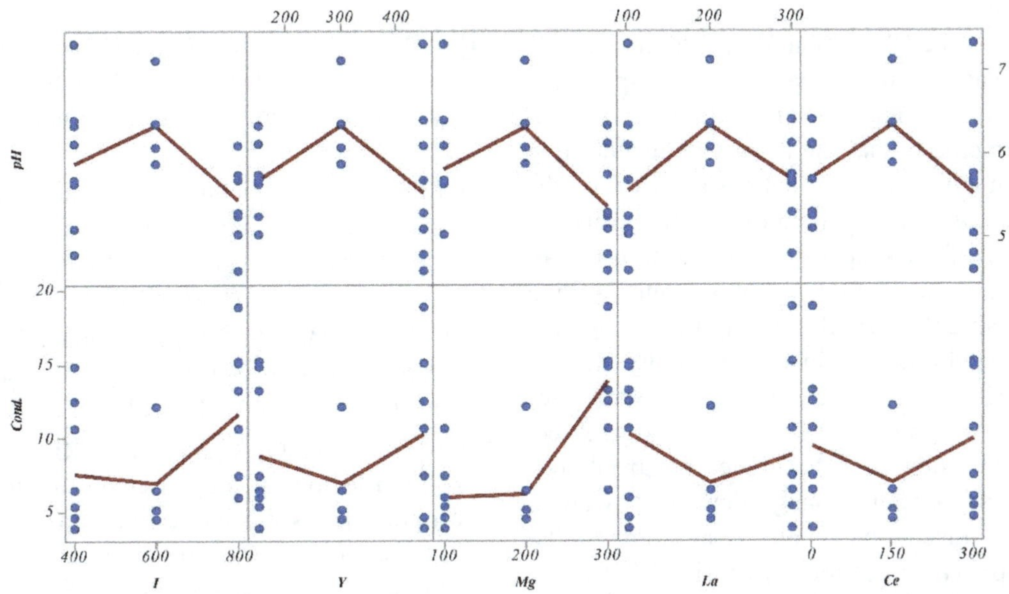

Fig. 9. Matrix plots of conductivity ($\mu S.cm^{-1}$) and pH vs. additive concentrations (ppm).

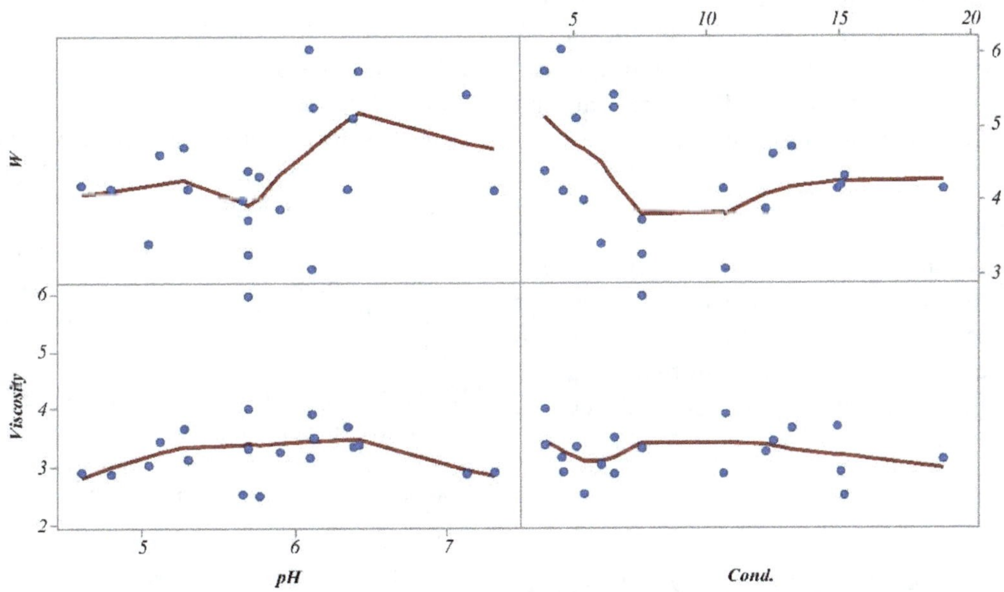

Fig. 10. Matrix plots of deposition weight (g) and viscosity (mPa.s) vs. pH and conductivity ($\mu S.cm^{-1}$).

4. Conclusions

This work described the preparation of homogeneous doped alumina nano powder ethanol-based suspensions using the cation salts additives and iodine. It was confirmed experimentally that the additive concentration is an important key parameter that has an effect on the viscosity of the suspension. According to this work the lower viscosity of the suspensions in addition to cation salts might be expected in optimum concentration which may enable the particles to arrange themselves at more optimized positions in suspension by electrostatic forces. This mechanism could be represented for higher deposited weight. When viscosity decreased, optimized suspension lead to higher electrophoretic mobility and simultaneously deposited weight.

Based on the discussion of electrical conductivity, viscosity and pH value distributions during EPD, at the initial stage of EPD, charged particles move toward anode by electrophoretic motion and M^{+x} ions that must be move toward a cathode deposited with them in anode.

Conductivity increased by increasing in I, Mg^{2+} and Y^{3+} cations concentrations. La^{3+} and Ce^{3+} cations concentrations have not significant effect and Mg^{2+} cation concentration has significant effect on the electrical conductivity. Maximum deposited weight can be achieved in pH around 6 and electrical conductivity around 5 μS/cm.

Acknowledgments

The authors would like to acknowledge the Iranian Research Organization for Science and Technology (IROST) for providing research facilities. This research did not receive any specific grant from funding agencies in the public, commercial, or not-for-profit sectors.

References

[1] S. Dey, S. Bhattacharjee, T.K. Rout, D.K. Sengupta, T. Uchikoshi, L. Besra, Effect of Electrode Reactions during Aqueous Electrophoretic Deposition on Bulk Suspension Properties and Deposition Quality, Key Eng. Mater. 654 (2015) 3–9.

[2] H.C. Hamaker, Formation of a deposit by electrophoresis, Trans. Faraday Soc. 35 (1940) 279–287.

[3] R. Moreno, B. Ferrari, Nanoparticles Dispersion and the Effect of Related Parameters in the EPD Kinetics, in: J.H. Dickerson, A.R. Boccaccini (Eds.), Electrophor. Depos. Nanomater., Springer, New York, 2012, pp. 73–128.

[4] L. Besra, P. Samantaray, S. Bhattacharjee, B.P. Singh, Electrophoretic deposition of alumina on stainless steel from non-aqueous suspension, J. Mater. Sci. 42 (2007) 5714–5721.

[5] T.S. Suzuki, T. Uchikoshi, Y. Sakka, Effect of sintering additive on crystallographic orientation in AlN prepared by slip casting in a strong magnetic field, J. Eur. Ceram. Soc. 29 (2009) 2627–2633.

[6] L. Jin, X. Mao, S. Wang, M. Dong, Optimization of the rheological properties of yttria suspensions, Ceram. Int. 35 (2009) 925–927.

[7] A. Tsetsekou, C. Agrafiotis, A. Milias, Optimization of the rheological properties of alumina slurries for ceramic processing applications Part I: Slip-casting, J. Eur. Ceram. Soc. 21 (2001) 363–373.

[8] M. Stuer, Z. Zhao, P. Bowen, Freeze granulation: Powder processing for transparent alumina applications, J. Eur. Ceram. Soc. 32 (2012) 2899–2908.

[9] S. Heydarian, Z. Ranjbar, S. Rastegar, Electrophoretic Deposition Behavior of Chitosan Biopolymer as a Function of Solvent Type, Polym.-Plast. Technol. Eng. 54 (2015) 1193–1200.

[10] A.M. Popa, J. Vleugels, J. Vermant, O. Van der Biest, Influence of surfactant addition sequence on the suspension properties and electrophoretic deposition behaviour of alumina and zirconia, J. Eur. Ceram. Soc. 26 (2006) 933–939.

[11] Z. Zhang, Y. Huang, Z. Jiang, Electrophoretic Deposition Forming of SiC-TZP Composites in a Nonaqueous Sol Media, J. Am. Ceram. Soc. 77 (1994) 1946–1949.

[12] M. Stuer, P. Bowen, Yield stress modelling of doped alumina suspensions for applications in freeze granulation: towards dry pressed transparent ceramics, Adv. Appl. Ceram. 111 (2012) 254–261.

[13] P. Biswas, M. Kiran Kumar, K. Rajeswari, R. Johnson, U.S. Hareesh, Transparent sub-micrometre alumina from lanthanum oxide doped common grade alumina powder, Ceram. Int. 39 (2013) 9415–9419.

[14] D.C. Montgomery, Design and Analysis of Experiments, 8th ed., John Wiley & Sons, Inc., New York, 2001.

[15] G. Wang, P. Sarkar, P.S. Nicholson, Influence of Acidity on the Electrostatic Stability of Alumina Suspensions in Ethanol, J. Am. Ceram. Soc. 80 (2005) 965–972.

[16] B. Workie, B.E. McCandless, Z. Gebeyehu, Electrophoretic Deposition of Aluminum Nitride from Its

Suspension in Acetylacetone Using Iodine as an Additive, J. Chem. 2013 (2013) 1–7.

[17] H. Idriss, Ethanol Reactions over the Surfaces of Noble Metal/Cerium Oxide Catalysts, Platin. Met. Rev. 48 (2004) 105–115.

[18] P. Kostetskyy, G. Mpourmpakis, Structure-activity relationships in the production of olefins from alcohols and ethers: a first-principles theoretical study, Catal. Sci. Technol. 5 (2015) 4547–4555.

[19] S. Novak, K. König, Fabrication of alumina parts by electrophoretic deposition from ethanol and aqueous suspensions, Ceram. Int. 35 (2009) 2823–2829.

[20] M. Milani, S.M. Zahraee, S.M. Mirkazemi, Influence of electrophoretic deposition parameters on pore size distribution of doped nano alumina plates, Ceram. Silik. 60 (2016) 299–307.

[21] L. Besra, M. Liu, A review on fundamentals and applications of electrophoretic deposition (EPD), Prog. Mater. Sci. 52 (2007) 1–61.

4

Synthesis of pure monoclinic zirconia nanoparticles using ultrasound cavitation technique

Maryam Ranjbar[1*], Mostafa Yousefi[1,2], Mahboobe Lahooti [3], S. Heydar Mahmoudi Najafi[1], Azim Malekzadeh[3]

[1] *Department of Chemical Technologies, Iranian Research Organization for Science and Technology (IROST), Tehran, Iran*

[2] *National Iranian Oil Products Distribution Company (NIOPDC), Zahedan Region, Zahedan, Iran*

[3] *Department of Chemistry, Faculty of Sciences, Damghan University, Damghan, Iran*

HIGHLIGHTS

- The sonochemical synthesis and characterization of a nano-sized zirconium(IV)-minoxidil complex have been investigated in two different solvents.
- The pure monoclinic zirconia nanoparticles were readily synthesized from thermal decomposition of the complex as a precursor in both solvents.
- The average particles size of zirconia were significantly reduced when monoethylene glycol was used as solvent.
- The sonochemical method can be successfully applied as an efficient method for the preparation of nano-scale materials.

GRAPHICAL ABSTRACT

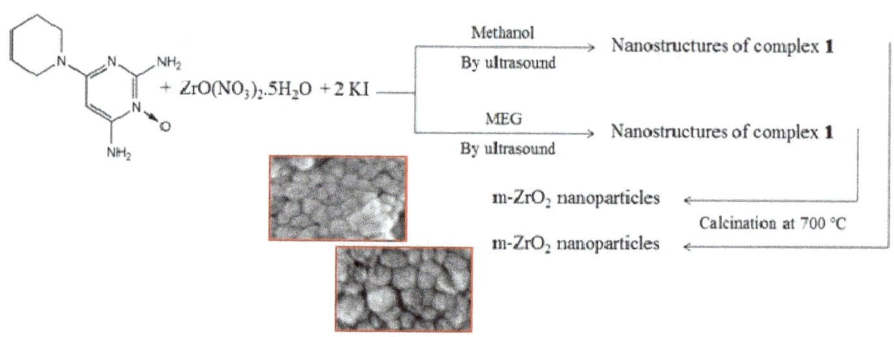

ARTICLE INFO

Keywords:
Zirconia nanoparticles
Monoclinic
Zr(IV) complex
Sonochemical method
Thermal decomposition

ABSTRACT

In the current study, synthesis and characterization of a new nano-structured zirconium(IV)-minoxidil complex (1), where minoxidil, ($C_9H_{15}N_5O$; 6-(1-Piperidinyl)-2,4-pyrimidinediamine 3-oxide), have been investigated in two different solvents. The compound 1 has been synthesized by sonochemical method in the presence of methanol and monoethylene glycol (MEG) as solvents and characterized by scanning electron microscopy (SEM), Fourier transform infrared (FT-IR) spectroscopy and elemental analysis. The thermal stability of the compound 1 has been studied by thermal gravimetric (TG) and differential thermal analyses (DTA). Pure monoclinic (m) zirconia (ZrO_2) nanoparticles were readily synthesized from thermal decomposition of the compound 1 as a new precursor in both solvents. The products were characterized by FT-IR, XRD, and SEM to exhibit the phase and morphology. The results showed that, pure zirconia was produced with particle size of 53 nm and the crystal system was monoclinic when methanol was used as solvent in complexation process. While the particles size of zirconia with the same structure were significantly reduced to 25 nm, using MEG as solvent. This study demonstrates that the coordination compounds may be suitable precursors for the simple one-pot preparation of nano-scale metal oxides with different morphologies.

* *Corresponding author.*
E-mail address: marandjbar@irost.ir

1. Introduction

Macroscopic properties of materials strongly depend on both the size and morphologies of the microscopic particles they are made up from. This is especially true for materials with morphological features smaller than a micron in at least one dimension, which are commonly called nano-scale materials, or simply nanomaterials [1]. During the last two decades, the rational design and syntheses of novel coordination supramolecular compounds, which involves self-assembly of organic ligands with appropriate functional groups and metal ions with specific directionality and functionality, have made considerable progress in the field of supramolecular chemistry and crystal engineering [2–4]. The considerable interest is driven by the impact on basic structural chemistry as well as by possible applications in a number of fields such as catalysis, molecular adsorption, luminescence, magnetism, nonlinear optics, and molecular sensing that are not found in mononuclear compounds [5]. By decreasing the size of coordination supramolecular compounds as polymers in nano-size, surface area would be increased. Hence making coordination supramolecular compounds in any form in nano-scale is certainly a major step forward toward the technological applications of these new materials [6]. Due to their possessing combination of unique properties such as excellent refractoriness and chemical resistance, good mechanical strength, high ionic conductivity, low thermal conductivity at high temperature along with relatively high thermal expansion coefficient and good thermal stability, ultrafine zirconia particles have attracted much interest recently [7]. Based on the above mentioned properties, they have found a broad industrial applications including: fuel-cell technology [8], as a catalyst or catalyst support [9], oxygen sensor [10], protective coating for optical mirrors and filters [11], nanoelectronic devices, thermal-barrier coatings [12], ceramic biomaterial [13], and thermo luminescence UV dosimeter [14]. Zirconia exhibits several crystalline modifications at different temperatures: monoclinic, tetragonal, and cubic. At very high temperatures (>2370 °C) the material has a cubic structure. At intermediate temperatures (1150–2370 °C) it has a tetragonal structure. At lower temperatures (below 1150 °C) the material transforms to the monoclinic structure which is a thermodynamically stable phase [15].

A variety of zirconium complexes have been prepared successfully and modified [16,17]. These precursors could be good candidates for the synthesis of cursors could be good candidates for the synthesis of ZrO_2 nanoparticles with different morphologies. So far, few studies has been taken place on using Zr(IV) supramolecules as precursor. We have been investigated the synthesis of monoclinic and tetragonal ZrO_2 nanoparticles from thermal decomposition of the $[ZrO(dmph)I_2]$, as a new precursor in methanol and monoethylene glycol as solvents, respectively [17]. Also, our group worked on the synthesis of pure tetragonal ZrO_2 nanoparticles from isophthalic acid-zirconium(IV) complex by thermal decomposition at 700 °C for 4 h [18]. Salavati-Niasari et al. have been reported the synthesis of single-phase cubic ZrO_2 nanoparticles from bis-aqua tris-salicylaldehydato zirconium(IV) nitrate; $[Zr(sal)_3(H_2O)_2](NO_3)$, as the new precursor [19]. This paper describes a facile and environmentally friendly synthetic sonochemical preparation of nano-sized Zr(IV) complex in the presence of methanol and separately in MEG as solvents, and its use as a new precursor for the preparation of pure m-ZrO_2 nanoparticles. To the best of our knowledge, this is the first report on the synthesis of a Zr(IV) complex with minoxidil as a ligand via sonochemical method and production of ZrO_2 nanoparticles with the same crystal systems by Zr(IV) complex as precursor. The results revealed that in comparison with traditional synthetic techniques, such as solvent diffusion, hydrothermal and solvothermal methods, the ultrasonic synthesis is a simple, efficient, low cost, and environmentally friendly approach to nano-scale coordination supramolecular compounds [20]. Sonochemical method can lead to homogeneous nucleation and a substantial reduction in crystallization time compared with conventional oven heating when nanomaterials are prepared [21]. These extreme conditions permit access to a range of chemical reaction spaces normally not accessible, for example many coordination supramolecular compounds have been synthesized via the sonochemical method, such as: La(III)-LH_2 [22], Zr(IV)-Neocuproine [17], Cu(I)-Neocuproine [1], Zr(IV)-LH_2 [23], Zr(IV)-Isophthalic acid [18], Cd(II)-Thioacetamide [24], Pb(II)-Minoxidil [25], Zn(II)-LH2 [26], and Pb(II)-LH_2 [27].

Minoxidil, ($C_9H_{15}N_5O$; 6-(1-Piperidinyl)-2,4-pyrimidinediamine 3-oxide), has been exploited as an anti-alopecia agent since the late 1980s. It has become one of the most popular anti-alopecia medicaments in the world with a market exceeding $100 million per year in the 1980s–1990s in the United States alone. In 1995 the world market was $125 million [28]. A review of the literature shows that very few metal complexes of minoxidil have been reported. After making a coordination polymer of minoxidil and Pb(II) in presence of KI [29], it was considered interesting to investigate the influence of the bromide counter ion (Br^-) in the polymerization process due to the expected structural diversity [25].

2. Experimental

2.1. Materials and physical measurements

All reagents and solvents were purchased from Merck Chemical Co. and used without further purification. Melting points were measured with a Branstead Electrothermal 9100 apparatus. FT-IR spectra were recorded on a Bruker tensor 27 spectrophotometer in the range 400–4000 cm^{-1} using the KBr disk technique. Elemental analyses (carbon, hydrogen, and nitrogen) were performed using a Heraeus CHN-O- Rapid analyzer. A multiwave ultrasonic generator (UP400S, Hielscher), equipped with a converter/transducer and titanium oscillator (horn), 12.5 mm in diameter, operating at 20 KHz with a maximum power output of 600 W, was used for ultrasonic irradiation. The ultrasonic generator automatically adjusts the power level. Thermogravimetric analysis (TGA) and differential thermal analyses (DTA) of the title compound were performed on a computer-controlled PL-STA 1500 apparatus. Powder sample of 1 was loaded into alumina pans and heated with a ramp rate of 10 °C/min from room temperature to 600 °C under argon atmosphere. X-ray powder diffraction (XRD) measurements were performed using an X'pert diffractometer of Philips Company with graphite monochromatic Cu Kα (λ= 1.5418 Å) radiation at room temperature in the 2 θ range of 20-90 °. The samples were characterized by a scanning electron microscope (Philips XL30) with gold coating.

2.2. Synthesis of nano-structured zirconium(IV)-minoxidil complex (1) by the sonochemical method

0.23 g (1 mmol) of zirconyl nitrate pentahydrate, $ZrO(NO_3)_2.5H_2O$, and 0.34 g (2 mmol) of potassium iodide (KI) were dissolved in 10 ml methanol. Then, 0.21 g (1 mmol) of minoxidil was dissolved in 10 ml methanol and both of solutions were mixed together and sonicated for 1 hour in an ultrasound vessel to obtain a homogeneous mixture with the rated output power of 600 W and frequency of 25 KHz. After 1 hour a bright yellow precipitate was formed. It was isolated by centrifugation (4000 rpm, 15 min), washed with water and acetone to remove residual impurities and finally dried in air (yield: 0.51 g, 65%). dec. p. ~280°C. Elemental analysis, found (%): Zr, 0.68; C, 13.91; H, 2.75; N, 11.48. FT-IR (cm-1) selected bands observed were: 3319(b), 2933(w), 2852(w), 1637(b), 1383(s), 1279(w), 1251(s), 1231(w), 1125(s), 1023(s), 981(s), 954(w), 825(w), 773(s), 670(w), and 507(s). The experiment was repeated by using MEG as the solvent. The resulting yellow powder of complex 1, (0.53 g) was obtained in 68% yield. dec. p. ~245°C. Elemental analysis, found (%): Zr, 0.56; C, 16.46; H, 4.20; N, 3.06. FT-IR (cm-1) selected bands observed were: 3470(b), 2940(w), 2375(w), 1625(w), 1540(s), 1375(w), 1340(s), 1280(w), 1140(w), 1070(s), 1035(w), 825(w), 810(w), 760(w), 655(w), 625(w), and 490(s).

2.3. Synthesis of ZrO_2 nanoparticles by thermal decomposition of compound 1

For synthesis of zirconia nanoparticles, the compound 1 were calcinated at 700 °C for 4 hour, under atmospheric air. The final products were collected and washed with water and acetone several times, dried in air, and kept for further characterization. The FT-IR spectra and powder XRD diffraction shows that the calcination was completed and the entire organic compound were decomposed. Also, the XRD patterns show the products using both precursors are m-ZrO_2.

3. Results and discussion

Reaction of minoxidil as ligand with zirconyl nitrate pentahydrate and potassium iodide under ultrasonic irradiation in methanol and separately in MEG as solvents, led to the formation of a nano-structured zirconium(IV)-minoxidil complex (1), in 65% and 68% yields, respectively. Scheme 1 gives an overview of the method used for the synthesis of compound 1 nanostructure. The FT-IR spectra for the free minoxidil ligand, Zr(IV) complex in methanol and monoethylene glycol and the as-prepared ZrO_2 nanoparticles in the frequency range from 400-4000 cm^{-1}, are compared in Figs. 1 and 2, respectively. By comparing the FT-IR spectra of minoxidil (Fig. 1a) and the Zr(IV) complex (Figs. 1b and 2a), it was found that minoxidil has been coordinated to Zr^{4+} ion, and forming new complex (1). The only difference among these characteristic peaks is either the peak intensity or a slight shift in the peak position. All distinct absorption peaks of minoxidil are illustrated in Table 1,

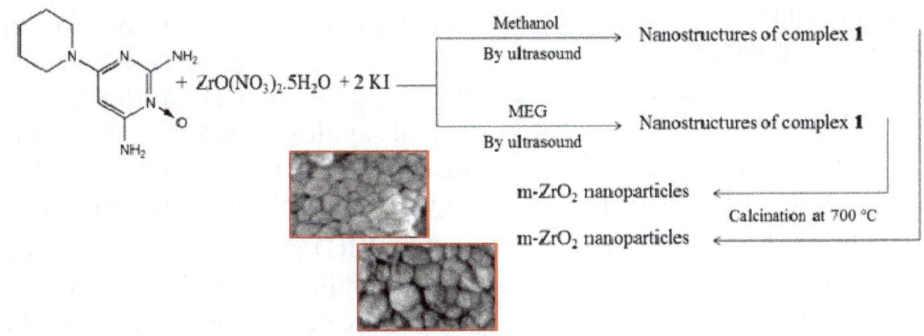

Scheme 1. Schematic diagram illustrating the formation of complex 1 and ZrO_2 nanopowders.

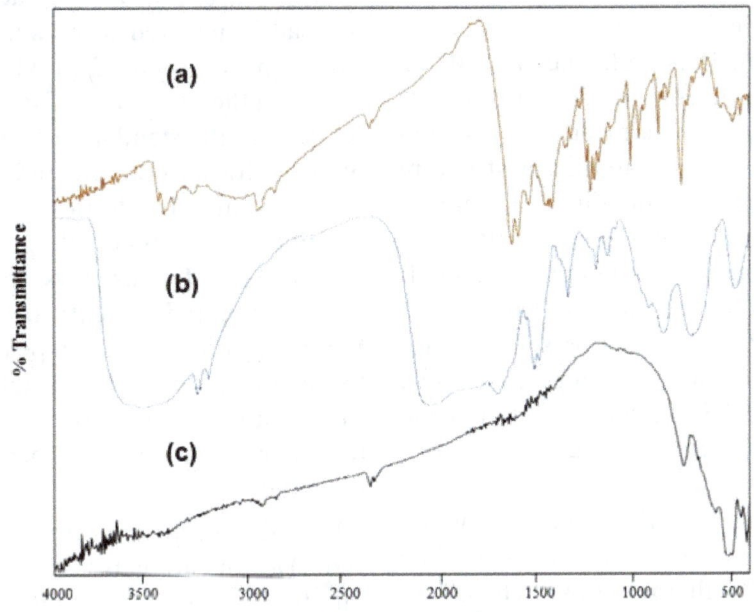

Fig. 1. FT-IR spectra of (a) minoxidil ligand, (b) nanoparticles of compound 1 synthesized by sonochemical method (using methanol), and (c) m ZrO_2 nanoparticles.

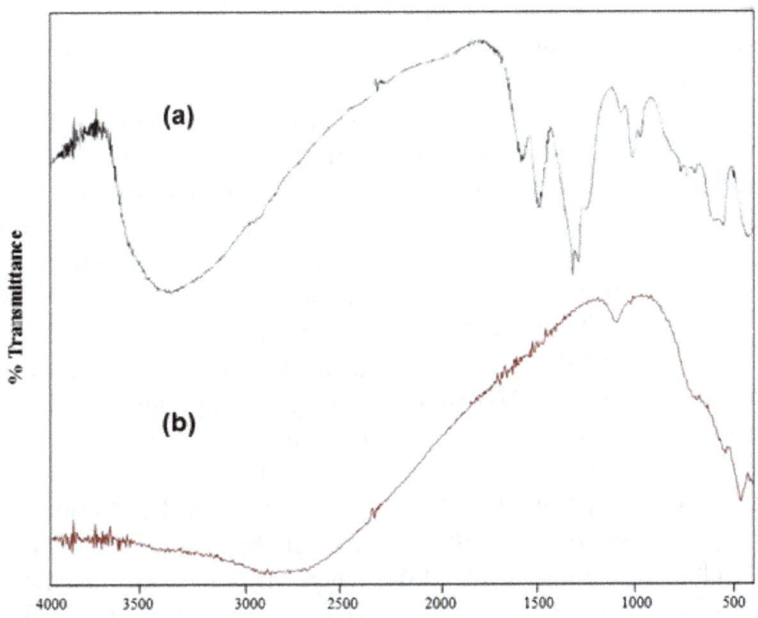

Fig. 2. FT-IR spectra of (a) nanoparticles of compound 1 synthesized by sonochemical method (using MEG), and (b) m-ZrO_2 nanoparticles.

Table 1.
FT-IR absorption peaks of minoxidil [30].

Wavenumber observed (cm^{-1})	Group identified
3432	N-H stretch
3289	H-bonded N-H
2950	Aromatic and aliphatic C-H stretch
1642	Aromatic C-H stretch
1560	Aromatic O-C stretch
1375	N-H bending
1232	Aromatic C-N stretch

and every absorption peak was assigned to corresponding vibration [30]. It is well known that the FT-IR spectra are very useful in determining the crystal phase for ZrO_2 [31]. The extended spectrum in the 400–4000 cm-1 region signified a structure with the surface of the sample containing H_2O molecules. As can be seen from Figs. 1c and 2b, all above bands disappeared when the compound 1 was calcinated. The band at 420-750 cm^{-1} is attributed to the vibration of Zr-O-Zr bond. The presence of adsorbed water molecule in zirconia samples is shown by the bands at around 2600-2000 and 1600-1000 cm-1 as stretching and bending frequencies, respectively.

To examine the thermal stability of the compound 1 nanostructure that obtained via sonochemical method (using methanol), thermal gravimetric (TG) and differential thermal analyses (DTA) were carried out between 20 and 600 °C in air (Fig. 3). As shown in Fig. 3, three weight loss steps were obtained in the TG curve. According to the TG curve, the first weight loss observed at 165 °C was attributed to removal of coordinated methanol, the second weight loss shown at 506 °C corresponds to removal of minoxidil and iodine atoms. At high temperature (up to 506 °C), the decomposition of the complex occurs to ultimately give solid that appears to be zirconia. Mass loss calculations of the end residue and the XRD pattern of the final decomposition product (Fig. 5a) show the formation of m-ZrO_2. There was no organic residue left after 506 °C, as confirmed by the FT-IR spectrum of the residual mass. The DTA curve displays two distinct exothermic peaks at 309 °C and 405 °C for the complex (Fig. 3).

The morphology and size of compound 1 nanostructure which prepared by sonochemical method in methanol and MEG were characterized by SEM and is shown in Fig. 4. Typical SEM images as shown in Figs. 4a and 4b, demonstrate nanoparticles with size about 66 nm and low uniformity. After using MEG as a solvent, the particles size decreased to about 29 nm (Figs. 4c and 4d). Furthermore, this micrograph reveals agglomerated particles with a homogeneous distribution. Also, quasi-spherical shaped morphology has been observed for the nanoparticles.

The XRD patterns shown in Figs. 5a and 5b are corresponding to the zirconia obtained by calcination of the compound 1 synthesized in methanol and separately in MEG at 700 °C for 4 hour, respectively. Fig. 5a shows all diffraction peaks which can be readily indexed to crystalline monoclinic phase ZrO_2 (space group P2/c) which is consistent to the values in the literature (JCPDS no. 001-0750). Compared with the standard diffraction pattern, no peaks of impurities were detected, indicating high purity of the product. In addition, the intense and sharp diffraction peaks suggest the crystallinity of the product. When MEG used as solvent, the same phase was formed under identical conditions, as shown in Fig. 5b. All the reflection peaks in this pattern could be readily indexed to crystalline monoclinic phase ZrO_2 with the same space group (JCPDS no. 007-0343), no characteristic peaks of impurities were detected.

In order to further investigation, the crystallite size (Dc) of products were calculated from the major diffraction peak of the corresponding zirconia using the Debye–Scherrer formula (1).

$$D_c = \frac{0.89\lambda}{\beta \cos\theta} \quad (1)$$

In Eq. (1), λ is the wavelength of X-ray (1.5418 Å) for Cu Kα radiation, β the full width at half maximum, FWHM of prominent intensity peak (using the 100% relative intensity peak), and θ the peak position. The average crystallite size (Dc) and lattice parameters for two samples are reported in Table 2. As seen from Table 2, the average size of zirconia prepared in MEG is smaller than the one obtained using methanol. Also, average size of the products is in good agreement with the SEM images. Figs. 5 (a,b) and (c,d) show the SEM photographs of ZrO_2 nanoparticles in methanol and MEG, respectively. It represents that the obtained nanoparticles have hexagonal shape with uniform dimensions. As a result, when MEG has been used as solvent, ZrO_2 particles sizes decreased of about 53 nm to 25 nm.

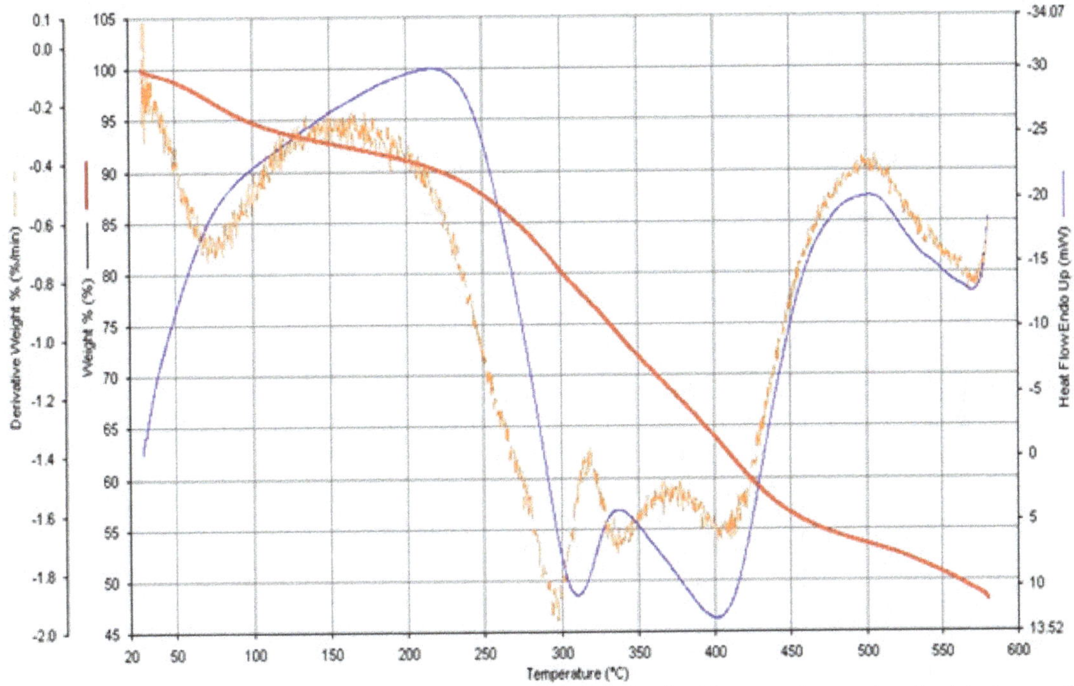

Fig. 3. TGA and DTA diagrams of the compound 1 nanoparticles synthesized by sonochemical method (using methanol).

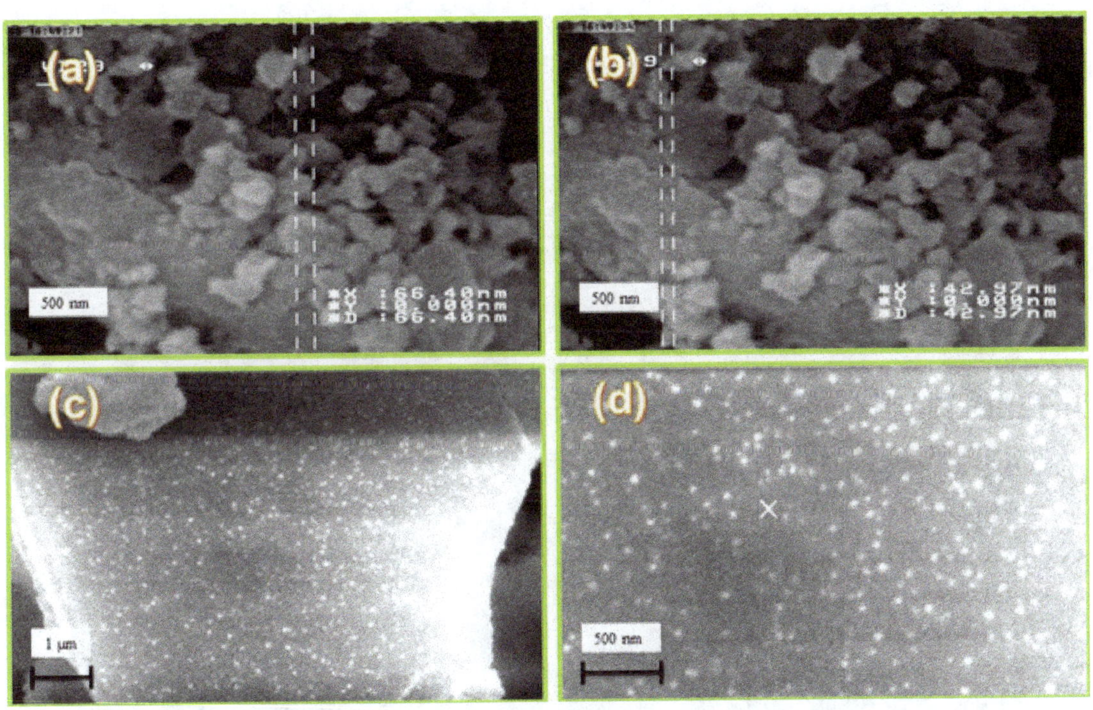

Fig. 4. SEM images of the compound 1 nanoparticles synthesized by sonochemical method in (a, b) methanol as solvent, and (c, d) MEG as solvent.

4. Conclusions

In this work we have successfully demonstrated the synthesis of nano-structured zirconium(IV)-minoxidil complex (1) in methanol and MEG as solvents by sonochemical method. To proceed, pure monoclinic zirconia nanoparticles were synthesized from calcination othe compound 1 at 700 °C. The XRD pattern indicates that the well-crystallized zirconia nanoparticles can be easily obtained under the current synthetic conditions. The results reveal that, when MEG was used as solvent, the produced m-ZrO_2 showed smaller particles size. In overall we can concluded that using nano-structured supramolecular

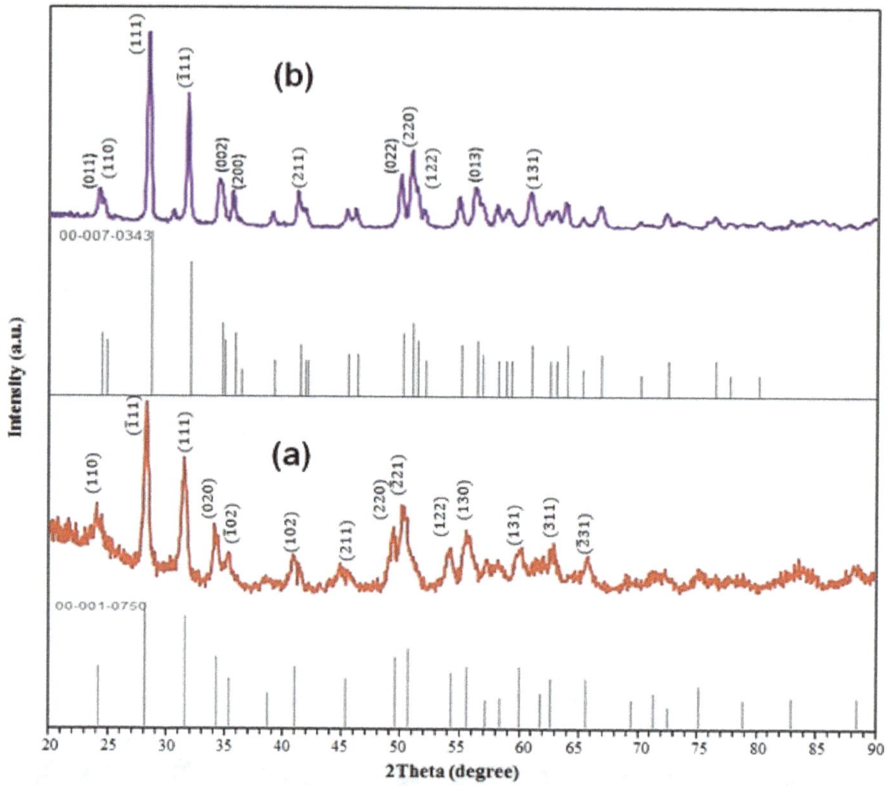

Fig. 5. XRD patterns of m-ZrO_2 nanoparticles synthesized by thermal decomposition of compound 1 in (a) methanol as solvent, and (b) MEG as solvent.

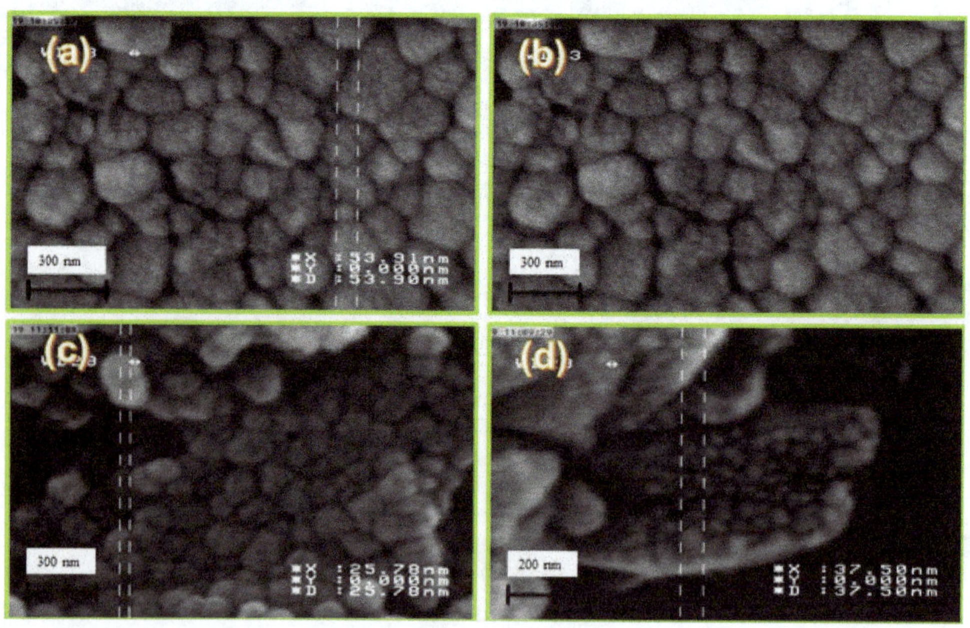

Fig. 6. SEM images of ZrO_2 nanoparticles in (a, b) methanol as solvent, and (c, d) MEG as solvent.

Table 2.

Lattice parameters and particle size of ZrO_2 nanoparticles synthesized by thermal decomposition of compound 1 at 700 °C

Name of the compound	Crystal structure	Solvent	Lattice parameters			Cell volume	Average size (nm)	
			a(Å)	b(Å)	c(Å)		XRD	SEM
zirconia	Monoclinic	Methanol	5.21	5.26	5.37	145.29	76	53
zirconia	Monoclinic	MEG	5.14	5.20	5.31	140.30	59	25

compounds can be appropriate precursors for production of nano-sized metal oxide materials with different and interesting morphologies. The study also establishes that sonochemical method can be successfully applied as an efficient method for the preparation of nano-scale materials.

Acknowledgment

The authors would like to express gratitude for support by Iranian Research Organization for Science and Technology (IROST), University of Damghan, and Nanotechnology Initiative Council.

References

[1] M. Ranjbar, M. Nabitabar, Ö. Çelik, M. Yousefi, Sonochemical synthesis and characterization of nanostructured copper(I) supramolecular compound as a precursor for the fabrication of pure-phase copper oxide nanoparticles, J. Iran. Chem. Soc. 12 (2015) 551-559.

[2] V. Safarifard, A. Morsali, Applications of ultrasound to the synthesis of nanoscale metal–organic coordination polymers, Coord. Chem. Rev. 292 (2015) 1-14.

[3] L. Aboutorabi, A. Morsali, Structural transformations and solid-state reactivity involving nano lead(II) coordination polymers via thermal, mechanochemical and photochemical approaches, Coord. Chem. Rev. 310 (2016) 116-130.

[4] M.Y. Masoomi, A. Morsali, Sonochemical synthesis of nanoplates of two Cd(II) based metal–organic frameworks and their applications as precursors for preparation of nano-materials, Ultrason. Sonochem. 28 (2016) 240-249.

[5] A.Y. Robin, K.M. Fromm, Coordination polymer networks with O- and N-donors: what they are, why and how they are made, Coordin. Chem. Rev. 250 (2006) 2127-2157.

[6] T. Uemura, S. Kitagawa, Nanocrystals of coordination polymers, Chem. Lett. 34 (2005) 132-137.

[7] M. Tahmasebpour, A.A. Babaluo, M.K. Razavi Aghjeh, Synthesis of zirconia nanopowders from various zirconium salts via polyacrylamide gel method, J. Eur. Ceram. Soc. 28 (2008) 773-778.

[8] S. Park, J.M. Vohs, R.J. Gorte, Direct oxidation of hydrocarbons in a solid-oxide fuel cell, Nature 404 (2000) 265-267.

[9] Y.W. Li, D.H. He, Z.X. Cheng, C.L. Su, J.R. Li, M.J. Zhu, Effect of calcium salts on isosynthesis over ZrO_2 catalysts, Mol. Catal. A 175 (2001) 267-275.

[10] E.C. Subbarao, H.S. Maiti, Science and technology of zirconia, Adv. Ceram. 24 (1988) 731.

[11] Q. Zhang, J. Shen, J. Wang, G. Wu, L. Chen, Sol–gel derived ZrO_2–SiO_2 highly reflective coatings, Int. J. Inorg. Mater. 2 (2000) 319-323.

[12] P.K. Wright, A.G. Evans, Mechanisms governing the performance of thermal barrier coatings, Curr. Opin. Solid State Mater. Sci. 4 (1999) 255-265.

[13] C. Piconi, G. Maccauro, Zirconia as a ceramic biomaterial, Biomaterials 20 (1999) 1-25.

[14] P. Salas, E.D. Rosa-Cruz, L.A. Diaz-Torres, V.M. Castano, R. Melendrez, M. Barboza-Flores, Monoclinic, ZrO_2 as a broad spectral response thermoluminescence UV dosemeter, Radiat. Meas. 37 (2003) 187-190.

[15] P. Gao, L.J. Meng, M.P. dos Santos, V. Teixeira, M. Andritschky, Study of ZrO_2–Y_2O_3 films prepared by rf magnetron reactive sputtering, Thin Solid Films 377 (2000) 32-36.

[16] N. Tayyebi Sabet Khomami, F. Heshmatpour, B. Neumuller, A novel dinuclear zirconium (IV) complex derived from [Zr(acac)$_4$] and a pentadentate Schiff base ligand: Synthesis, characterization and catalytic performance in synthesis of indole derivatives, Inorg. Chem. Commun. 41 (2014) 14-18.

[17] M. Ranjbar, M. Lahooti, M. Yousefi, A. Malekzadeh, Sonochemical synthesis and characterization of nano-sized zirconium(IV) complex: new precursor for the preparation of pure monoclinic and tetragonal zirconia nanoparticles, J. Iran. Chem. Soc. 11 (2014) 1257-1264.

[18] M. Ranjbar, M. Yousefi, M. Lahooti, A. Malekzadeh, Preparation and characterization of tetragonal zirconium oxide nanocrystals from isophthalic acid-zirconium(IV) nanocomposite as a new precursor, Int. J. Nanosci. Nanotechnol. 8 (2012) 191-196.

[19] M. Salavati-Niasari, M. Dadkhah, M.R. Nourani, A. Amini Fazl, Synthesis and characterization of single-phase cubic ZrO_2 spherical nanocrystals by decomposition route, J. Clust. Sci. 23 (2012) 1011-1017.

[20] J. H. Bang, K.S. Suslick, Applications of ultrasound to the synthesis of nanostructured materials, Adv. Mater. 22 (2010) 1039-1059.

[21] W. J. Son, J. Kim, J. Kim, W. S. Ahn, Sonochemical synthesis of MOF-5, Chem. Commun. (2008) 6336-6338.

[22] M. Ranjbar, M. Yousefi, Synthesis and characterization of lanthanum oxide nanoparticles from thermolysis of nano-sized lanthanum(III) supramolecule as a novel precursor, J. Inorg. Organomet. Polym. 24 (2014) 652-655.

[23] M. Ranjbar, M. Yousefi, Facile preparation of zirconia nanostructures by new method: nano-scale zirconium (IV) coordination supramolecular compound as precursor, Iran. J. Sci. Technol. Trans. Sci. (2016). doi:10.1007/s40995-016-0069-9.

[24] M. Ranjbar, M. Yousefi, R. Nozari, S. Sheshmani, Synthesis and characterization of cadmium-thioacetamide nanocomposites using a facile sonochemical approach: a precursor for producing CdS nanoparticles via thermal decomposition, Int. J. Nanosci. Nanotechnol. 9 (2013) 203-212.

[25] M. Ranjbar, Ö. Çelik, S. H. Mahmoudi Najafi, S. Sheshmani, N. Akbari Mobarakeh, Synthesis of lead(II) minoxidil coordination polymer: a new precursor for lead(II) oxide and lead(II) hydroxyl bromide, J. Inorg. Organomet. Polym. 22 (2012) 837-844.

[26] M. Ranjbar, S.A. Mozaffari, E. Kouhestanian. H. Salar Amoli, Sonochemical synthesis and characterization of a Zn(II) supramolecule, bis(2,6 diaminopyridinium)bis(pyridine-2,6-dicarboxylato)zincate(II), as a novel precursor for the ZnO-based dye sensitizer solar cell, J. Photochem. Photobiol. A: Chem. 321 (2016) 110-121.

[27] M. Ranjbar, M. Yousefi, Sonochemical synthesis and characterization of a nano-sized lead (II) coordination polymer; a new precursor for the preparation of PbO nanoparticles, Int. J. Nanosci. Nanotechnol. 12 (2016) 109-118.

[28] A. Aslani, A. Morsali, M. Zeller, Dynamic crystal-to-crystal conversion of a 3D–3D coordination polymer by de- and re-hydration, Dalton Trans. 14 (2008) 5173-5177.

[29] M. Ranjbar, S. H. Mahmoudi Najafi, S. W. Ng, catena-Poly[lead(II)-[μ-2,4-di-amino-6-(piperidin-1-yl) pyrimidine N-oxide-κ^2O:O]di-μ-iodido], Acta Cryst. E65 (2009) m749-m749.

[30] N. B. Date, A.G. Purohit, S. S. Pimple, P.D. Chaudhari, Formulation and evaluation of coated microneedles for the treatment of hairloss, Int. J. Res. Rev. Pharm. Appl. Sci. 4 (2014) 1083-1101.

[31] H. Klug, L. Alexander, X-Ray Diffraction Procedures, John Wiley and Sons, New York, 1962.

Effects of temperature and particle size distribution on barite reduction by carbon monoxide gas

Saeed Vakilpour, Ahmad Ghaderi Hamidi*

Department of Metallurgy and Materials Engineering, Hamedan University of Technology, Hamedan, Iran.

HIGHLIGHTS

- Reduction of barite particles with different particle sizes by CO gas.
- Study of a technical grade barite with relatively low $BaSO_4$ content (82%).
- Kinetical relations of reduction of $BaSO_4$ to BaS.

GRAPHICAL ABSTRACT

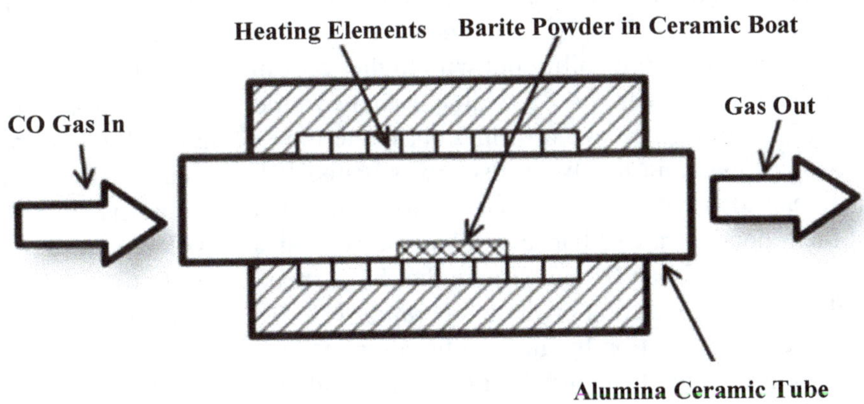

ARTICLE INFO

Keywords:
Barite
Carbon monoxide
Activation energy

ABSTRACT

In this research, a mineral barite powder was reduced by carbon monoxide gas and the effects of reduction time and temperature was investigated as well as barite particle size. Results showed that the best result would be feasible when the barite particle sizes are between (-70 +100) in mesh scale. The barite reduction could reach the maximum level (98%) after reduction by carbon monoxide at 850°C for one hour. Meanwhile, the same amount of reduction could have been achieved in shorter time intervals at higher temperatures. The kinetics model for reduction process was also determined and activation energy was calculated.

* *Corresponding author.*
E-mail address: ghaderi@hut.ac.ir

1. Introduction

Barite is the most abundant mineral of barium. Crude barite is widely used in drilling fluids for oil and gas exploration. Barite with quite high purity can be raw material for synthesis of barium and its compounds [1]. Barite (mainly barium sulfate) reduction to barium sulfide that is dissolvable in water is an imperative step in synthesis of barium compounds [2]. In industrial plants, charcoal is usually used to reduce barite [3]. Nevertheless the high consumption of charcoal besides producing impure barium sulfide with longer reduction times and higher rate of pollution makes the gases more favorable reduction agents [4]. Carbon monoxide, hydrogen or reformed methane can be used in the process of reduction by gases. Some researchers investigated the possibilities of using hydrogen gas for barite reduction in temperature ranges of 800 to 1100°C [5, 6]. The hydrogen has shown strong reducing ability but with higher cost in comparison with others [7, 8]. Although there would be a difficulty for methane reformation, reduction of barite with methane has been also investigated [9]. Barite can also be reduced by CO bearing nitrogen and the reduction rate of barite in temperature ranges of 850 to 1000°C can be increased either by increasing the CO concentration or the temperature [10]. In the process of reduction by gas, the burning gas also can provide needed heat and fluid bed. These advantages decrease the reduction temperature to lower than 1000 °C and time to lower than one hour. This also is favorable energetically but still using gas as a reduction agent for barite has not been industrialized [11].

There is no report about reduction of barite by pure CO. there is only one article that have used nitrogen gas with partially CO (carbon monoxide fractions of 2.4–9.6%) [10]. In this research the effects of some parameters of reduction in CO atmosphere has investigated. The aim of this research is to determine the start temperature for barite reduction by carbon monoxide gas, fulfilling the conditions for lowering the temperature and energy consumption and keeping the reduction rate at the highest simultaneously.

2. Materials and Methods

Milled mineral barite was used as raw material. Chemical analysis of the barite that was determined by X-ray florescence (XRF) is shown in Table 1. X-ray diffraction results (Fig.1) also shows that barium sulfate is dominant phase in primary specimen.

The barite samples with specific particle size distributions were used for gas reduction. Reduction cycles were examined at temperature range of 600 to 1000°C. Heating rate was 15 °C/min and 99.9% purity carbon monoxide gas flow was kept during the reduction cycle with flow rate of 0.1 L/min.

Samples with weight of 1.000 g were reduced in a tube furnace and weight reductions were measured

Table 1.
Chemical analysis of mineral barite

Compound	$BaSO_4$	$SrSO_4$	Fe	CaO	SiO_2	Al_2O_3
Wt.%	81.52	2.05	0.28	3.75	7.91	4.47

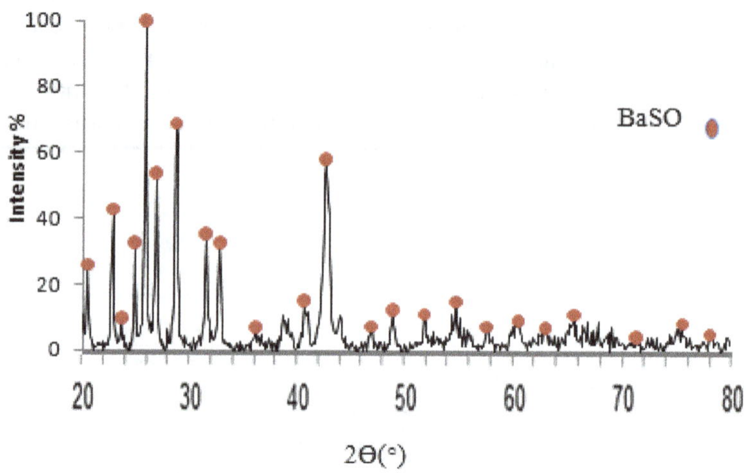

Fig. 1. XRD patterns of milled primary barite.

tube furnace and weight reductions were measured as the criterion for reduction. Using equations 1 and 2, it is possible to calculate the reduction levels. The parameters in equation 1 and 2 are defined in Table 2.

$$\Delta m_{BaSO4\text{-}red} = \Delta m_{tot} - (\Delta m_{SrSO4\text{-}red} + \Delta m_{BaSO4\text{-}vol} + \Delta m_{BaSO4\text{-}moi}) \tag{1}$$

$$\text{Reduction Percentage} = (\Delta m_{BaSO4\text{-}red} / \Delta m_{Sto} - \Delta m_{BaSO4\text{-}red}) * 100\% \tag{2}$$

3. Results and discussion

3. 1. Thermodynamics of barite reduction by CO

Reaction 3 demonstrates the barite reduction by CO gas. The Gibbs free energy of supposed reaction versus temperature is shown in Fig.2.

$$BaSO_{4(s)} + 4CO_{(g)} \rightarrow BaS_{(s)} + 4CO_{2(g)} \tag{3}$$

As is evident in Fig.2, barite reduction by CO gas should be possible from 500 °C but experimental results are not in agreement with the above thermodynamic predictions. As is demonstrated in Fig.3, barite isothermal reduction started form 600°C and reached to a significant amount at 850°C. The increasing of reduction temperature above 950°C did not lead to further reduction. Therefore temperature range of 850-950°C as optimized temperature was selected for study of other parameters.

3.2. Effect of barite particle size

Three different ranges of particle size of milled barite as: (-20, +50), (-70, +100) and (-140, +170) were chosen to be reduced at 900 °C. The reduction times were also set at 15, 30, and 45 minutes. The reduction results versus time are demonstrated in Fig.4.

Results showed that barite samples with coarse, medium and fine particle size revealed the lowest, the highest, and the medium reduction rates, respectively. The finest particles showed the lowest amount of reduction rate. The fine particle powder mass had narrow channels for gas diffusion and has exposed more resistance against gas exchange. Although the fine powder had the highest surface/mass ratio but deficiency of gas transportation decelerated the reductin process. On the other hand, the coarse particles had the lowest surface/mass ratio therefore the

Table 2.
Defining the parameters of equations 1 and 2

Calculated weight reduction due to stoichiometric reduction of whole $BaSO_4$ to BaS	Δm_{Sto}
weight reduction due to barite reductlon	$\Delta m_{BaSO4\text{-}red}$
Weight reduction due to volatiles elimination	$\Delta m_{BaSO4\text{-}vol}$
Weight reduction due to moisture elimination	$\Delta m_{BaSO4\text{-}moi}$
Weight reduction due to full reduction of $SrSO_4$ to SrS	$\Delta m_{SrSO4\text{-}red}$
Measured total weight reduction of specimen	Δm_{tot}

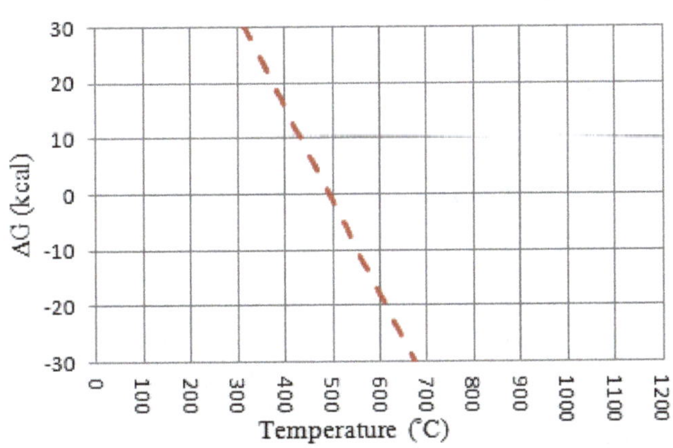

Fig. 2. Gibbs free energy changes of barite reduction reaction versus temperature [12].

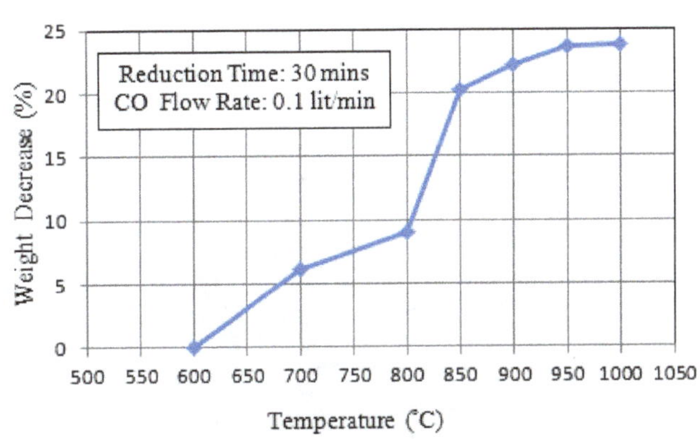

Fig.3. Weight decrease due to barite reduction by CO at different temperatures for 30 min.

reducing agent (CO gas) needed to diffuse across a thick reacted surface layer (BaS) to compelete reaction of the reduction. The specimen of medium particle size barite (-70 +100) showed the highest reduction rate and a beter balance between two abow effective parameters has been established. Therefore it is predictable that barite powders with particles bigger than tested coarse powder and smaller than tested fine powder would have lower reduction rate. XRD results for primary barite and the reduced sample are shown in Fig.5 and demonstrates that BaS (barium sulfide) is dominant phase of reduced sample.

3.3. Time effect on barite gas reduction

To investigate the effect of time on reduction process, four different timings was set at 15, 30, 45, and 60 minutes. The reduction temperatures were arranged at 850, 900, and 950°C, other parameters were also kept constant. Results are demonstrated in Fig.6.

According to Fig. 6, at reduction temperature of 850°C, reduction is more sensitive to time changes in comparison to others. Results are suggesting that the biggest portion of reduction process has been

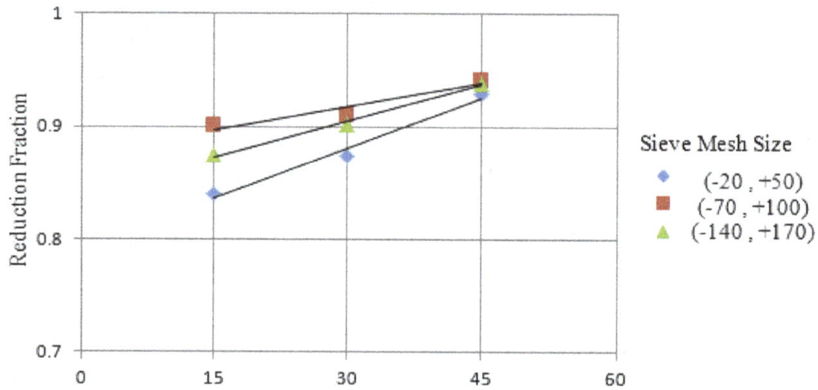

Fig. 4. Effect of barite particle size on reduced mass fraction at 900°C.

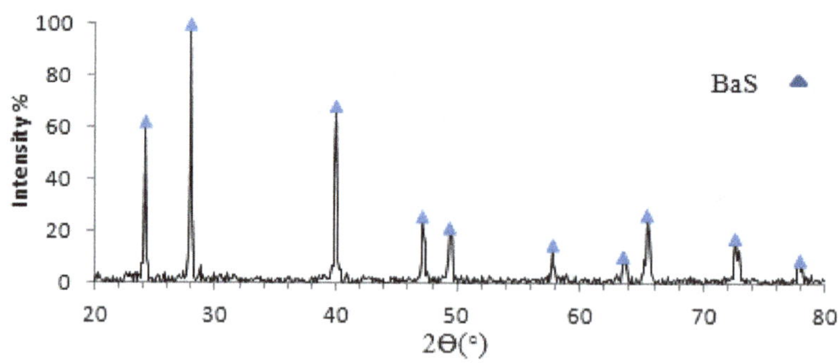

Fig. 5. XRD patterns of reduced barite specimen.

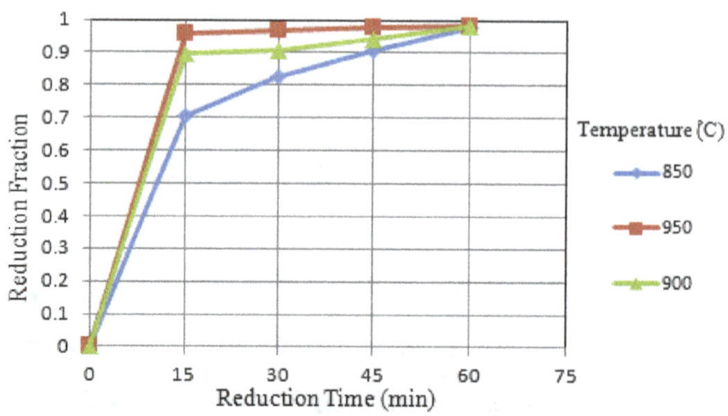

Fig. 6. Reduced barite mass fraction versus of isothermal holding time.

completed in the first time period (less than 15 minutes), specially for the higher temperatures as 950°C. Therefore, it is worth mentioning that the reduction rate will change slightly at higher temperatures. For different temperatures, the maximum levels of reduction were almost same.

3.4. The kinetics of barite reduction by gas

3.4.1. The kinetics model

Reduction progress at all three temperatures i.e. 850, 900 and 950°C were considered as the first order reactions as are proven in Fig.7. Results showed that theoretical and experimental data are highly matched when the assumption is based on first order reactions. Therefore, it would be safe to say that the barite reductions in all three temperatures were according to first order reaction [13].

As mentioned in section 3.1, the barite reduction with CO gas is done according to equation 3. Therefore the reduction products were the Barium sulfide and CO_2 gas. It means reacting a solid with a gas produced another solid and another gas. In this condition and by assuming that the barite particles were spherical, the properties of reduction product formed on primary barite particles (BaS) is determining of the reduction progression. There would be two kinetic models which could be offered to investigate the reaction mechanism [14,15].

To determine the kinetic model, reduced (dimensionless) time method has used. This is an applicable method to determine the speed controlling mechanism of a reaction [16]. In this method, the general equation is g(x)=kt, where; x is the reaction fraction, k is the speed constant and t is the time. As mentioned before, for this reduction reaction, two types of equations are suggested as equations 4a and 4b:

$$1-(1-x)^{1/3} = kt \tag{4a}$$

$$1-\left(\tfrac{2}{3}\right)x-(1-x)^{2/3} = kt \tag{4b}$$

For the relation 4a, the reduction rate is controlled by the rate of reduction chemical reaction and for relation 4b, the diffusion rate across reduction products is speed controller. Results showed that experimental data is matched with kinetics model b (98.45%) comparing to another model. Consequently, the diffusion rate of gaseous species (CO&CO_2) through reduction products (BaS layer) had been the controlling mechanism of reduction rate. Fig.8 shows full compliance between the kinetics model and experimental data.

3.4.2. Activation energy calculation of reduction reaction

For calculating the activation energy, relation 5 exponential equation is used where, the k is speed constant, k0 is the frequency factor, R is the gases constant, T is the temperature and E is activation energy [16]:

$$k = k_0 \exp(-E/RT) \tag{5}$$

First, the speed constant for model 4.b at 850, 900, 950, and 1000 °C has determined and then Fig. 9 has been draw according to above exponential equation. Results from Fig.9 show that (E/R) ration is 1.9943 and activation energy is 165.81 kJ/mol.

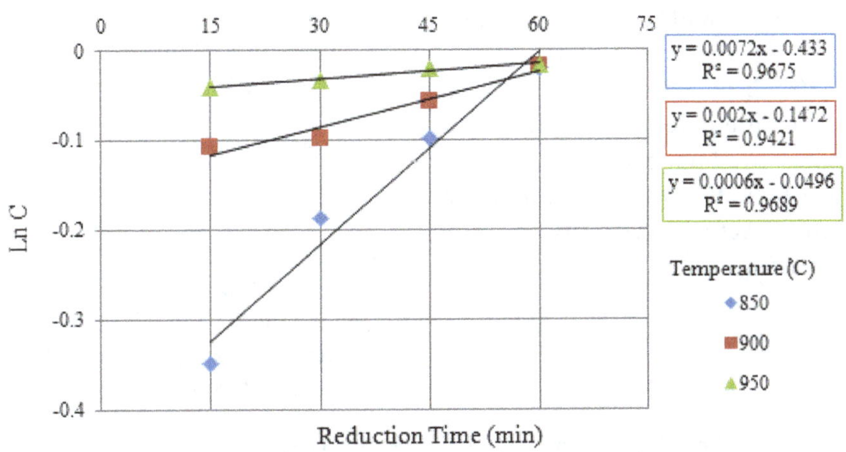

Fig. 7. Determining the barite reduction reaction order.

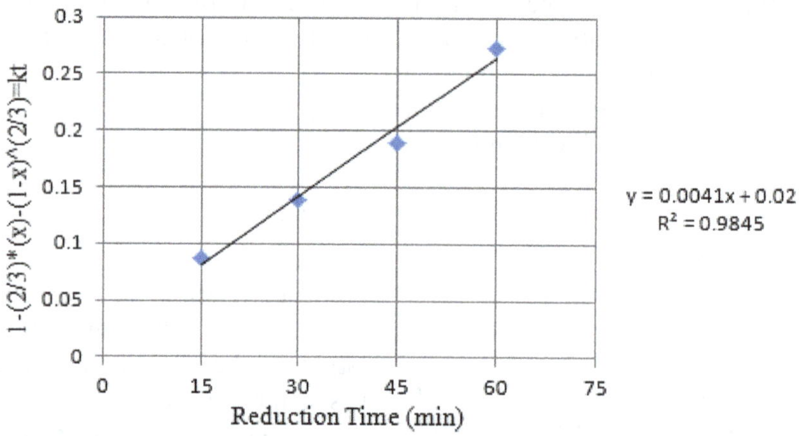

Fig. 8. Kinetics model of barite reduction process with gas at 850 °C.

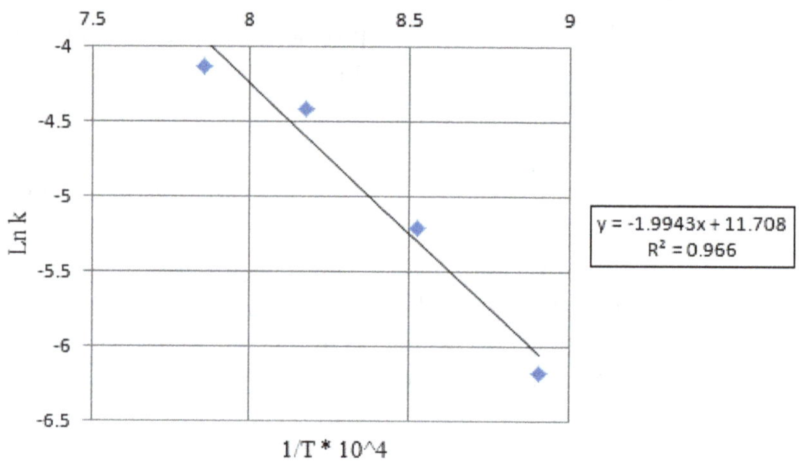

Fig. 9. Calculating activation energy for barite reduction with gas.

4. Conclusions

The barite reduction with CO gas should have been thermodynamically feasible from 500 °C, but practically it started at 600 °C because at lower temperatures the reaction was very slow. The reduction rate was improved considerably at 850 °C and it reduction was completed in one hour. By increasing the reduction temperature to 950 °C, reduction rate continued to increase and the maximum reduction level was achievable in 15 minutes. Medium particle size barite (-70 +100) has highest reduction rate in comparison to fine and coarse powder. Activation energy of barite reduction by CO was 165.81 kJ/mol.

References

[1] A. Salem, Y. T. Osgouei, The effect of particle size distribution on barite reduction, Mater. Res. Bull. 44 (2009) 1489–1493.

[2] A. Salem, S. Jamshidi, Effect of paste humidity on kinetics of carbothermal reduction of extruded barite and coke mixture, Solid State Sci. 14 (2012) 1012–1017.

[3] D. Guzmán, J. Fernández, S. Ordoñez, C. Aguilar, P. A. Rojas, D. Serafini, Effect of mechanical activation on the barite carbothermic reduction, Int. J. Miner. Process. 102–103 (2012) 124–129.

[4] A. Salem, Y. T. Osgouei, and S. Jamshidi, Kinetic Study of Barite Carbothermic Reduction in Presence of Sodium Carbonate as Catalyst, Iran. J. Chem. Eng. 7 (2010).

[5] Y. Pelovski, K. Ninova, I. Gruncharov, I. Dombalov, Isothermal reduction of barite with hydrogen, J. Therm. Anal. 36 (1990) 2037–2043.

[6] R. V. Culver, C. J. Hamdrof, E. C. R. Spooner, The reduction of barytes with hydrogen. i. kinetic studies in a differential reactor, J. Appl. Chem. 3 (1958) 810–819.

[7] W. V. Schulmeyer, H. M. Ortner, Mechanisms of the hydrogen reduction of molybdenum oxides, Int. J. Refract. Met. Hard Mater. 20 (2002) 261–269.

[8] E. Jamshidi, H. A. Ebrahim, A new clean process for barium carbonate preparation by barite reduction with methane," Chem. Eng. Process. Process Intensif. 47 (2008) 1567–1577.
[9] N. G. Gallegos, J. M. P. Lopez, Kinetic study of cobalt oxides reduction by hydrogen, Mater. Chem. Phys. 19 (1988) 431–446.
[10] P. S. Hlabela, H. W. J. P. Neomagus, F. B. Waanders, O. S. L. Bruinsma, Thermal reduction of barium sulphate with carbon monoxide-A thermogravimetric study, Thermochim. Acta 498 (2010) 67–70.
[11] S. Jamshidi, A. Salem, Role of extrusion process on kinetic of carbothermal reduction of barite, Thermochim. Acta 503–504 (2010) 108–114.
[12] H. G. Lee, Chemical Thermodynamic of Metals and Materials. Imperial College Press, 1999.
[13] M. J. Pilling and P. W. Seakins, Reaction kinetics., 2nd ed. Oxford Science Publications, 1996.
[14] J. H. Espenson, Chemical Kinetics and Reaction Mechanisms. 2002.
[15] F. Habashi, Kinetics of Metallurgical Processes. Métallurgie Extractive Québec, 1999.
[16] H.S.Ray, Kinetics of metallurgical reactions. Oxford & IBH Publishing, 1993.

6

Effect of Sn doping on structural and optical properties of 2D α-MoO_3 nanostructures

Amin Eftekhari[1], Mohammad Bagher Rahmani[1,*], Fatemeh Masdarolomoor[2]

[1]*Department of Physics, Shahrood University of Technology, Shahrood, Iran*

[2]*Department of Chemistry, Shahrood University of Technology, Shahrood, Iran*

HIGHLIGHTS

- The simple and low cost method of spray pyrolysis was utilized to synthesize 2D α-MoO_3 thin films.
- Sn doped MoO_3 thin films with different amounts of impurity were deposited with success
- Undoped and Tin doped MoO_3 thin films were characterized structurally and optically using various techniques.

GRAPHICAL ABSTRACT

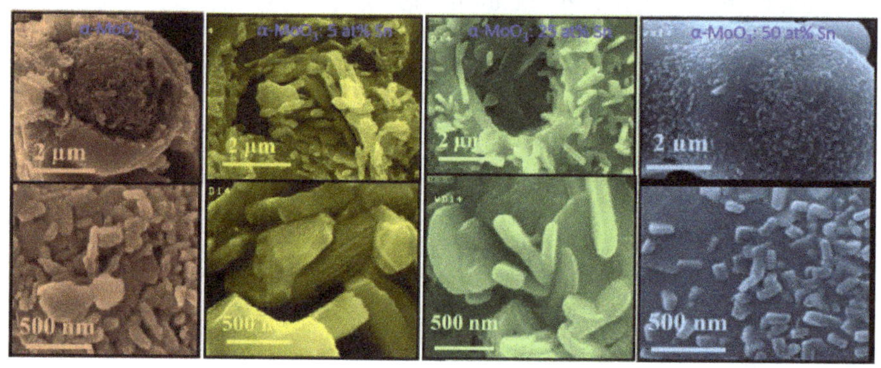

ARTICLE INFO

Keywords:
Crystal structure
Molybdenum trioxide
Micro-spheres
Nano-platelets
Spray pyrolysis

ABSTRACT

Undoped and tin (Sn) doped molybdenum trioxide (α-MoO_3) nanostructured thin films (which has lamellar (2D) structure) have been prepared using a simple and cost effective technique of spray pyrolysis on glass substrates at 450°C. Surface morphology, optical and structural properties of samples have been investigated using FESEM, UV-Vis spectroscopy and XRD analysis techniques, respectively. FESEM images showed the formation of some discrete micro-spheres on the surface, which with the increasing in the amount of dopant homogenous and dense nano-platelets was grown on top of these micro-spheres. The XRD pattern analysis shows that all samples have been grown in orthorhombic (α-MoO_3) crystal structure and except for the sample doped with 50 at% Sn which had a weak peak of SnO_2, no peak have been observed corresponding to the incorporation of Sn. By increasing the amount of impurity, optical transmittance of samples were increased from ~27 to 50%. Also, the band gap of samples were calculated using transmission data. An increasing of band gap from 3.34 to 3.89 eV was observed with increasing in the amount of doping.

* *Corresponding author.*
E-mail address:mbrahmani@shahroodut.ac.ir ; mbrahmani@yahoo.com

1. Introduction

In the past decade, nanostructures of metal oxides have attracted the attention of many scientists. These structures have inimitable electrical, optical, magnetic and catalytic properties [1]. Among them, molybdenum trioxide (MoO_3), recently due to the fascinating properties and applications of nanostructures on large areas are taken into consideration [2]. Because of its layered structure, MoO_3 has been recently focused due to its electrochromic and photochromic properties, which makes it a good candidate for use in important applications such as smart windows and display devices [3]. MoO_3 is a n-type semiconductor with a wide band gap of about 2.9 -3.5eV and due to these mentioned properties its applications as catalyzers, gas sensors and solar cells have been reported [4, 5]. Generally MoO_3 has three forms: the well-known thermodynamically stable orthorhombic α-MoO_3, metastable monoclinic β-MoO_3, and hexagonal h-MoO_3. Between these three forms, α-MoO_3 has attracted most interests because of its diverse properties as in supercapacitors [6]. Using different techniques for growing thin films of this material leads to diverse morphologies and structures and hence to distinct properties. Thin films preparation of MoO_3 have been reported by different deposition techniques such as thermal evaporation [4], spray pyrolysis [5, 7], hydrothermal [8], physical vapor deposition [1], sol-gel [9], spin coating [10], sputtering [11] and electron beam evaporation [12]. Among these techniques spray pyrolysis has attracted researchers' attention because it's a simple and cost effective method. Also, it can be implemented for depositing thin films on large areas. However, it is involved with a large number of interrelated variables like flow rate, carrier gas pressure, distance between the substrate and spraying nozzle and etc., which are all to be optimized to obtain a thin film with the desired quality [13]. Control of growth parameters with aiming to achieve well-defined nanostructured morphologies and also, doping with a metal or metal oxide impurities may open new opportunities for exploring exclusive physical and chemical properties and more potential uses [14-17]. There are several reports on doping of MoO_3 with different elements as impurities. M. Kovendhan et al. have deposited undoped and lithium (Li) (1–5 wt%) doped MoO_3 thin films on ITO coated glass substrates using spray pyrolysis at a substrate temperature of 325 °C. In their report, Li doping has induced modifications in physical and chemical properties of their samples [13]. Niobium (Nb) doped molybdenum trioxide (MoO_3) thin films have been synthesized using spray pyrolysis deposition technique, by S.S. Mahajan et al. [18]. Their report indicates that by increasing the dopant concentration, the structure of MoO_3 undergoes a phase transformation from orthorhombic to amorphous. In another case, J. Kaur et el. have studied the properties of SnO_2 thin films prepared by a sol–gel spin coating process using different percentages (1, 5, 10 and 20 wt. %) of MoO_3 as impurity on glass substrates. They have observed that with an increase in doping concentration the reflectance decreases because of an increase in roughness and a decrease in the number of free electrons [19].

In this paper, we have studied the effect of Sn doping (Sn/Mo atomic ratio =0.05, 0.10, 0.15, 0.25, 0.50 in precursor solution) on structural and optical properties and surface morphology of MoO_3 thin films on glass substrate prepared using the simple route of spray pyrolysis. The importance of the work that has been done in this manuscript is due to the fabrication of Sn doped 2D α-MoO_3 nano-platelets using the simple and low cost technique of spray pyrolysis. Two-dimensional (2D) materials can show unique electronic and optical properties when the number of planes are reduced, due to changes in the electronic band structure. Recently, fabrication of biosensors and high-performance field effect transistors have been reported using 2D nature of α-MoO_3 [20-22].

2. Experimental

2.1. Materials and methods

In order to deposition of thin films, firstly, glass substrates were washed by deionized water and soap, then substrates ultrasonically cleaned in a mixture of acetone and ethanol. Finally, the substrates were washed with deionized water and dried with a mild stream of clean and dry air.

For synthesis of undoped and Sn doped MoO_3 thin films, ammonium molybdate tetrahydrate ($(NH_4)_6Mo_7O_{24}.4H_2O$ extra pure (Merck)) was utilized as starting material and deionized water as solvent. Tin (II) chloride dehydrate ($SnCl_2.2H_2O$ (Merck)) was taken as the source of impurity. In order to prepare spray solution, firstly, 3.089 gr of $(NH_4)_6Mo_7O_{24}.4H_2O$ was dissolved in 20 mL deionized water and different amounts of $SnCl_2.2H_2O$ (Sn/Mo atomic ratio =0.05, 0.10, 0.15, 0.25, 0.50 in precursor solution) were dissolved in 30 ml deionized water and ethanol (1:1 ratio) by magnetic stirring. Then prepared solutions were stirred using a magnetic stirring heater at 40 °C for 2 hours until a clear and absolutely transparent solutions were obtained. Lastly, the obtained solutions were sprayed on top of glass substrates. Deposition parameters for spraying all samples were controlled as following: Substrate temperature: 450 °C, Nozzle to spray distance: 35cm, Dry and clean air pressure: 2 mbar, Precursor solution spray rate: 5 mL/min. Also, the heater holding substrates was spinning with 20 rev/min to make sure that we get uniform thin films on glass substrates.

2.2 Characterization Techniques

Structural properties of prepared thin films were characterized using X-ray diffractometer (XRD), (Unisance S300) equipped with a graphitic monochromator of Cu Kα radiation (λ=0.154056 nm) over the 2θ scan range of 20-70°. To study the surface morphology and optical properties of samples, field emission scanning electron microscope (FESEM), (Hitachi s.4160) and UV-Vis optical spectrometer (300-1100 nm), (Shimadzu UV-Vis 1800) equipment were utilized, respectively.

3.Results and Discussion

3.1. XRD analysis

Fig. 1 shows the X-ray diffraction (XRD) patterns of undoped and tin doped MoO_3 thin films deposited on glass substrates by spray pyrolysis method. The XRD pattern of the undoped sample shows the polycrystalline nature with orthorhombic (α-MoO_3) crystal structure with lattice parameters of a=3.9620 Å, b=13.8580 Å, c=3.6970 Å (according to JCPDS card No. 005-0508). The strong intensity of the reflection peaks of (0 4 0) and (0 6 0) proves the existence of the lamellar structure [4]. Also, Fig. 1 Exhibits that Sn doping of MoO_3 caused attenuation of strong α-MoO_3 peaks, and this continues with increasing in the amount of impurity. However, no peaks have been observed corresponding to the formation of secondary phase regarding SnO_2. Just for the sample with the highest percentage of doping (i.e. Sn/Mo = 0.5), a weak peak of (2 0 2) was observed which can be attributed to the preliminary formation of orthorhombic crystal structure for SnO_2 (according to JCPDS card No. 029-1484). This may be because of atomic substitution of Sn in Mo sites for low concentrations of impurity and beginning the formation of second phase of SnO_2 at high concentrations of doping.

The crystallite size (D (nm)) of prepared thin films was assessed using the famous Scherrer formula [6]:

$$D = \frac{S\lambda}{\beta \cos\theta} \tag{1}$$

where S is the shape factor (~ 0.9), λ is the wavelength of X-rays (1.5406 Å), β (rad) is the full width at half maxima (FWHM), and θ is the diffracting angle (rad). Also, the strain (ε (no dimension)) and dislocation density (δ $(nm)^{-2}$) of the crystallites determined using the following formulas:

$$\varepsilon = \frac{\beta \cos\theta}{4} \tag{2}$$

$$\delta = \frac{1}{D^2} \tag{3}$$

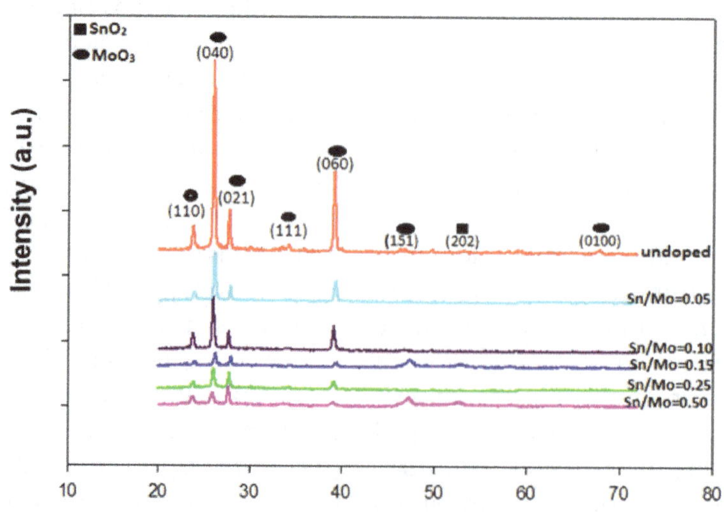

Fig. 1. XRD patterns of undoped and 5, 10, 15, 25 and 50 at% Sn doped MoO_3 thin films.

The crystallite size, strain and dislocation density of crystallites corresponding to the (0 4 0) plane for all samples were calculated and presented in Table 1. Analysis of the data indicates that with increasing the amount of impurity concentration in the precursor solution the crystallite size decreased from about 78 to 23 nm, for undoped and the highly tin doped (Sn/Mo = 0.50) samples, respectively. Also, doping intensification caused increasing the strain and dislocation density of crystallites, which this confirms degradation in crystal quality. This result is also in correlation with FESEM images of samples.

3.2 Surface morphology

In order to observe the surface morphology changes of samples due to the addition of impurities, field emission scanning electron microscopy (FESEM) of thin films were utilized. Before imaging all samples were coated with a thin layer of Au, using DC sputter coater in order to get clear and high contrast pictures. Fig. 2 shows the FESEM images of undoped and tin doped MoO_3 thin films prepared by spray pyrolysis method. Fig. 2(a) reveals that for undoped sample FE-SEM images show the formation of some dispersed micro-spheres on the surface. Insets of Fig. 2(a) shows that shell of every micro-sphere has been composed of stacked thin layers and the sphere has been filled by semi-rectangular nano-platelets. Figures 2 (b) - (d) show the surface morphology of tin doped MoO_3 samples with 5, 25 and 50 at% of impurity, respectively. Increasing the dopant amount causes formation of nucleation sites on every platelets and then creation of rectangular nano-plates which can be attributed to the formation of the new SnO_2 orthorhombic crystal structure which is formerly confirmed by XRD pattern analysis.

3.3 Optical properties

The optical properties of all samples were stud-

Table 1.
The crystallite size (D), FWHM (β), strain (ε) and dislocation density (δ) of undoped and 5,10,15,25 and 50 at% tin MoO_3 thin films along (0 4 0) plane.

sample	FWHM ($\times10^{-3}$rad)	D(nm)	ε ($\times10^{-4}$)	δ($\times10^{-4}$ (nm)$^{-2}$)
Undoped	1.7898	78.88	4.35	1.60
Sn/Mo=0.05	1.9850	71.71	4.83	1.94
Sn/Mo=0.10	2.0925	68.00	5.09	2.16
Sn/Mo=0.15	4.1933	33.95	10.21	8.67
Sn/Mo=0.25	4.6786	30.41	11.39	10.80
Sn/Mo=0.50	6.1200	23.24	14.91	18.50

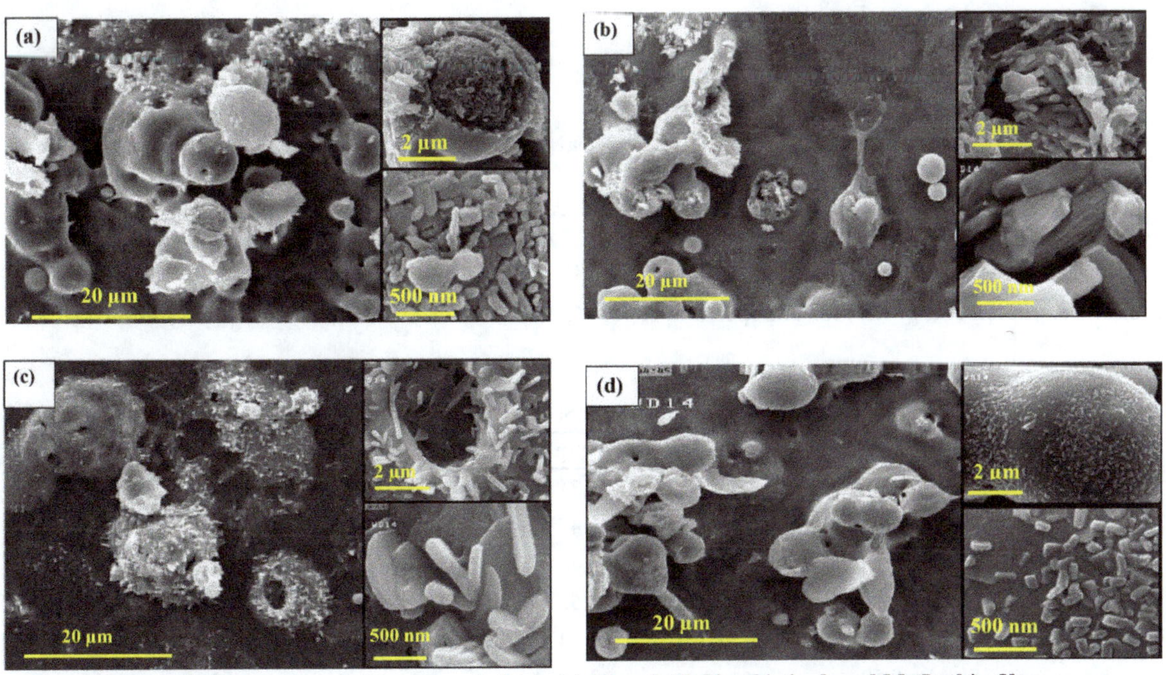

Fig. 2. FESEM images of: (a) undoped, (b) 5, (c) 25 and (d) 50 at% tin doped MoO_3 thin films

ied in the wavelength range of 300 -1100 nm. Fig. 3 shows the optical transmittance spectra of thin films on glass substrates and indicates that the transmittance has increased from about 28 to 51% (in the average wavelength of visible region, i.e. 550 nm), with increasing the percentage of tin as dopant from 0 to 50 at% in precursor solution, respectively. This causes the shift of absorption edges toward lower wavelengths, and can be attributed to the occupation of all states close to the conduction band and the optical band gap increases with increasing the doping amount of Sn according to Burstein-Moss effect [23].

The optical band gap of thin films was obtained by extrapolating of linear region of the curve $(\alpha h\upsilon)^2$ versus photon energy ($h\upsilon$) as shown in Fig. 4. Optical band gaps of samples are given in Table 2. The results indicate that the optical band gap increased with increasing the percentage of tin doping and is in agreement with the changes in the crystallite size. Besides Burstein-Moss effect on band gap broadening, the well-known quantum confinement phenomenon causes the increase of band gap by decreasing the crystallite size when the percentage of dopant rises [6].

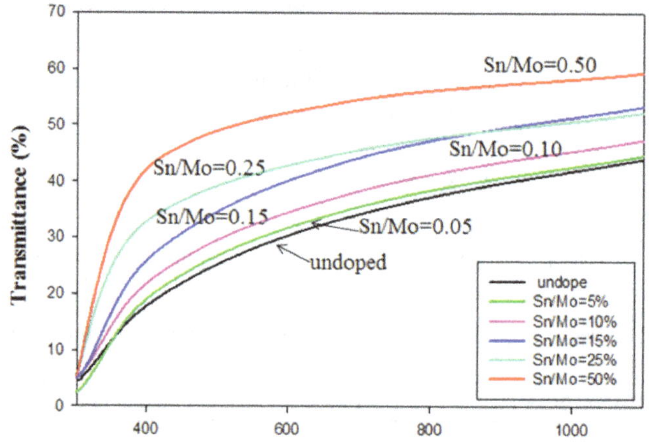

Fig. 3. The optical transmittance spectra of undoped and 5, 10, 15, 25 and 50 at% tin doped MoO_3 thin films.

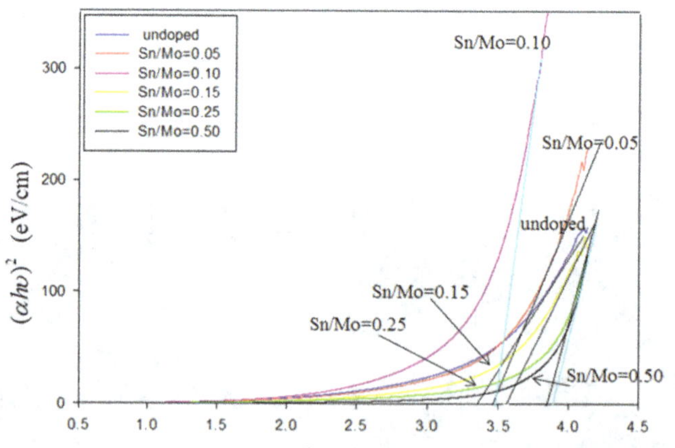

Fig. 4. $(\alpha h\upsilon)^2$ vs. photon energy ($h\upsilon$) for undoped and 5, 10, 15, 25 and 50 at% tin doped MoO_3 thin films.

Table 2. Optical band gap and transmittance of undoped and 5, 10, 15, 25 and 50 at% tin doped MoO_3 thin films.

sample	Band gap (eV)	Transmittance (%)
Undoped	3.34	27
Sn/Mo=0.05	3.47	29
Sn/Mo=0.10	3.48	32
Sn/Mo=0.15	3.56	38
Sn/Mo=0.25	3.84	41
Sn/Mo=0.50	3.89	51

4. Conclusions

In this work, we have studied the effect of Sn as a dopant on structural and optical properties of MoO_3 thin films deposited on glass substrates at 450 °C, using the simple and cost effective route of spray pyrolysis. XRD pattern analysis revealed that undoped samples have polycrystalline nature with α-MoO_3 orthorhombic phase which has lamellar (2D) structure. Only one peak was observed for the highly Sn doped MoO_3 thin films, which corresponds to the formation of new second phase of SnO_2 with orthorhombic lattice structure and other samples with lower percentages of impurity just show decreasing in the intensity of MoO_3 main peaks, which is due to incorporation of Sn in Mo sites. The measured transmittance spectra in the average wavelength of 550 nm for samples increased from about 28 to 51% with increasing in the percentage of tin dopant from 0 to 50 at% in precursor solution, respectively. Also, same increasing was observed in the calculated band gap of samples, which was 3.34 to 3.89 eV for samples from 0 to 50% impurity, respectively.

References

[1] K.-K. Wang, F.-X. Wang, Y.-D. Liu, G.-B. Pan, Vapor growth and photoconductive property of single-crystalline MoO_3 nanosheets, Mater. Lett., 102–103 (2013) 8-11.

[2] S. Balakumar, R.A. Rakkesh, A.K. Prasad, S. Dash, A.K. Tyagi, Nanoplatelet Structures of MoO_3 for H2 Gas Sensors, IEEE, (2011) 514-517.

[3] M.C. Rao, K. Ravindranadh, A. Kasturi, M.S. Shekhawat, Structural Stoichiometry and Phase Transitions of MoO_3 Thin Films for Solid State Microbatteries, Res. J. Recent Sci., 2(4) (2013) 67-73.

[4] M.B. Rahmani, S.H. Keshmiri, J.Yu, A.Z. Sadek, L. Al-Mashat, A. Moafi, K. Latham, Y.X. Lie, W. Wlodarski, K. Kalantar-zadeh, Gas sensing properties of thermally evaporated lamellar MoO_3, Sens. Actuators, B, 145 (2010) 13-19.

[5] H.M. Martínez, J. Torres, L.D.L. Carreño, M.E. Rodríguez-García, Effect of the substrate temperature on the physical properties of molybdenum tri-oxide thin films obtained through the spray pyrolysis technique, Mater. Charact., 7 5 (2013) 184 - 193.

[6] J. Li, X. Liu, Preparation and characterization of α-MoO_3 nanobelt and its application in supercapacitor, Materials Letters 112 (2013) 39-42.

[7] B. Kannan, R. Pandeeswari, B.G. Jeyaprakash, Influence of precursor solution volume on the properties of spray deposited α-MoO_3 thin film, Ceram. Int., 40 (2014) 5817-5823.

[8] H. Sinaim, A. Phuruangrat, S. Thongtem, T. Thongtem, Synthesis and characterization of heteronanostructured Ag nanoparticles MoO_3 nanobelts composites, Mater. Chem. Phys., 132 (2012) 358-363.

[9] S.-Y. Lin, C.-M. Wang, K.-S. Kao, Y.-C. Chen, C.-C. Liu, Electrochromic properties of MoO_3 thin films derived by a sol–gel process, J. Sol-Gel Sci. Technol., 53 (2010) 51-58.

[10] G. Wang, T. Jiu, P. Li, J. Li, C. Sun, F. Lu, J. Fang, Preparation and characterization of MoO_3 hole-injection layer for organic solar cell fabrication and optimization, Sol. Energy Mater. Sol. Cells, 120 (2014) 603-609.

[11] H.W. Choi, N.D. Theodore, T.L. Alford, ZnO–Ag–MoO_3 transparent composite electrode for ITO-free, PEDOT: PSS-free bulk-heterojunction organic solar cells, Sol. Energy Mater. Sol. Cells, 117 (2013) 446-450.

[12] R. Sivakumar, R. Gopalakrishnan, M. Jayachandran, C. Sanjeeviraja, Characterization on electron beam evaporated α-MoO_3 thin films by the influence of substrate temperature, Curr. Appl. Phys., 7 (2007) 51-59.

[13] M. Kovendhan, D.P. Joseph, P. Manimuthu, S. Sambasivam, S.N. Karthick, K. Vijayarangamuthu, A. Sendilkumar, K. Asokan, H.J. Kim, B.C. Choi, C. Venkateswaran, R. Mohan, 'Li' doping induced physicochemical property modifications of MoO_3 thin film, Appl. Surf. Sci., 284 (2013) 624-633.

[14] A.A. Firooz, T. Hyodo, A.R. Mahjoub, A.A. Khodadadi, Y. Shimizu, Synthesis and gas-sensing properties of nano- and meso-porous MoO_3-doped SnO_2, Sens. Actuators, B, 147 (2010) 554-560.

[15] P. Tyagi, A. Sharma, M. Tomar, V. Gupta, Metal oxide catalyst assisted SnO_2 thin film based SO_2 gas sensor, Sens. Actuators, B, 224 (2016) 282-289.

[16] V. Galstyan, E. Comini, C. Baratto, G. Faglia, G. Sberveglieri, Nanostructured ZnO chemical gas sensors, Ceram. Int., 41 (2015) 14239-14244.

[17] L.-l. Sui, Y.-M. Xu, X.-F. Zhang, X.-L. Cheng, S. Gao, H. Zhao, Z. Cai, L.-H. Huo, Construction of three-dimensional flower-like α-MoO_3 with hierarchical structure for highly selective triethylamine sensor, Sens. Actuators, B, 208 (2015) 406-414.

[18] S.S. Mahajan, S.H. Mujawar, P.S. Shinde, A.I. Inamdar, P.S. Patil, Structural, optical and electrochromic properties of Nb-doped MoO_3 thin film, Appl. Surf. Sci., 254 (2008) 5895-5898.

[19] J. Kaur, V.D. Vankar, M.C. Bhatnagar, Effect of MoO_3 addition on the NO_2 sensing properties of SnO_2 thin film, Sens. Actuators, B, 133 (2008) 650-655.

[20] M.M.Y.A. Alsaif, M.R. Field, T. Daeneke, A.F.

Chrimes, W. Zhang, B.J. Carey, K.J. Berean, S. Walia, J. van Embden, B. Zhang, K. Latham, K. Kalantar-zadeh, J.Z. Ou, Exfoliation Solvent Dependent Plasmon Resonances in Two-Dimensional Sub-Stoichiometric Molybdenum Oxide Nanoflakes, ACS Applied Materials & Interfaces, 8 (2016) 3482-3493.

[21] S. Balendhran, S. Walia, M. Alsaif, E.P. Nguyen, J.Z. Ou, S. Zhuiykov, S. Sriram, M. Bhaskaran, K. Kalantar-zadeh, Field Effect Biosensing Platform Based on 2D α-MoO_3, ACS Nano, 7 (2013) 9753-9760.

[22] M.M.Y.A. Alsaif, A.F. Chrimes, T. Daeneke, S. Balendhran, D.O. Bellisario, Y. Son, M.R. Field, W. Zhang, H. Nili, E.P. Nguyen, K. Latham, J. van Embden, M.S. Strano, J.Z. Ou, K. Kalantar-zadeh, High-Performance Field Effect Transistors Using Electronic Inks of 2D Molybdenum Oxide Nanoflakes, Adv. Func. Mater., 26 (2016) 91-100.

[23] A.H. Omran Alkhayatt, S.K. Hussian, Fluorine highly doped nanocrystalline SnO_2 thin films prepared by SPD technique, Mater. Lett., 155 (2015) 109-113.

7

Energy modeling and simulation including particle technologies within single and double pass solar air heaters

Fahimeh Ebrahim Asghari, Hossein Ghadamian*, Mohammad Aminy

Department of Energy, Materials and Energy Research Center (MERC), Tehran, Iran

HIGHLIGHTS

- Quantitative compare of the efficiencies in one and double pass solar air heaters.
- Target optimum performance & decide to select number of passes in solar air heaters.
- Double pass efficiency is higher in more inlet air temperature & lower solar irradiation.

GRAPHICAL ABSTRACT

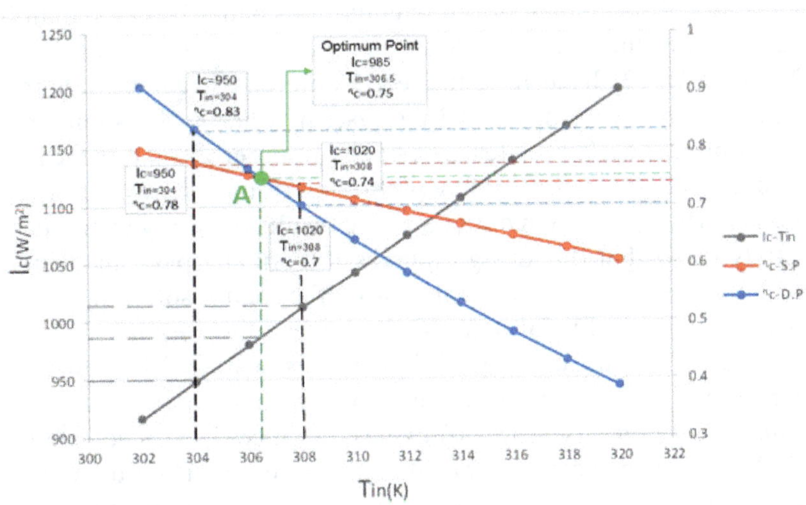

ARTICLE INFO

Keywords:
Energy modeling
Energy performance
Particles parameters
Heat transfer

ABSTRACT

In order to obtain the best performance of the solar air heaters, it is necessary to find optimum performance conditions. The aim of this research paper is to achieve optimum conditions, by comparing single and double pass solar air heaters. Also, a brief review study of various related research works of all scenarios for a single and double pass and packed bed (including particle technologies) solar air heaters was carried out to observe the challenges of the mentioned systems. Energy modeling and simulation with EES and MATLAB open source code software indicated significant results in efficiency. According to the obtained results, it can be explained that double pass duct not necessarily always increases the overall system energy efficiency. Results of this work indicate, higher ambient air temperature (inlet air temperature) and lower solar irradiation can increase overall energy efficiency of solar double pass systems. More precisely at solar irradiation of 916 W/m^2 and inlet air temperature of 302 K, the system achieves the targeted optimum value in energy efficiency, approximately 90%, which is considerably more than the 65% as an average value.

* *Corresponding author.*
E-mail address: h.ghadamian@merc.ac.ir

1. Introduction

One method to develop and improve solar air heaters is to increase the heat transfer rate by increasing the number of air channels (Pass). The mentioned method has its advantages and disadvantages. Although improvement in heat transfer due to increasing surface area assumes an advantage but pressure drop and therefore momentum energy reduction is a disadvantage of this method. Therefore, there is an optimized range that is being addressed in this study [1].

In the field of double and single pass solar air heaters, many studies have been carried out, where they compare solar air heaters in different conditions and features, some of them are as follow;

C.D. Ho et al. [2] have investigated double pass solar air heaters by using recycling system. This model of double pass air heater is designed in order to increase inlet air temperature by premixing of inlet air and partially recycled outlet air. The results of their study show that this design increases thermal driving force and subsequently, increases the heat loss which is not useful for solar air heater energy efficiency at all. On the other hand, it increases heat transfer coefficient in forced convection approach that causes to improve energy efficiency. As mentioned, this design can cause a considerable increase in efficiency.

In research work of Prashant Dhiman et al. [3], in order to analyze the system, a Parallel Flow Packed Bed Solar Air Heater (abbreviated PFPBSAH) has been considered. That consist of two glass covers, two plates, and two air channels, above and between the plates with porous media, which is above the absorber plate. A packed bed is a volume of porous media obtained by packing particles of selected material into a duct. Also, PFPBSAH is found to perform more efficiently than the conventional non-porous double flow solar air heaters with 10%~20% increase in its thermal efficiency. Significant improvement in the packed bed solar air heater performance can be obtained through a careful choice of packed bed particles parameters and the fraction of mass flow rate [4].

A.A. El-Sebaii et al. [5] investigated double pass air heaters with two glass cover in two scenarios, First one with packing only in lower air ducts, the other in both upper and lower air ducts. Investigated parameters in this paper include: outlet air temperature, outlet heat power, pressure drop and hydraulic efficiency. Mass flow rate of air and filled porous materials are effective in the energy efficiency of the system. B.M. Ramani et al. [6] have implemented and simulated double pass solar air heater with parallel flow. In this research, not any comparison between single and double pass solar air heaters has not been done, but Ravi Kant Ravi et al. [7] and Ghadimi, M., et al. [8] have investigated a variety of different techniques to improve energy efficiency of double pass solar air heaters such as using packed-bed, extended surfaces and corrugated absorbent surfaces. These works include cost analysis, thermo-hydraulic properties and design of these type of air heaters. They used developed correlations for comparison purposes among energy efficiency of different designs. Besides, double pass air heater with one added glass cover (air heater with two glass covers) was investigated in their studies.

In summary, many research works, double pass solar air heaters with various features are compared. However, as a challenge of comparison, it is necessary to compare the performance of single and double pass air heaters by energy modeling and numerical solution, in terms of temperature (thermal) and energy.

The new opinion of this study is quantitative comparison and optimization between single and double pass solar air heaters, as a challenge which has not been done up to now. The result of this research issued that in a certain range of performance and application, each of single and double pass can have better efficiency. This viewpoint (kind of comparison) is not available in the previously mentioned investigations. In fact, innovative results of this work can provide the ability to predict optimum solar air heater type according to the required solar irradiation and inlet temperature range.

2. System energy modeling

The overall Figures of single and double pass solar air heaters are shown in Fig. 1 & 2. Double pass air heaters can have different designs, the model shown in Fig 2, has been applied in the coding procedure [9].

To simplify and solve the systems equations, the following assumptions are considered [10].

1. The air heater operates under steady state conditions;
2. The capacitance of the absorber plate is negligible;
3. The temperature of air varies only in the direction of the flow (x direction);
4. The area of the absorber plate is equal to the aperture area of the air heater;

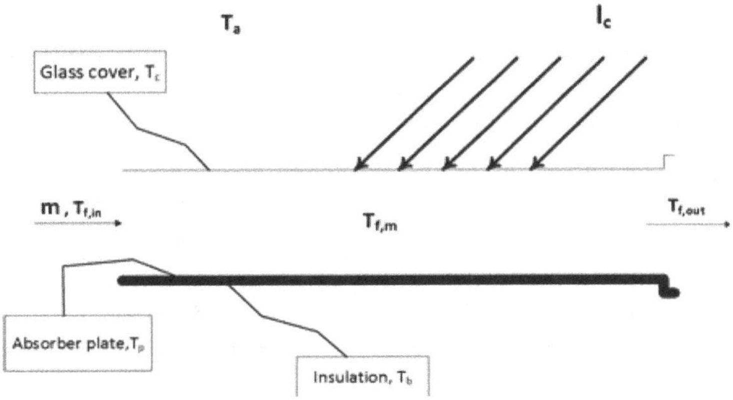

Fig. 1. Single pass solar air heater

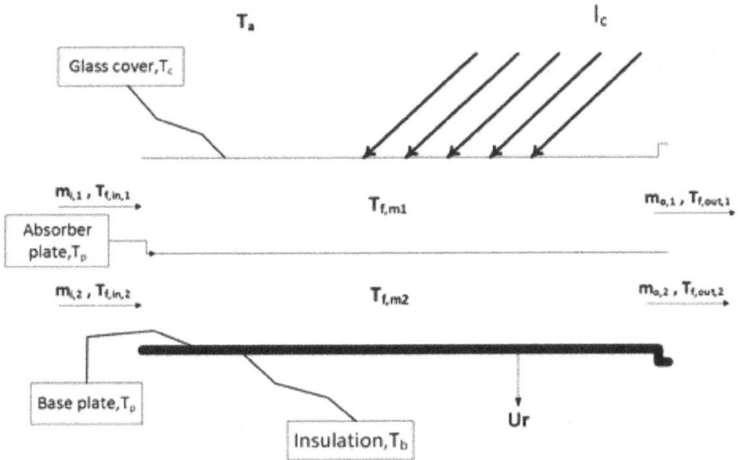

Fig. 2. Double pass solar air heater

5. The heat flow through the glazing and the base plate is one dimensional and in the y-direction.

To peruse the aim of energy modeling of single pass solar air heater, the extracted equations from resources were coded in EES software. Simultaneously, the energy modeling of double pass air heater is done in MATLAB software. Both of software are open source codes type. But double pass energy modeling has been concluded in MATLAB software because of limitations in EES software. Furthermore, the governing equations that used in energy modeling, for example, convection and radiation heat transfer coefficients (Equations 1~3), are presented in Table 1 [10-12].

Table 1.
Heat Transfer Coefficient Equations

$h_w = k(75+0.42\,\mathrm{Re}^{0.6})/X$	(1)
$h_{r,cs} = \sigma\varepsilon_c(T_c^2+T_s^2)(T_c+T_s)$	(2)
$h_{r,pc} = \sigma(T_p^2+T_c^2)(T_p+T_c)/\{1/\varepsilon_p+1/\varepsilon_i-1\}$	(3)

Heat loss coefficients (Equations 4 & 5) are used to calculate the waste heat from the air heater.

Table 2.
Heat Loss Coefficient Equations

$U_e = (k/x_1)\times s/W$	(4)
$U_b = \dfrac{1}{(x_1/k)+(1/h_b)}$	(5)

Energy and heat balances in different parts of solar air heaters are applied to execute Equations 6~10 in Table 3[10-12].

Energy efficiency in the solar air heater is obtained from the Equation 12. Furthermore, Equation 11 is applied to calculate mass flow inlet of each channel in double pass solar air heater [10-12].

Regarding the assumptions, simplified equations are solved in the mentioned program, based on an algorithm that is shown in Fig. 3.

Table 3.

Heat and Energy balances

$\bar{S}_c + h_{r,pc}(T_p - T_c) - h_1(T_f - T_c) = h_w(T_c - T_a) + h_{r,cs}(T_c - T_s)$	(6)
$\bar{S}_p = h_3(T_p - T_{f,2}) + h_2(T_p - T_{f,1}) + h_{r,pc}(T_p - T_c) + h_{r,pb}(T_p - T_b)$	(7)
$h_2 \times A_c \times (T_p - T_f) = h_1 \times A_c \times (T_f - T_c) + \dot{m} \times C_{pa} \times (T_{f1,out} - T_{f1,in}) + U_e \times L \times s \times (T_f - T_a)$	(8)
$h_3 A_c (T_p - T_{f,2}) = h_4 A_c (T_{f,2} - T_b) + \dot{m}_2 C_{pa} (T_{f2,out} - T_{f2,in}) + U_{e,2} s_2 L (T_{f,2} - T_a)$	(9)
$h_4 (T_{f,2} - T_b) + h_{r,pb} (T_p - T_b) = U_b (T_b - T_a)$	(10)

Table 4.

Energy efficiency and Mass flow rate

$\dot{m}_i = \rho_i W_{s_i} C_{V,i} F_{V,i} \sqrt{2\left\{233.9\rho_a h\left(\frac{T_{f,out} - T_a}{T_{f,out} + \frac{7}{13}T_a}\right)\right\}}$	(11)
$\eta_c = \frac{\dot{m}_1 C_{p,1}(T_{f,1out} - T_{f,1in}) + \dot{m}_2 C_{p,2}(T_{f,2out} - T_{f,2in})}{I_c \times A_c}$	(12)

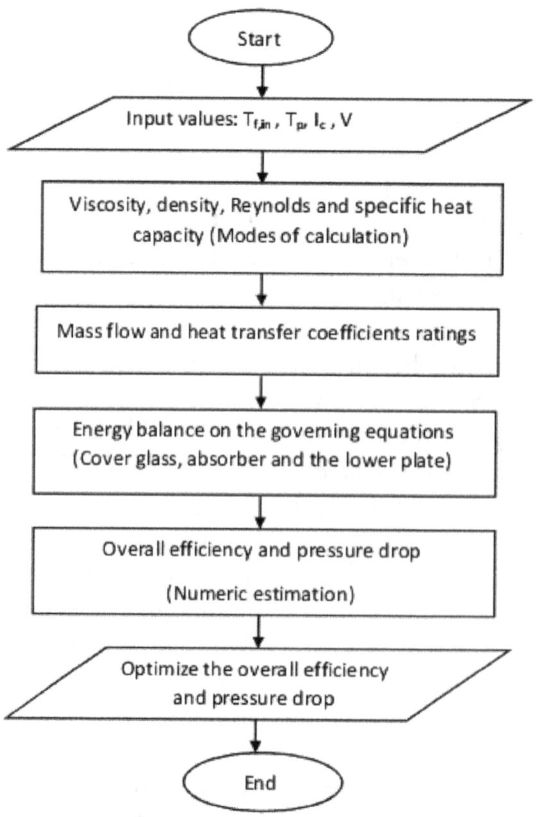

Fig. 3. The algorithm of simulation

3. Results and discussion

According to the algorithm in Fig. 3, coding procedure has been carried out in EES and MATLAB software, the obtained results are shown in Table 2. Regarding the studies, the parameters that have the most effect on final energy efficiency and outlet temperature are inlet air temperature (T_{in}) and inlet solar irradiation (I_c) to the solar air heater. Therefore, graphs are depicted based on these two parameters [13]. Fig. 4, 5 & 6 are demonstrating diagrams that concluded by numerical solution of single and double pass studied systems. Fig. 4 shows energy efficiency diagram according to inlet solar irradiation and air temperature to the single pass air heater. In Fig. 5, outlet air temperature versus inlet solar irradiation and air temperature in double pass air heater is depicted in EES software. In Fig. 6, the energy efficiency curve versus inlet irradiation and air temperature is shown for double pass air heater in MATLAB software.

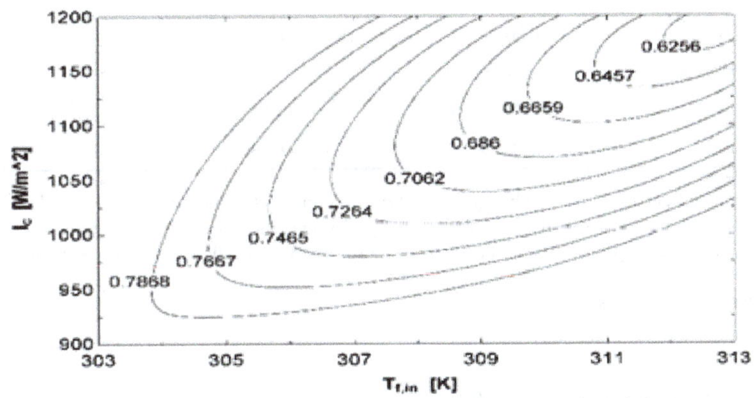

Fig. 4. Thermal efficiency diagram of one pass solar air heater

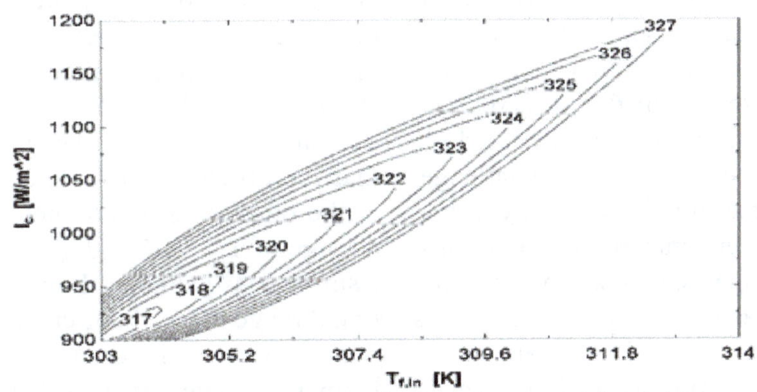

Fig. 5. Outlet air temperature diagram of one pass solar air heater

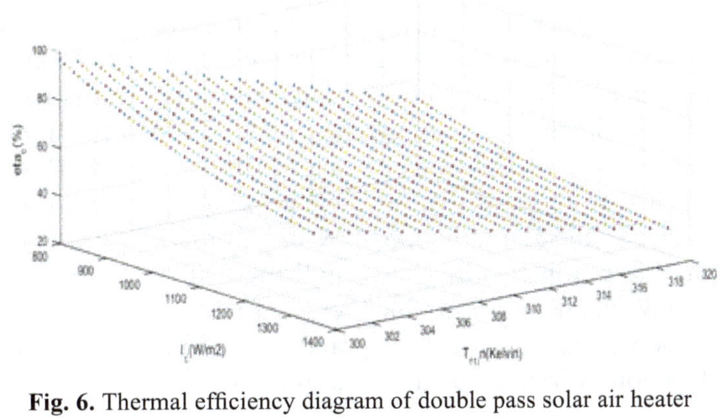

Fig. 6. Thermal efficiency diagram of double pass solar air heater

The results of calculations are given in the following Table 5.

Table 5.
Results of energy modeling

Inletairtemperature (K)	Inlet Irradiation (W/m^2)	Single pass air heater thermal efficiency	Double pass air heater thermal efficiency	Double pass packed bed (with particles) air heater overall efficiency [3]
300	888	0.8174	0.9106	0.48*1
302	916	0.7964	0.9083	N.A.*2
304	948	0.7752	0.8345	0.50*1
306	980	0.7539	0.7656	0.41*1
308	1012	0.7327	0.7011	N.A.*2
310	1042	0.7115	0.6417	0.44*1
312	1074	0.6903	0.5845	N.A.*2
314	1106	0.6691	0.5307	N.A.*2
316	1388	0.6478	0.4799	N.A.*2
318	1168	0.6266	0.4326	N.A.*2
320	1200	0.6054	0.3869	N.A.*2

*1: As a challenge, the generally overall efficiency of packed bed particles type, in comparison with none-packed is lower than in rate, that is because of more pressure drop taking place in bed type. As an issue, the bed type is always followed is thermal efficiency instead of overall efficiency, which assist them to ignore the important term of pressure drop, that affects the significant drop in overall efficiency. This research article considers all terms for purpose of comparison. [3]
*2: Not Available data based on previously referred studies. [3]

According to the Fig. 4, 5 & 6 in both single & double systems, rising in temperature and solar irradiation, negatively affects the efficiencies. It is observed that the effect of decreasing in energy efficiency, the solar irradiation is more than the inlet temperature value. According to the results of this research and the implemented model, generally double passed solar air heaters are better options, but as a challenge, it should be considered that this is valid only at low solar irradiation and higher temperature conditions. In this situation, the system shows a more rational energy efficiency; Such as the Iran northern regions and the tropics. While in the single pass air heater, any increase in solar irradiation as well as inlet air temperature results in lower inlet and outlet temperature difference. Also in low irradiation intensity and inlet air temperature, we observe an increase in energy efficiency. This suggests that outlet air temperature rising is not necessarily reflect the increasing energy efficiency of the single pass solar air heaters, But the appropriate range of both can increase energy efficiency.

Optimizing process for air heater contains two independent variables, temperature and solar radiation, besides, one dependent variable is outlet efficiency (outlet temperature). Efficiency functions for double and single pass systems are non-linear and those are strongly depended on endogenous variables contain geometry, an absorber plate and etc. The regression of the performance functions of one and double passes system, and finding their cross (optimum) point A, (I_c=965 W/m^2, T_{in}=306.5 K, η_c=0.75) is obtained and demonstrated in Fig. 7. The results of the study show that the efficiency of double pass system is higher in the range of T_{in}<306 and/or I_c <985, whereas for a range of T_{in} >306 and/or I_c >985, the efficiency of a single pass is higher. It is understood, by the functional range in Fig. 7, often, double passes systems is not effective in higher inlet solar irradiations and air temperatures ranges. This fact is because of absorber inherent opposite behavior. By these means, in this mentioned conditions, that works like thermal shield rather than absorber.

and single pass systems are non-linear and those are strongly depended on endogenous variables contain

According to the results of Table 5, for the double pass air heater performance, appropriate temperature range and solar irradiation for optimum performance are under 306 K and 1000 W/m^2, respectively. As it shown in Fig. 7, The Best energy efficiency in double pass solar air heater is marked at point A (I_c=965 W/m^2, T_{in}=306.5 K, η_c=0.75). The quantitative results indicate that the double pass air heater has better energy efficiency in areas with higher air temperature and lower irradiation. While temperature and irradiation rising has less impact on single pass air heater.

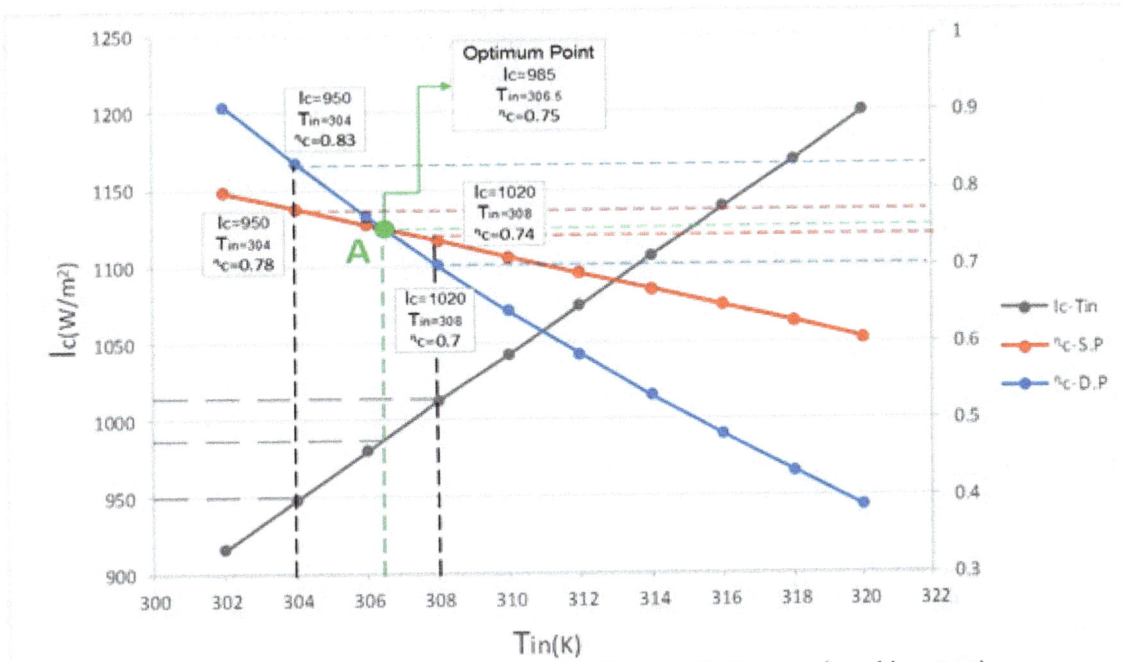

Fig. 7 .Single & double pass comparison in efficiency (Optimum point achievement)

heater. It should be noted that the energy efficiency does not fall extremely with sharp steep, such as double pass system.

4. Conclusions

The following conclusions are obtained that the foregoing theoretical studies and numerical calculations:

- Based on the above results, the double passed system does not necessarily increase the energy efficiency and it strongly depends on inlet situations.
- Double pass air heater in areas with higher inlet temperature and lower irradiations, shows better in performance.
- With the inlet air temperature of 302 K and inlet solar irradiation 916 W/m², the system energy efficiency of 90% is achieved in double pass systems as a considerable result.
- With lower solar irradiations & inlet air temperature values, the double pass packed bed type solar air heater shows lower overall efficiency compare with none packed bed types, based on critical obligatory pressure drop taking place inside the bed.

Nomenclature

A_c	(m²)	absorber surface area
Δp	(pa)	pressure drop
Re		Reynolds number
V	(m/s)	air velocity
I_c	(W/m²)	energy incident per unit area per unit time
T_a	(K)	ambient temperature
$T_{f,in}$	(K)	inlet air temperature
$T_{f,out}$	(K)	outlet air temperature
T_p	(K)	absorber plate temperature
T_b	(K)	base plate temperature
T_c	(K)	glass cover temperature
T_s	(K)	sky temperature
m_i	(kg/s)	mass flow rate inlet to each channel
h_w	(W/m².K)	Wind heat transfer coefficient
$h_{r,cs}$	(W/m².K)	Radiant heat transfer from solar air heater cover to sky
$h_{r,pc}$	(W/m².K)	Radiation heat transfer coefficient from air heater absorber plate to cover
$h_{r,pb}$	(W/m².K)	Radiation heat transfer coefficient from air heater absorber plate to base plate
U	(W/m².K)	heat lose coefficient
η_c		energy efficiency
σ	(kW/m².K⁴)	Stefan-Boltzman constant
ε		emittance
k	(W/m K)	thermal conductivity
ρ	(kg/m³)	density of air
s	(m)	the depth of air heater channel
W	(m)	the width of the air heater
$\overline{S}$	(W/m²)	monthly average absorbed solar irradiation per unit area
C_p	(J/kg K)	specific heat capacity of air

Subscripts

1	top (upper) channel

2	bottom (lower) channel
a	ambient
b	base plate
c	glass cover
p	absorber plate
s	sky
in	inlet
out	outlet

References

[1] F. Esmailie, H. Ghadamian, and M. Aminy, Modeling and simulation of a solar flat plate collector as an air heater considering energy efficiency, Mech. Ind., 15 (2014) 455–464.

[2] C. D. Ho, H. M. Yeh, and R. C. Wang, Heat-transfer enhancement in double-pass flat-plate solar air heaters with recycle, Energy, 30 (2005) 2796–2817.

[3] P. Dhiman, N. S. Thakur, A. Kumar, and S. Singh, An analytical model to predict the thermal performance of a novel parallel flow packed bed solar air heater, Appl. Energy, 88 (2011) 2157–2167.

[4] S. Singh and P. Dhiman, Thermal performance of double pass packed bed solar air heaters - A comprehensive review, Renew. Sustain. Energy Rev., 53 (2016) 1010–1031.

[5] a. a. El-Sebaii, S. Aboul-Enein, M. R. I. Ramadan, and E. El-Bialy, Year round performance of double pass solar air heater with packed bed, Energy Convers. Manag., 48 (2007) 990–1003.

[6] W. Colborne and W. Moffat, Fundamental analysis of chimney performance. ASHRAE J1(3). 1959.

[7] R. K. Ravi and R. P. Saini, A review on different techniques used for performance enhancement of double pass solar air heaters, Renew. Sustain. Energy Rev., 56 (2016) 941–952.

[8] M. Ghadimi, H. Ghadamian, A. Hamidi, F. Fazelpour, and M. Behghadam, Analysis of free and forced convection in airflow windows using numerical simulation of heat transfer, Int. J. Energy Environ. Eng., 3 (2012) 1–10.

[9] S. Suman, M. K. Khan, and M. Pathak, Performance enhancement of solar collectors - A review, Renew. Sustain. Energy Rev., 49 (2015) 192–210.

[10] H. Rajakaruna, F. K. Forson, and M. a. a. Nazha, Experimental and simulation studies on a single pass, double duct solar air heater, Energy Convers. Manag., 44 (2003) 1209–1227.

[11] F. Esmailie, M. Aminy, and H. Ghadamian, Energy Intensity Diagnostics Contributed to Solar Dryers Energy Challenges, J. Clean Energy Technol., 3 (2015) 388-392.

[12] H. Salarian, H. Ghadamian, and M. K. Assadi, An experimental and modeling study of a dehumidification tower, Int. J. Phys. Sci., 6 (2011) 2852–2860.

[13] B. M. Ramani, A. Gupta, and R. Kumar, Performance of a double pass solar air collector, Sol. Energy, 84 (2010) 1929–1937.

8

Thin film nanocomposite forward osmosis membrane prepared by graphene oxide embedded PSf substrate

Saharnaz Tajik[1], Omid Moini Jazani[1*], Soheila Shokrollahzadeh[2], Seyed Mahdi Latifi[2]**

[1]*Department of Chemical Engineering, Faculty of Engineering, University of Isfahan, Isfahan, Iran*

[2]*Department of Chemical Technologies, Iranian Research Organization for Science and Technology (IROST), Tehran, Iran*

HIGHLIGHTS

- Novel polysulfone-graphene oxide (PSf-GO) nanocomposite substrates were prepared via phase inversion method for fabrication of thin film nanocomposite forward osmosis (TFN-FO) membranes.
- The existence of functional groups in GO nanoplates increases the substrate hydrophilicity which leads to increase in water permeability of the TFN membranes
- The substrate embedded with 0.5 wt% of GO nanoplates exhibited highest water flux which is the optimum concentration of GO loading.
- Addition of GO nanoplates to the PSf substrate is an effective method to control ICP in FO membranes.

ARTICLE INFO

Keywords:
Forward Osmosis (FO)
Nanocomposite Substrates
Graphene Oxide nanoplates
Internal Concentration Polarization (ICP)

GRAPHICAL ABSTRACT

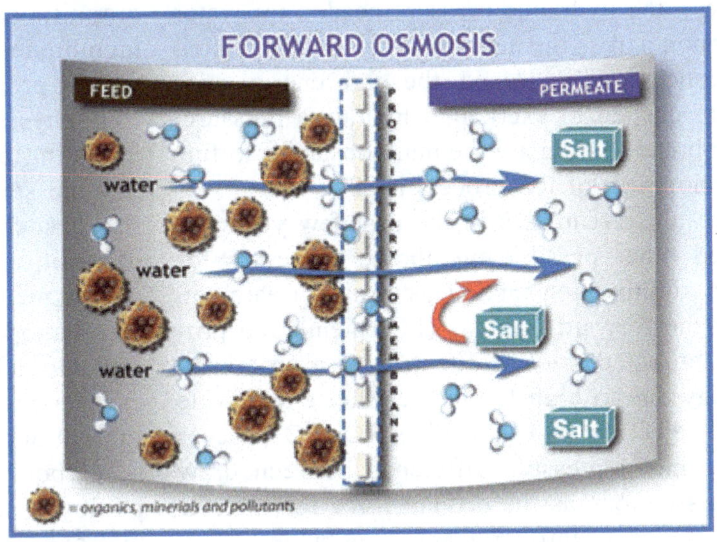

ABSTRACT

One of the limiting factors in good performance of forward osmosis (FO) membranes is the internal concentration polarization (ICP). To reduce ICP, thin film nanocomposite forward osmosis (TFN-FO) membranes were fabricated by adding different amounts of graphene oxide (GO) nanoplates (0-1 wt. %) to polymer matrix of polysulfone (PSf) substrate. The prepared nanocomposite membranes exhibited both hydrophilicity and porosity higher than that of neat PSf counterpart. An optimum amount of 0.5 wt. % was obtained for GO addition into the membranes. The corresponding fabricated thin film nanocomposite (TFN) membrane (TFNG0.5) revealed a water permeability of 2.44 L/m^{2}h bar which is 66% higher compared to an in-house made composite membrane. The FO performance of TFN was assessed by DI water as feed solution and 1 M NaCl as draw solution in AL-DS orientation. The water flux of the synthesized FO membranes increased upon adding of GO nanoplates and reached to a maximum water flux of 37.74 (L/m^{2}h) for TFNG0.5 membrane. This flux is about 3 times higher than TFC membranes without significant changes in their salt rejection. The higher water flux of the TFN membranes can be attributed to ICP decrease originating from reduction of structural parameter of the membranes.

* *Corresponding author.*
E-mail address: o.moini@eng.ui.ac.ir
** *Corresponding author.*
E-mail address: shokrollahzadeh@yahoo.com

1. Introduction

One of the main problems of human society is the shortage of potable water sources that is occurred as a result of rapid population growth and economic development [1]. Membrane technology was introduced as a solution to tackle the crisis of water scarcity through desalination of brackish/salty water with low cost [2-4]. The membrane processes such as reverse osmosis (RO) and nanofiltration (NF) are the most widely methods for seawater desalination and wastewater treatment to product potable water of high quality [5, 6]. The RO is a membrane process in which pure water is obtained by applying hydraulic pressure to the salt solution [7]. In the reverse osmosis process, the hydraulic pressure needed for water transmission across the membrane is directly related to salt concentration. Hence, the high energy consumption, needed to exert high hydraulic pressure, and membrane fouling are the main obstacles to further development of RO process [7, 8].

As a result, researchers in the past few years investigated other possible membrane processes for seawater desalination and wastewater treatment, amongst the forward osmosis (FO) membrane process has attracted a lot of attention as a promising technology due to high fouling resistance and wide range of rejection [9-11]. Unlike RO, FO process utilizes osmotic pressure difference between draw and feed solutions as the driving force for transfer of water across a semi-permeable membrane [2, 12, 13]. In comparison with the pressure-driven membrane processes, it is claimed that FO has potential advantages including lower energy consumption [14, 15], higher water recovery [16, 17], lower fouling tendency and consequently easier cleaning [18-20]. Nevertheless, one of the main problems in FO process is internal concentration polarization (ICP), an unfavorable phenomenon which limits the membrane performance by decreasing the osmotic driving force [21-23]. The ICP is inevitable phenomenon in osmotic driven membrane processes which is based on the membrane orientation. Both concentrative ICP and dilutive ICP can happen in FO processes [24, 25]. In active layer facing the draw solution (AL-DS orientation), a concentrative ICP happens in which the rejected feed solutes accumulate in the support layer of the membrane [26, 27]. In other orientation (active layer facing the feed solution, AL-FS), a dilutive ICP occurs through dilution of the draw solution in support layer by water permeation from feed solution side and aggregation of the feed solutes in the vicinity of the active layer [1, 28]. Therefore, for ICP reduction, the support layer with minimal structural parameter value (S, thickness tortuosity/porosity) is preferred in FO process which leads to a better membrane performance (higher water flux and better solute rejection) [29, 30].

FO membranes with good performance are developed by tailoring their structure [31-34]. In this regard, addition of nanomaterials into the polyamide layer as the membrane substrate is introduced as an effective way to modify the properties and structure of the membrane [27, 28, 35, 36]. By addition of hydrophilic nanomaterials such as titanium dioxide [37], zeolite [38], silica [39], multiwalled carbon nanotubes [40] and graphene oxide modified graphitic carbon nitride [41] into the substrate, FO membranes show improved performance. Due to unique properties, nanomaterials may cause changes in physical-structural properties of the membrane including porosity, hydrophilicity, tensile strength, thermal stability and ICP [42-45].

Emadzadeh et al. [46, 47] investigated the effect of addition of TiO2 nanoparticles into PSf matrix on the membrane performance. They reported that TiO_2 nanoparticles played an important role in water flux due to ICP reduction. Wang et al. [40] fabricated nanocomposite substrate using carbon nanotube for FO applications. They showed that the adding of carbon nanotubes caused changes in the morphology and membrane properties such as porosity, hydrophilicity and structural parameters. Their prepared nanocomposite membranes showed salt rejection over 90%. The presence of zeolite nanoparticles in substrate of FO membranes was also investigated by Ma et al. [38]. The nanocomposite substrates exhibited lower structural parameter and minimal ICP which led to increase in the water flux. Graphene oxide (GO) nanoplates have attracted considerable attention as a filler due to their unique properties such as (1) high specific surface area which creates better interaction with the polymeric support layer; (2) two dimensional nanostructure; (3) high chemical-mechanical stability and (4) excellent hydrophilicity due to existence of oxygenous functional hydrophilic groups such as hydroxyl, carboxyl and carbonyl groups [48-50]. A literature review shows that researchers have utilized GO in preparation of the membranes for different applications. It is confirmed that incorporation of GO improves water flux, enhances salt rejection, increases hydrophilicity and decreases fouling [51-53].

In this study, novel polysulfone-graphene oxide

(PSf-GO) nanocomposite substrates were prepared via phase inversion method for fabrication of thin film nanocomposite forward osmosis (TFN-FO) membranes. The membrane performance is evaluated in terms of water flux, salt rejection, and water/salt permeability. The effects of GO addition on the morphology and structure of the prepared membranes are analyzed by FESEM, AFM and contact angle measurements.

2. Experimental

2.1. Materials

The polysulfone (PSf, Ultrason S 6010) as the polymeric substrate was purchased from BASF Co., Germany. N-methyl-pyrrolidone (NMP, 99.5%, Daejung) and polyvinyl pyrrolidone (PVP-K30, M_w=40,000 g/mol, Daejung) were used as a solvent and a pore former in preparation of the casting solution. The graphene oxide (GO) nanosheets prepared by Staudenmaier method [54] were used for the modification of the substrate properties. M-phenylenediamine (MPD, ≥99%, Acros organics), trimesoyl chloride (TMC, Merck) and n-hexane (>95%, Daejung) were employed to fabricate the PA layer of the membranes. Sodium chloride (NaCl, 99.9%, Pars Namak Co.) was used for the FO tests.

2.2 Preparation of nanocomposite substrates

The flat sheet PSf/GO nanocomposite substrates were prepared using conventional phase inversion method [55]. The casting solution was composed of PSf (12 wt%), PVP (1 wt%) and accurate amounts of GO nanoplates in NMP solvent. The substrates based on their composition are listed in Table 1. Accurate amounts of GO nanoplates (0.1, 0.5 and 1 wt%) were dispersed in NMP using ultrasonication (360 W, 20 kHz, misonix, New York) for 30-45 min (depending on the concentration of GO). Then, PSf and PVP were dissolved in the prepared solution and stirred at 60 °C for 24 h to obtain a homogeneous solution. The polymer solution was degassed degassed to remove air bubbles followed by casting on a clean glass plate using a casting knife with the gate height of 100 μm. The casted polymer solution was immediately immersed in the tap water bath. After few minutes, the formed membrane was separated from the surface glass and washed with deionized water. Finally, the membranes were stored in deionized water bath for 24 h to remove the residual solvent. The membranes were placed between two paper filters for 24 h at room temperature to dry.

Table 1.
The composition of PSf substrate matrix.

Substrate type	PSf (wt. %)	PVP (wt. %)	GO nanoplates (wt. %)
G0	12.0	1.0	-
G0.1	12.0	1.0	0.1
G0.5	12.0	1.0	0.5
G1	12.0	1.0	1.0

2.3. Fabrication of polyamide rejection layer by interfacial polymerization

The PA rejection layer was fabricated on top surface of the PSf and PSf/GO substrates by interfacial polymerization reaction of MPD and TMC monomers. The dried substrate was taped to a glass plate and was immersed in a 2 wt% MPD aqueous solution for 2 min and the excess MPD solution was removed from the substrate surface by a paper filter. The MPD saturated substrate was then soaked in a 0.1 wt. % TMC/n-hexane solution for 1 min. This step is the start of interfacial polymerization reaction which leads to formation of ultrathin PA layer. Finally, the fabricated membranes were washed and stored in deionized water bath at 5 °C. The fabricated membranes with G0, G0.1, G0.5 and G1 substrates were named as TFC, TFNG0.1, TFNG0.5 and TFNG1, respectively.

2.4. Characterization

2.4.1. Characterization of the membranes morphology

The morphologies of the top surface and the cross section of both substrates and TFN membranes were observed by a Field Emission Scanning Electron Microscope (FESEM, JSM-7610F, JEOL). In order to scan cross section, the samples were fractured in liquid nitrogen and were dried for 24 h at room temperature. For imaging, the samples were coated with an ultrathin layer of gold by sputter coater. Atomic Force Microscopy (AFM, DME DualScope 95, Denmark) was used to examine surface morphology and roughness of the membranes in non-contact mode.

2.4.2. Measurement of contact angle

To investigate the hydrophilicity of the substrate as a function of GO loading, the contact angle of deionized (DI) water placed on the substrate surface was measured by contact angle measuring instrument (G10, KRUSS, Germany). Using sessile drop method, the contact angle (CA) for each sample was measured in 5 random locations and the average value was reported.

2.4.3. Determination of substrate porosity

The overall porosity (ε) of the substrate membrane was calculated by gravimetric method. Initially, dried samples (2 2 cm) were weighed (m_d , g). Then, the samples were immersed in water bath for 24 h. The residual water on the surface of the wet substrate was quickly removed by a clean tissue and the samples were immediately re-weighed (m_w, g). The porosity was calculated by the following equation [34]:

$$\varepsilon = \frac{(m_w - m_d)/\rho_{H_2O}}{(m_w - m_d)/\rho_{H_2O} + m_d/\rho_p} \tag{1}$$

Where ρ_{H_2O} (1.0 g/cm^3) and ρ_p (1.23 g/cm^3) are the densities of water and PSf, respectively.

2.5. Determination of water and salt permeability of the membranes

The water and salt permeability of the fabricated membranes were determined using a dead end filtration unit [30, 32]. The dead end stirred cell volume and effective area were 400 cm^3 and 19.6 cm^2, respectivley. The feed reservoir was pressurized using nitrogen gas. The water permeability (A, L/m^2 h bar) was calculated from the pure water permeation fluxes under pressure of 3 bar.

$$A = \frac{J_w}{\Delta P} \tag{2}$$

Where J_w is the permeate water flux (L/m^2 h) and (bar) is the applied hydraulic pressure [56].

A 2000 mg/L NaCl solution was used for determination of salt permeability. The salt permeability (B, L/m^2 h or m/s) was calculated from the following solution-diffusion theory, where in the salt rejection (R) was determined by conductivity measurement of the both permeate and feed solutions while a conductivity-concentration calibration curve was used. [57].

$$\frac{1-R}{R} = \frac{B}{A(\Delta P - \Delta \pi)} \tag{3}$$

Where A (L/m^2 h bar) is water permeability,ΔP (bar) and Δπ (bar) are pressure difference and osmotic pressure difference, respectively. Also, the salt rejection was calculated as follows:

$$R(\%) = \left(1 - \frac{C_p}{C_f}\right) \times 100 \tag{4}$$

Where C_f (mol/L) and C_p (mol/L) are the salt concentration of the feed and the permeate solutions, respectively [56].

2.6. FO performance of the membranes

The separation performance of the fabricated FO membranes (water flux and salt reverse diffusion) was evaluated using FO lab-scale setup [58] with an effective membrane area of 15.8 cm2. In all experiments, DI water and 1.0 M NaCl solutions were used as the feed and draw solutions, respectively. The membranes were tested in (AL-DS) orientation as active layer faces with the draw solution. The water flux (J_v , L/m^2 h), from the feed solution to the draw solution was calculated (Equation 5) by measuring the volume change of the draw solution during the tests.

$$J_v = \frac{\Delta V_{draw}}{A_m \times \Delta t} \tag{5}$$

In this equation, ΔV_{draw} (L) is the volume change of the draw solution, Δt (h) refers to the time interval and A_m (m^2) is the effective membrane area.

Using a conductometer (LF96, WTW) and applying the conductivity calibration curve, the salt concentration in the feed solution was measured. The salt reverse diffusion (J_s , g/m^2 h) is the salt leakage rate from the draw solution to the feed solution which was calculated by the change of salt concentration in the feed solution.

$$J_s = \frac{V_t.C_t - V_0.C_0}{A_m.\Delta t} \tag{6}$$

The salt leakage rate is calculated through equation 6 where V_0 (L) and V_t (L) are the initial and final volumes of the feed solution, respectively; C_0(mol/L) and C_t(mol/L) are the initial and final salt concentrations of the feed solution, respectively.

2.7. Determination of structural parameter

The structural parameter can be predicated based on the classical model developed by Loeb et al. [26], by the following equations.

In the AL-FS mode (the active layer is facing the feed solution)

$$S = \frac{D}{J_v} \ln \frac{B + A\pi_{D,b}}{B + A\pi_{F,b} + J_v} \quad (7)$$

In the AL-DS mode (the active layer is facing the draw solution)

$$S = \frac{D}{J_v} \ln \frac{B + A\pi_{D,b} - J_v}{B + A\pi_{F,b}} \quad (8)$$

In the aforementioned equations, D (m^2/s) is the solute diffusion coefficient; $\pi_{D,b}$ (bar) and $\pi_{F,b}$ (bar) are the osmotic pressures of the bulk draw and feed solutions, respectively. J_V (L/m^2 h) is also the FO water flux.

3. Results and discussion

3.1. Characterization of the nanocomposite substrates

The effect of adding GO nanoplates on PSf matrix and also its loading on properties and morphology of the fabricated substrates were investigated. The results were compared to the properties and morphology of the pure PSf substrate.

3.1.1. Influence of adding GO nanoplates on the substrate properties

The overall porosity of the composite and nanocomposite substrates prepared in this study was measured and listed in Table 2. As it can be seen, the PSf/GO substrates except G1 have a higher porosity than PSf substrate having no GO nanoplates. Nevertheless, the porosity of all the substrates is in a suitable range (between 74% and 82%), which can be attributed to the low concentration of PSf and PVP additive [59]. The higher porosity of the nanocomposite substrates is due to perch of nanoplates between the intertwined polymer chains which reduces their compaction. Also, the existence of GO hydrophilic nanoplates into the polymer matrix accelerates the phase inversion process and finally leads to thickness decrease of the substrate [60]. However, the results indicate that in high concentrations of GO (more than 0.5 wt%) the substrate porosity is reduced. It is probably due to increased viscosity of the casting solution in high concentrations of GO. As GO concentration exceeds 0.5%, hydrophilicity of the membranes improves significantly and thickness of the substrate increases due to reduced solvent/non-solvent exchange rate. [60, 61]. Zinadini et al. [50] observed the similar behavior for PES/GO membrane.

The hydrophilicity of the membranes is one of the important factors which affects the ICP and water flux, and is evaluated based on water contact angle with the membrane surface [30]. Reducing the contact angle means the hydrophilicity improvement of the prepared membranes. As summarized in Table 2, the contact angle of the substrates first decreases upon adding GO nanoplates to the polymer solution and then increases. Previous studies also demonstrated that adding of hydrophilic nanomaterials into a PSf matrix as fillers increases the substrate hydrophilicity [37, 38, 40, 47, 50, 52]. Pure PSf substrate showed higher contact angle (86°), while the contact angle decreased to 75.1° and 69.8° with adding 0.1 and 0.5 wt% of GO nanoplates, respectively. The increase in hydrophilicity coupled with the decrease in thickness of the nanocomposite substrates (Table 2) may facilitate passage of water molecules through the membranes showing less resistance. This process results in improvement of the membrane performance [37, 62]. Nevertheless, it was observed that the contact angle in high concentration of GO nanoplates (1 wt.%) increased to 72.4° which signifies that high amount of GO addition into the polymer solution does not affect the substrate hydrophilicity positively. It is believed that at high concentrations, GO nanoplates are agglomerated on the surface of the substrate, which leads to reduction in effective surface of the nanoplates as well as reduces population of the functional groups existing on the substrate surface [50, 63].

3.1.2. Influence of GO nanoplates on the substrate morphology

The FESEM images of the top surface and cross section of both composite and nanocomposite substrates are illustrated in Fig. 1. The images of the top surface of the membranes show that the nanocomposite substrates (G0.5 and G1 in Fig. 1(b) and (c))

Table 2
Properties of the synthesized substrates with different concentrations of GO nanoplates.

Substrate	Thickness (μ m)	Porosity (%)	Contact angle (°)
G0	87.19	74.51	86.0
G0.1	85.43	77.91	75.1
G0.5	84.80	82.92	69.8
G1	85.91	75.60	72.4

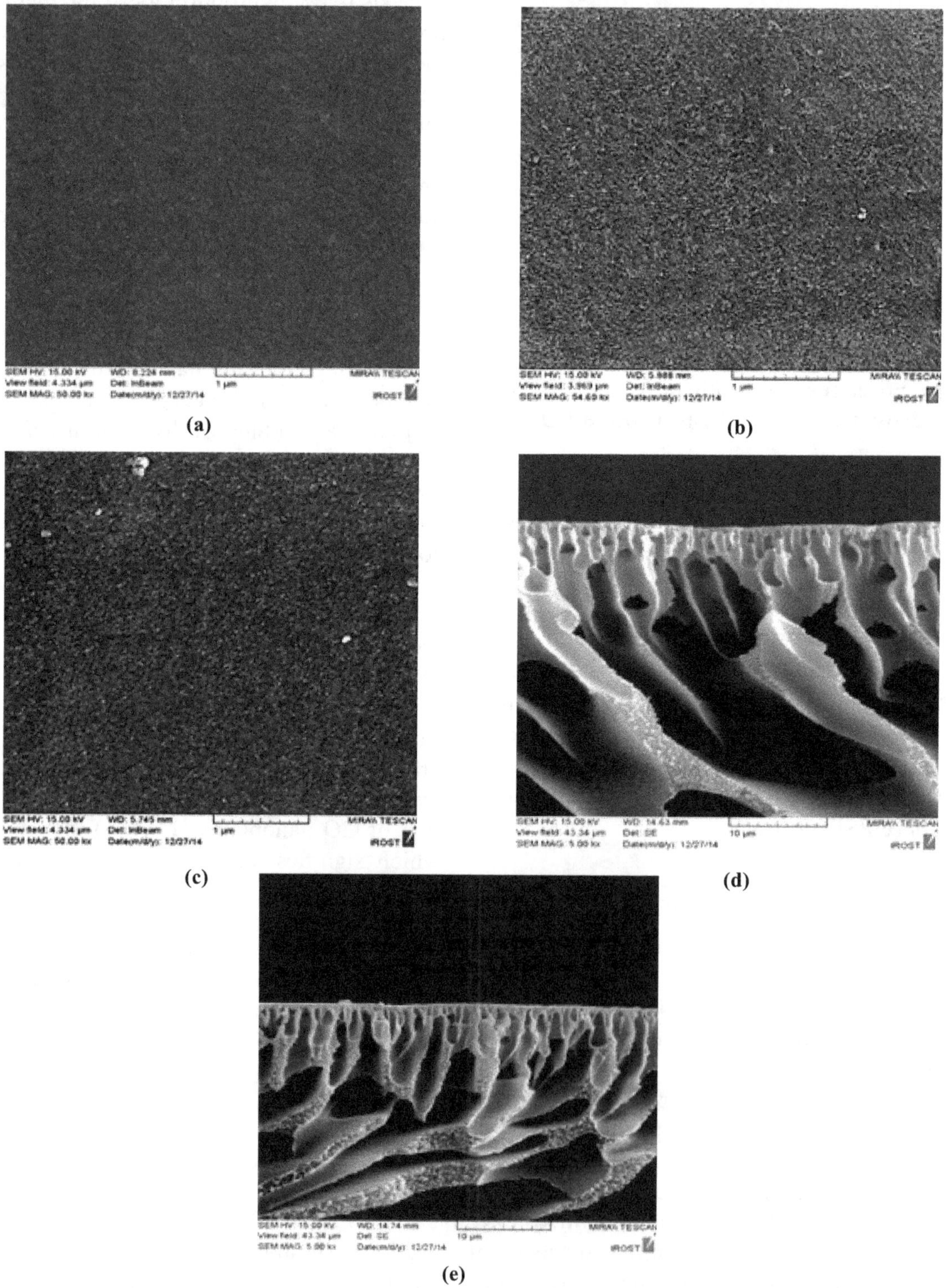

Fig. 1. The FESEM images of PSf substrate and PSf/GO substrates. (a) top surface of the substrate having no GO; (b) top surface of G0.5 substrate; (c) top surface of G1 substrate; (d) cross section of the substrate having no GO, and (e) cross section of G0.5 substrate.

are more porous than PSf substrate having no GO nanoplates. This results are in agreement with the porosity measurements listed in Table 2. The substrates exhibited an asymmetric structure containing a dense top layer and a sublayer with macro voids. The FESEM images show that embedding GO nanoplates in PSf matrix may lead to important changes in the substrate microstructure. As can be seen, the nanocomposite substrates presented longer finger-like structure compared to neat PSf substrate (Fig. 1(d) and (e)) [46].

The difference in morphology of the nanocomposite substrates with the substrate having no GO nanoplates can be originated from increasing rates of solvent (NMP) exit and non-solvent (water) arrival during phase inversion process [51]. It may be expressed that the GO nanoplates, having hydrophilic nature, increases transfer rate of water molecules from water bath to the substrate which ultimately leads to the formation of more finger-like pores and higher overall porosity [41, 64].

Fig. 2 shows digital photos taken from the top and bottom surfaces of G0.5 substrate. As these photos demonstrate, the color of top surface of the nanocomposite substrate is darker than its bottom surface. The reasonable explanation for this phenomenon is that during phase inversion process, GO nanoplates because of high hydrophilicity move toward top surface of the membrane where is the most vulnerable to water [50, 51]. Immigration of GO nanoplates into top surface of the membrane leads to improvement in the substrate hydrophilicity [64].The data obtained from the contact angle measurements (Table 2) support this hypothesis. These results are consistent with previous reports [50, 51, 65]. Also, Vatanpour et al. reported the similar results for carbon nanotube incorporated PES membrane [60, 66].

3.2. Characterization of the fabricated TFC and TFN membranes

3.2.1. Morphology of the TFC/TFN membranes

The FESEM images of the top surface of the TFC and TFNG0.5 membranes are depicted in Fig. 3. The TFC and TFNG0.5 membranes exhibited the ridge-valley surface morphology, which signifies formation of polyamide (PA) layer (through interfacial polymerization between MPD and TMC monomers) on the substrate membrane [36, 62]. Also, comparison of top surface of the TFC and TFNG0.5 membranes (Fig. 3(a) and (b)) displayed enhancing nodular surface of TFN membrane due to presence of GO nanoplates on the surface substrate [39]. According to the cross section images of the membranes (Fig. 3), it seems that polyamide layer formed on the nanocomposite substrates has better adhesion than neat PSf substrate (strong evidences exist in higher magnifications). This is probably due to higher hydrophilicity of the substrates nanocomposite [40]. Hydrophilic groups facilitate better absorption of MPD on the surface and improve its diffusion into pores of the substrates due to existence of interaction between the hydrophilic groups and MPD monomers [40, 62]. Therefore, thinner and deeper polyamide layer is formed within the substrate pores. Based on these results and other research reports, it is offered that the modification of substrate hydrophilicity can be effective method to achieve a PA dense layer from interfacial polymerization, which leads to the improvement of membrane performance [62, 67, 68].

Fig. 2. Digital photographs of the top and bottom surfaces of PSf/GO substrate.

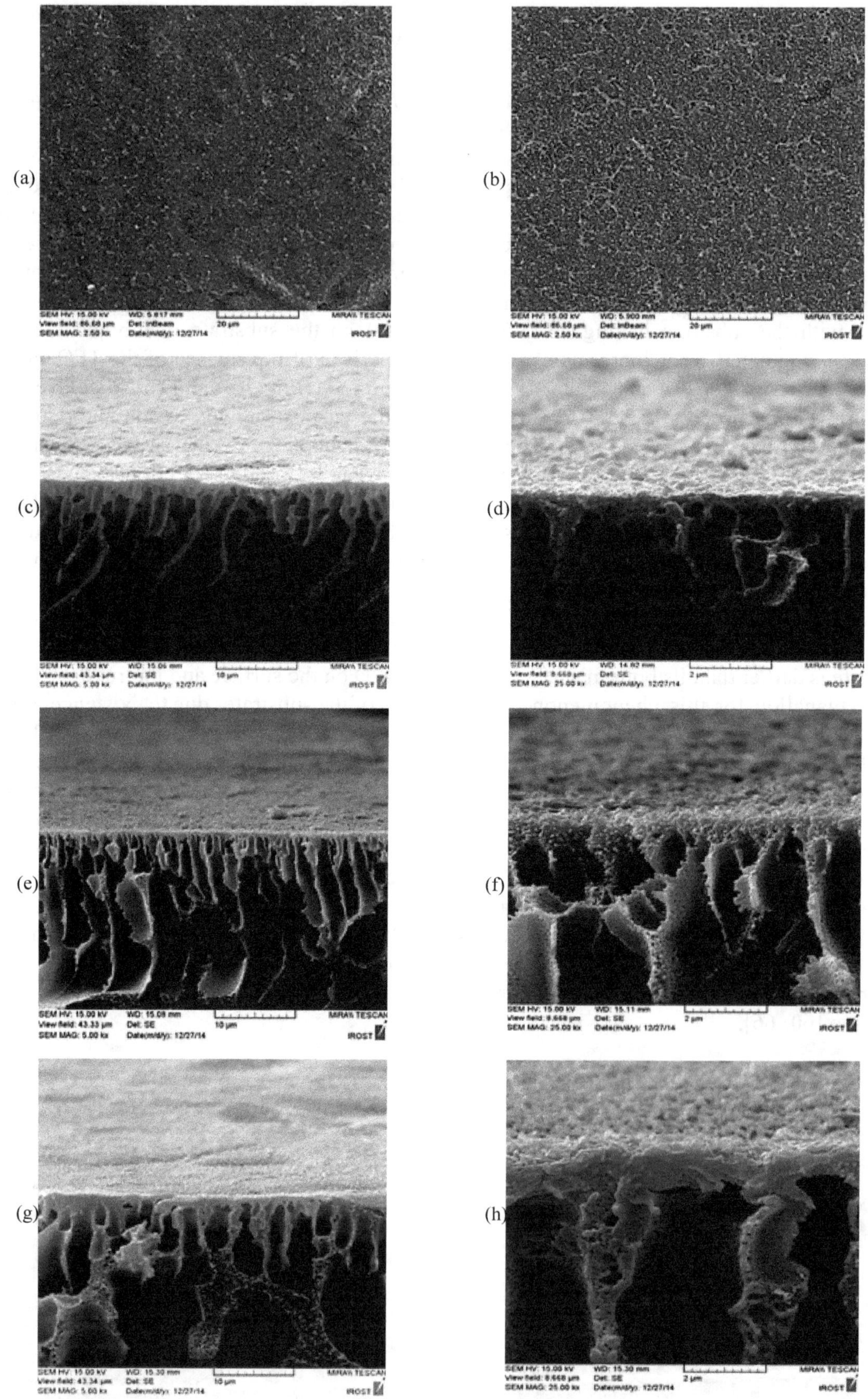

Fig. 3. FESEM images of the TFC/TFN membranes. Top views of (a) TFC; and (b) TFNG0.5 membranes. Cross section views of TFC membrane at low (c) and high (d) magnification; TFNG0.5 membrane at low (e) and high (f) magnification; TFNG1 membrane at low (g) and high (h) magnification.

The previous studies confirmed that physical-chemical characteristics of the substrate membrane and its morphology affect structure and roughness of PA layer prepared by interfacial polymerization [67, 69-71]. Therefore, the surface roughness of all the prepared membranes was examined by AFM images. The three dimensional AFM images taken from PA layer of the nanocomposite and composite membranes are illustrated in Fig. 4. In topographical images, the brightest region reveals the noontide of the membrane surface and the dark zones demonstrate the valleys or membrane pores. The roughness parameters of the membranes surface including the mean roughness (R_a), the root mean square of z valus (R_{ms}) and the maximum vertical distance between the highest data points (R_z) which were calculated using AFM images and SPM DME software in scanning area of 5 5 , are demonstrated in Table 3. It is observed that the surface roughness of the fabricated TFN membranes on the nanocomposite substrates is higher than the TFC membrane. So that, the highest value of roughness parameters belongs to the TFNG0.5 membrane with R_a=36.3 nm, R_{ms}=44.4 nm and R_z=198 nm. The higher surface roughness of the nanocomposite membranes can be ascribed to the nodular of GO nanoplates incorporated in the substrate. These results are in a good agreement with Emadzadeh et al. findings [46]. Also, Ganesh et al. reported that it may occur due to the rapid exchange of solvent and non-solvent during phase inversion process because of the presence of hydrophilic GO nanoplates [51]. However, the subsequent decrease in roughness could be due to the agglomeration of GO nanoplates at high concentrations (> 0.5 wt. %). The increase in surface roughness of TFN membranes significantly facilitates increase of water flux through expansion of the membrane effective area. Hirose et al. reported that a semi-linear relationship exists between water flux and the membrane surface roughness [72].

3.2.2. Water and salt permeability of the prepared membranes

A common method to evaluate the performance of the membranes is the use of a dead end filtration system. By using this system, the water and salt permeability of the prepared membranes were measured and the obtained results are presented in Table 4. As can be observed, all the TFN membranes except TFNG1 exhibited higher water permeability compared to the TFC membrane. So that, the highest value of water permeability was obtained for the TFNG0.5 membrane equal to 2.44 L/m^2 h bar, which is about 1.7 times higher than TFC membrane. The increase of the water permeability of the membranes as a functional of GO loading is due to improvement of hydrophilicity coupled with increase in porosity of the nanocomposite substrates which reduces water diffusion resistance across nanocomposite membrane [38, 40]. Meanwhile, reducing water permeability for TFNG1 membrane can be explained similarly. The results suggest that improving of the substrate properties is an effective method for increasing water permeability of the membrane.

The salt permeability values for TFNG0.1, TFNG0.5 and TFNG1 membranes were reported as 6.31×10^{-7} m/s, 5.81×10^{-7}m/s and 5.97×10^{-7} m/s, respectively. These values are lower than the ones reported for the TFC membranes. This phenomenon can be attributed to the formation of an effective PA layer especially at TFNG0.5 membrane which leads to the increase of salt rejection of the membranes and results in reduced salt permeability. In FO membranes, the lower B/A (water permeability/salt permeability) ratio is favorable, which is indicative of less solute reverse diffusion from draw solution to the feed solution [27, 65]. The TFNG0.5 membrane exhibited lowest B/A value (85.81 kPa). Hence, this membrane has shown maximum separation efficiency.

Table 3
.Surface roughness parameters of TFC and TFN membranes

Membrane	Roughness parameters		
	R_a (nm)	R_{ms} (nm)	R_z (nm)
TFC	20.5	26.0	132
TFNG0.1	26.4	30.0	150
TFNG0.5	36.3	44.4	198
TFNG1	28.3	34.3	161

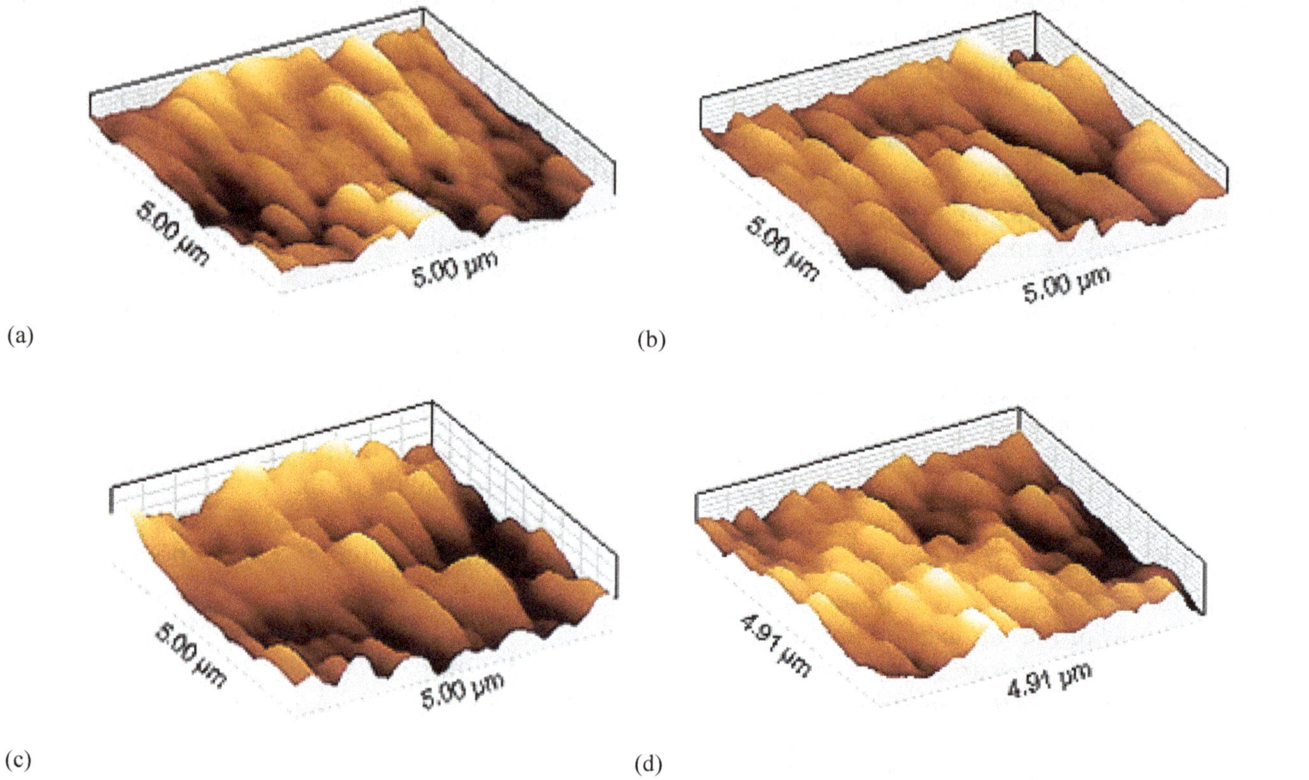

Fig. 4. AFM images of the membranes with different concentrations of GO nanoplates (a) TFC, (b) TFNG0.1, (c) TFNG0.5 and **(d)** TFNG1.

Table 4
Separation properties of prepared TFC and TFN FO membranes.

Membrane	Water permeability A ($L/m^2 h\,bar$)	Water permeability A (m/s Pa)	Salt permeability B (m/s)	B/A (kPa)
TFC	1.49	4.13	8.59	207.99
TFNG0.1	1.89	5.25	6.31	120.19
TFNG0.5	2.44	6.77	5.81	85.81
TFNG1	1.97	5.47	5.97	109.14

3.2.3. FO performance of the TFC/TFN membranes

The influence of the nanocomposite substrates containing different amount of GO nanoplates on FO performance of the prepared membranes was investigated and the results were depicted in Fig. 5. The water flux and salt rejection of the membranes were determined using DI water as the feed solution and 1 M NaCl as the draw solution in AL-DS orientation. Clearly, the water flux for all of TFN membranes is higher than the in-house made TFC membrane. This finding is completely consistent with previous reports wherein the prepared TFN membranes containing nanocomposite substrates (containing various nanomaterials such as zeolite, carbon nanotube and titanium dioxide) exhibited higher water flux [37, 38, 40, 46]. The increased water flux indicates that suitable structural changes in the substrate have affected FO performance of the membranes positively [39, 65]. The previous studies proved that the internal concentration polarization (ICP) is extremely effective on FO water flux as it can decrease the water flux to more than 80% during FO process [1, 30, 38]. In current study, the increase in water flux of TFN membranes verifies that substrate modification with GO nanoplates has reduced ICP. This claim is confirmed based on the structural parameter (S) values shown in Fig. 6. Since the structural parameter has an exponential relationship with the ICP, reducing structural parameter leads to the decrease of the ICP and results in better FO membrane performance [73]. Reduction in structural parameter with the increase of GO loading is related to improvement of the properties and structure of nanocomposite substrates [40]. The results of this study show that the substrate embedded with 0.5 wt% of GO nanoplates exhibited lowest S value (S=0.33 mm) which is the optimum concentration of GO loading. The water flux has shown the compatible trend with S value, when the GO changes. So that, the highest water flux of 37.74 L/m^2 h was achieved for the prepared membrane on G0.5 substrate that is 214% higher than of water flux for TFC membrane. Overall, following reasons can be presented for justifying why TFN membranes compared to TFC ones show higher water flux. (i) The improvement in the morphology and the properties of the substrates upon adding of GO nanoplates, shortens the water diffusion paths and leads to more comfortable transfer of water molecules through membrane and as a result increases water flux [47, 65]. (ii) As GO concentration increases, S value decreases and thereby negative effects arising from ICP are diminished. This phenomenon affects water transfer across the membranes and contributes to an increase of water flux [23, 74]. (iii) The surface roughness enhancement of the PA layer of TFN membranes significantly increases the water flux [72].

The salt rejection of the membranes is also shown in Fig. 5. The salt rejection of the TFN membranes (above 97%) is higher compared to the TFC membrane. The increased salt rejection from 95.63% for TFC membrane to 98.6% for the TFNG0.5 membrane is due to the more adhesive and more perfect formation of PA layer [40]. Nonetheless, a slight decline was observed at salt rejection of the TFNG1 membrane. This result may be due to the agglomeration of GO nanoplates in some locations of the substrate surface which can limit the interfacial polymerization reaction between MPD and TMC monomers and weakens the formation of PA layer and reduces the salt rejection [27, 75].

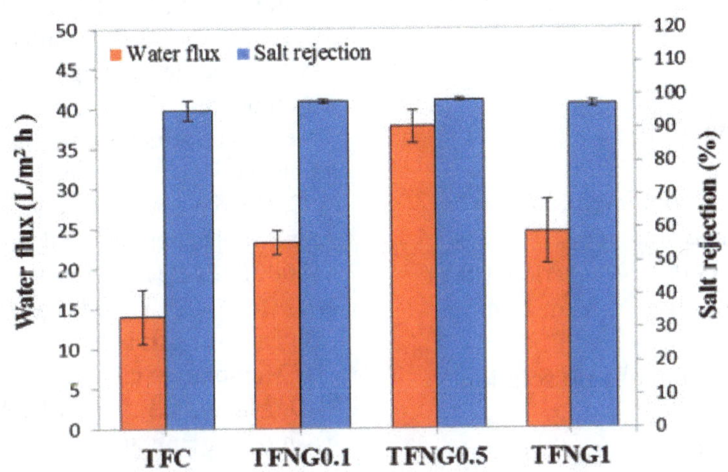

Fig. 5. Comparison of FO water flux and salt rejection of the prepared TFC and TFN membranes.

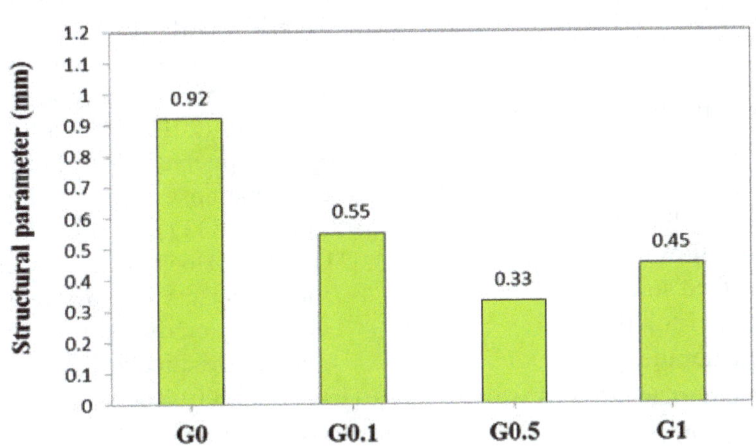

Fig. 6. The structural parameter of the prepared substrates with different loading of GO nanoplates.

4. Conclusions

In the current study, the forward osmosis nanocomposite membranes were prepared therein GO nanoplates were embedded. The effects of GO addition on morphology and performance of the fabricated nanocomposite membranes were investigated. The results demonstrated that the existence of functional groups including hydroxyl, carboxyl and epoxy in GO nanoplates increases the substrate hydrophilicity which leads to increase in water permeability of the TFN membranes. The TFN membranes exhibited higher water flux than TFC one. So that, the water flux increased from 14.09 L/m^2 h for TFC membrane to 37.74 L/m^2 h for TFNG0.5 membrane (i.e. 214% increase). Regarding the achieved results, it was suggested that the optimal concentration of GO nanoplates addition is 0.5 wt. %. The measurements of the substrate structural parameter indicated that with GO loading increment, the S value decreases. This means that addition of GO nanoplates to the PSf substrate is an effective method to control ICP in FO membranes.

Nomenclature & Abbreviations

A	Water permeability coefficient, L/m^2 h bar
A_m	Effective membrane area, m^2
B	Salt permeability coefficient, L/m^2 h (m/s)
C_0	Initial salt concentration of feed solution, mol/L
C_t	Final salt concentration of feed solution, mol/L
C_f	Salt concentration in feed solution, mol/L
C_p	Salt concentration in the permeate solution, mol/L
D	Solute diffusion coefficient, m^2/s
J_s	Salt reverse flux, g/m^2 h
J_w	Water flux in pressure driven process, L/m^2 h
J_v	Water flux in osmotic process, L/m2 h
P	Hydraulic pressure, bar
R	Salt rejection, %
S	Membrane structural parameter, mm
ΔV_{draw}	Volume change of draw solution, L
V_0	Initial volume of feed solution, L
V_t	Final volume of feed solution, L
m_d	Weight of dry membrane, g
m_w	Weight of wet membrane, g
Δt	Operation time interval, h

Greek letters

ε	Membrane porosity, %
π	Osmotic pressure, bar
ρ	Material density, g/cm^3

References

[1] S. Zhao, L. Zou, C.Y. Tang, D. Mulcahy, Recent developments in forward osmosis: Opportunities and challenges, Journal of Membrane Science, 396 (2012) 1-21.

[2] T.-S. Chung, S. Zhang, K.Y. Wang, J. Su, M.M. Ling, Forward osmosis processes: Yesterday, today and tomorrow, Desalination, 287 (2012) 78-81.

[3] H.-S.L.G.M. Geise, D.J. Miller, B.D. Freeman, J.E. Mcgrath, D.R. Paul, Water Purification by Membranes: The Role of Polymer Science, Journal of Polymer Science Part B: Polymer Physics, 48 (2010) 1685-1718.

[4] M.A. Shannon, P.W. Bohn, M. Elimelech, J.G. Georgiadis, B.J. Marinas, A.M. Mayes, Science and technology for water purification in the coming decades, Nature, 452 (2008) 301-310.

[5] M.M. Motsa, B.B. Mamba, A.D. Haese, E.M.V. Hoek, A.R.D. Verliefde, Organic fouling in forward osmosis membranes: The role of feed solution chemistry and membrane structural properties, Journal of Membrane Science, 460 (2014) 99-109.

[6] C. Charcosset, A review of membrane processes and renewable energies for desalination, Desalination, 245 (2009) 214-231.

[7] J. Dong, L. Li, T.M. Neoff, R. Lee, Desalination by reverse osmosis using MFI zeolite membranes, journal of Membrane Science, 243 (2004) 401-404.

[8] D. Mattai, K.P. Lee, T.C. Arnot, A review of reverse osmosis membrane materials for desalination – development to date and future potential, Journal of Membrane Science, 370 (2011) 1-22.

[9] O.A. Bamaga, A. Yokochi, B. Zabara, A.S. Babaqi, Hybrid FO/RO desalination system: preliminary assessment of osmotic energy recovery and designs of new FO membrane module configuration, Desalination, 268 (2011) 163-169.

[10] M. Elimelech, W.A. Phillip, The future of seawater desalination: energy, technology, and the environment, Science, 333 (2011) 712-717.

[11] E.M.V. Hoek, M.C.Y. Wong, K. Martinez, G.Z. Ramon, Impacts of operating conditions and solution chemistry on osmotic membrane structure and performance, Desalination, 287 (2012) 340-349.

[12] J.R. McCutcheon, R.L. McGinnis, M. Elimelech, Desalination by ammonia–carbon dioxide forward osmosis: Influence of draw and feed solution concentrations on process performance, Journal of Membrane

Science, 278 (2006) 114-123.

[13] C. Suh, S. Lee, Modeling reverse draw solute flux in forward osmosis with external concentration polarization in both sides of the draw and feed solution, Journal of Membrane Science 427 (2013) 114-123.

[14] Z. Liu, H. Bai, J. Lee, D.D. Sun, A low-energy forward osmosis process to produce drinking water, Energy & Environmental Science 4(2011) 2582.

[15] R.L. McGinnis, M. Elimelech, Energy requirements of ammonia–carbon dioxide forward osmosis desalination, Desalination, 207 (2007) 370-382.

[16] C.R. Martinetti, A.E. Childress, T.Y. Cath, High recovery of concentrated RO brines using forward osmosis and membrane distillation, Journal of Membrane Science, 331 (2009) 31-39.

[17] S. Zhao, L. Zou, Effects of working temperature on separation performance, membrane scaling and cleaning in forward osmosis desalination, Desalination, 278 (2011) 157-164.

[18] A. Achilli, T.Y. Cath, E.A. Marchand, A.E. Childress, The forward osmosis membrane bioreactor: a low fouling alternative to MBR processes, Desalination, 239 (2009) 10-21.

[19] S. Lee, C. Boo, M. Elimelech, S. Hong, Comparison of fouling behavior in forward osmosis (FO) and reverse osmosis (RO), Journal of Membrane Science 365 (2010) 34-39.

[20] B. Mi, M. Elimelech, Organic fouling of forward osmosis membranes: Fouling reversibility and cleaning without chemical reagents, Journal of Membrane Science, 348 (2010) 337-345.

[21] Y. Gao, Y.-N. Wang, W. Li, C.Y. Tang, Characterization of internal and external concentration polarizations during forward osmosis processes, Desalination, 338 (2014) 65-73.

[22] J.R. McCutcheon, R.L. McGinnis, M. Elimelech, A novel ammonia-carbon dioxide forward (direct) osmosis desalination process, Desalination, 174 (2005) 1-11.

[23] C.Y. Tang, Q. She, W.C.L. Lay, R. Wang, A.G. Fane, Coupled effects of internal concentration polarization and fouling on flux behavior of forwar dosmosis membranes during humic acid filtration, Journal of Membrane Science 354 (2010) 123-133.

[24] G.T. Gray, J.R. McCutcheon, M. Elimelech, Internal concentration polarization in forward osmosis" role of membrane orientation, Desalination, 197 (2006) 1-8.

[25] J.R. McCutcheon, M. Elimelech, Influence of concentrative and dilutive internal concentration polarization on flux behavior in forward osmosis, Journal of Membrane Science, 284 (2006) 237-247.

[26] S. Loeb, L. Titelman, E. Korngold, J. Freiman, Effect of porous support fabric on osmosis through a Loeb-Sourirajan type asymmetric membrane, Journal of Membrane Science, 129 (1997) 243-249.

[27] N. Niksefat, M. Jahanshahi, A. Rahimpour, The effect of SiO_2 nanoparticles on morphology and performance of thin film composite membranes for forward osmosis application, Desalination, 343 (2014) 140-146.

[28] M. Amini, M. Jahanshahi, A. Rahimpour, Synthesis of Novel Thin Film Nanocomposite (TFN) Forward Osmosis Membranes Using Functionalized Multi-walled Carbon Nanotubes, Journal of Membrane Science, 435 (2013) 233-241.

[29] N.-N. Bui, M.L. Lind, E.M.V. Hoek, J.R. McCutcheon, Electrospun nanofiber supported thin film composite membranes for engineered osmosis, Journal of Membrane Science, 385-386 (2011) 10-19.

[30] J.M.C. Puguan, H.-S. Kim, K.-J. Lee, H. Kim, Low internal concentration polarization in forward osmosis membranes with hydrophilic crosslinked PVA nanofibers as porous support layer, Desalination, 336 (2014) 24-31.

[31] Y.H. Cho, J. Han, S. Han, M.D. Guiver, H.B. Park, Polyamide Thin-Film Composite Membranes Based on Carboxylated Polysulfone Microporous Support Membranes for Forward Osmosis, Journal of Membrane Science, 445 (2013) 220-227.

[32] G. Han, S. Zhang, X. Li, N. Widjojo, T.-S. Chung, Thin film composite forward osmosis membranes based on polydopamine modified polysulfone substrates with enhancements in both water flux and salt rejection, Chemical Engineering Science, 80 (2012) 219-231.

[33] N. Widjojo, T.-S. Chung, M. Weber, C. Maletzko, V. Warzelhan, The role of sulphonated polymer and macrovoid-free structure in the support layer for thin-film composite (TFC) forward osmosis (FO) membranes, Journal of Membrane Science, 383 (2011) 214-223.

[34] N. Widjojo, T.-S. Chung, M. Weber, C. Maletzko, V. Warzelhan, A sulfonated polyphenylenesulfone (sPPSU) as the supporting substrate in thin film composite (TFC) membranes with enhanced performance for forward osmosis (FO), Chemical Engineering Journal, 220 (2013) 15-23.

[35] G.L. Jadav, P.S. Singh, Synthesis of novel silica-polyamide nanocomposite membrane with enhanced properties, Journal of Membrane Science, 328 (2009) 257-267.

[36] N. Ma, J. Wei, R. Liao, C.Y. Tang, Zeolite-polyamide thin film nanocomposite membranes: Towards enhanced performance for forward osmosis, Journal of Membrane Science, 405-406 (2012) 149-157.

[37] D. Emadzadeh, W.J. Lau, T. Matsuura, M. Rahbari-Sisakht, A.F. Ismail, A novel thin film composite

forward osmosis membrane prepared from PSf–TiO2 nanocomposite substrate for water desalination, Chemical Engineering Journal, 237 (2014) 70-80.

[38] N. Ma, J. Wei, S. Qi, Y. Zhao, Y. Gao, C.Y. Tang, Nanocomposite substrates for controlling internal concentration polarization in forward osmosis membranes, Journal of Membrane Science, 441 (2013) 54-62.

[39] X. Liu, H.Y. Ng, Fabrication of layered silica–polysulfone mixed matrix substrate membrane for enhancing performance of thin-film composite forward osmosis membrane, journal of Membrane Science, 481 (2015) 148-163.

[40] Y. Wang, R. Ou, Q. Ge, H. Wang, T. Xu, Preparation of polyethersulfone/carbon nanotube substrate for high-performance forward osmosis membrane, Desalination, 330 (2013) 70-78.

[41] Y. Wang, R. Ou, H. Wang, T. Xu, Graphene oxide modified graphitic carbon nitride as a modifier for thin film composite forward osmosis membrane, Journal of Membrane Science, (2014).

[42] J.H. Choi, J. Jegal, W.N. Kim, Fabrication and characterization of multi-walled carbon nanotubes/polymer blend membranes, Journal of Membrane Science, 284 (2006) 406-415.

[43] Y. Jia, H. Li, M. Wang, L. Wu, Y. Hu, Carbon nanotube: Possible candidate for forward osmosis, Separation and Purification Technology, 75 (2010) 55-60.

[44] E.S. Kim, G. Hwang, M.G. El-Din, Y. Liu, Development of nanosilver and multiwalled carbon nanotubes thin-film nanocomposite membrane for enhanced water treatment, Journal of Membrane Science, 394–395 (2011) 37–48.

[45] A. Rahimpour, M. Jahanshahi, S. Khalili, A. Mollahosseini, A. Zirepour, B. Rajaeian, Novel functionalized carbon nanotubes for improving the surface properties and performance of polyethersulfone (PES) membrane, Desalination, 286 (2011) 99–107.

[46] D. Emadzadeh, W.J. Lau, A.F. Ismail, Synthesis of thin film nanocomposite forward osmosis membrane with enhancement in water flux without sacrificing salt rejection, Desalination 330 (2013) 90-99.

[47] D. Emadzadeh, W.J. Lau, T. Matsuura, A.F. Ismail, M. Rahbari-Sisakht, Synthesis and characterization of thin film nanocomposite forward osmosis membrane with hydrophilic nanocomposite support to reduce internal concentration polarization, Journal of Membrane Science, 449 (2014) 74-85.

[48] C. Lee, X.D. Wei, J.W. Kysar, J. Hone, Measurement of the elastic properties and intrinsic strength of monolayer graphene, Science, 321 (2008) 385-388.

[49] H. Wu, B. Tang, P. Wu, Development of novel SiO_2-GO nanohybrid/polysulfone membrane with enhanced performance Journal of Membrane Science, 451 (2014) 94-102.

[50] S. Zinadini, A.A. Zinatizadeh, M. Rahimi, V. Vatanpour, H. Zangeneh, Preparation of a novel antifouling mixed matrix PES membrane by embedding graphene oxide nanoplates, Journal of Membrane Science, 453 (2014) 292-301.

[51] B.M. Ganesh, A.M. Isloor, A.F. Ismail, Enhanced hydrophilicity and salt rejection study of graphene oxide-polysulfone mixed matrix membrane, Desalination, 313 (2013) 199-207.

[52] Z. Xu, J. Zhang, M. Shan, Y. Li, B. Li, J. Niu, B. Zhou, X. Qian, Organosilane-functionalized graphene oxide for enhanced antifouling and mechanical properties of polyvinylidene fluoride ultrafiltration membranes, Journal of Membrane Science, 458 (2014) 1-13.

[53] H. Zhao, L. Wu, Z. Zhou, L. Zhang, H. Chen, Improving the antifouling property of polysulfone ultrafiltration membrane by incorporation of isocyanate-treated graphene oxide, Physical chemistry chemical physics : PCCP, 15 (2013) 9084-9092.

[54] S. Sheshmani, M.A. Fashapoyeh, Suitable chemical methods for preparation of graphene oxide, graphene and surface functionalized graphene nanosheets, Acta Chimica Slovenica, 60 (2014) 813-825.

[55] B. Chakrabarty, A.K. Ghoshal, M.K. Purkait, Preparation, characterization and performance studies of polysulfone membranes using PVP as an additive, Journal of Membrane Science, 315 (2008) 36-47.

[56] T.Y. Cath, A.E. Childress, M. Elimelech, Forward osmosis: principles, applications, and recent developments, Journal of Membrane Science, 281 (2006) 70-87.

[57] S. Loeb, G.D. Mehta, A two-coefficient water transport equation for pressure-retarded osmosis, Journal of Membrane Science, 4 (1979) 351-362.

[58] E.R. Cornelissen, D. Harmsen, K.F.d. Korte, C.J. Ruiken, J.-J. Qin, H. Oo, L.P. Wessels, Membrane fouling and process performance of forward osmosis membranes on activated sludge, Journal of Membrane Science, 319 (2008) 158-168.

[59] E. Yuliwati, A.F. Ismail, Effect of additives concentration on the surface properties and performance of PVDF ultrafiltration membranes for refinery produced wastewater treatment, Desalination, 273 (2011) 226-234.

[60] V. Vatanpour, S.S. Madaeni, R. Moradian, S. Zinadini, B. Astinchap, Fabrication and characterization of novel antifouling nanofiltration membrane prepared from oxidized multiwalled carbon nanotube/polyethersulfone nanocomposite, Journal of Membrane Science, 375 (2011) 284-294.

[61] S. Qiu, L. Wu, X. Pan, L. Zhang, H. Chen, C. Gao,

Preparation and properties of functionalized carbon nanotube/PSF blend ultrafiltration membranes, Journal of Membrane Science, 342 (2009) 165-172.
[62] M. Fathizadeh, A. Aroujalian, A. Raisi, Effect of lag time in interfacial polymerization on polyamide composite membrane with different hydrophilic sub layers, Desalination, 284 (2012) 32-41.
[63] E. Celik, H. Park, H. Choi, H. Choi, Carbon nanotube blended polyethersulfone membranes for fouling control in water treatment, Water Research, 45 (2011) 274-282.
[64] J. Lee, H.-R. Chae, Y.J. Won, K. Lee, C.-H. Lee, H.H. Lee, I.-C. Kim, J.-m. Lee, Graphene oxide nanoplatelets composite membrane with hydrophilic and antifouling properties for wastewater treatment, Journal of Membrane Science, 448 (2013) 223-230.
[65] M.J. Park, S. Phunstho, T. He, G.M. Nisola, L.D. Tijing, X.-M. Li, G. Chen, W.-J. Chung, H.K. Shon, Graphene oxide incorporated polysulfone substrate for the fabrication of flat sheet thin-film composite forward osmosis membranes.
[66] V. Vatanpour, S.S. Madaeni, R. Moradian, S. Zinadini, B. Astinchap, Novel antibifouling nanofiltration polyethersulfone membrane fabricated from embedding TiO2 coated multiwalled carbon nanotubes, Separation and Purification Technology, 90 (2012) 69-82.
[67] A.K. Ghosh, E.M.V. Hoek, Impacts of support membrane structure and chemistry on polyamide–polysulfone interfacial composite membranes, Journal of Membrane Science, 336 (2009) 140-148.
[68] H.I. Kim, S.S. Kim, Plasma treatment of polypropylene and polysulfone supports for thin film composite reverse osmosismembrane, journal of Membrane Science, 286 (2006) 193-201.
[69] L. Shi, S.R. Chou, R. Wang, W.X. Fang, C.Y. Tang, A.G. Fane, Effect of substrate structure on the perfor mance of thin-film composite forward osmosis hol low fiber membranes, Journal of Membrane Science, 382 (2011) 116-123.
[70] P.S. Singh, S.V. Joshi, J.J. Trivedi, C.V. Devmurari, A.P. Rao, P.K. Ghosh, Probing the structural varia tions of thin film composite RO membranes obtained by coating polyamide over polysulfone membranes of different pore dimensions, Journal of Membrane Science, 278 (2006) 19-25.
[71] X. Liu, H.Y. Ng, Double-blade casting technique for optimizing substrate membrane in thin-film composite forward osmosis membrane fabrication, Journal of Membrane Science, 469 (2014) 112-126.
[72] M. Hirose, H. Ito, Y. Kamiyama, Effect of skin layer surface structures on the flux behaviour of RO membranes, Journal of Membrane Science, 121 (1996) 209-215.
[73] W.C.L. Lay, J. Zhang, C. Tang, R. Wang, Y. Liu, A.G. Fane, Factors affecting flux performance of forward osmosis systems, Journal of Membrane Science, 394 (2012) 151-168.
[74] K.L. Lee, R.W. Baker, H.K. Lonsdale, Membranes for power generationby pressure-retarded osmosis, Journal of Membrane Science, 8 (1981) 141-171.
[75] J. Jegal, S.G. Min, K.-H. Lee, Factors affecting the interfacial polymerization of polyamide active layers for the formation of polyamide composite membranes, Journal of Applied Polymer Science 86 (2002) 2781-2787.

9

Improvement in photocatalysts and photocatalytic reactors for water and wastewater treatment

Hamed Baniamerian, Soheila Shokrollahzadeh*

Department of Chemical Technologies, Iranian Research Organization for Science and Technology (IROST), Tehran, Iran

HIGHLIGHTS

- Principles of semiconductors and photocatalysts were investigated due to their applications in the advanced oxidation processes.
- Various improvement strategies in photocatalysts were discussed.
- Recent developments in photocatalytic reactors in wastewater treatment were investigated.

GRAPHICAL ABSTRACT

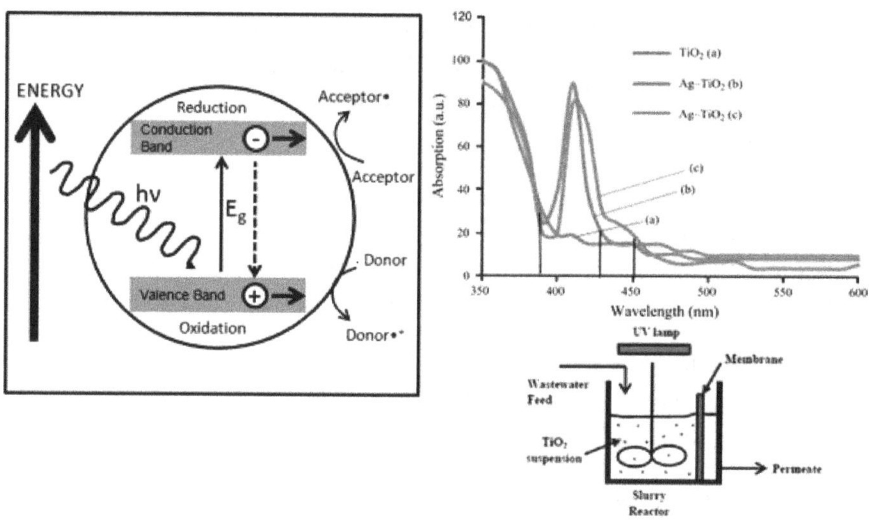

ARTICLE INFO

Keywords:
Semiconductor
Photocatalyst
Photocatalytic reactor
Water and wastewater treatment
Pollutants

ABSTRACT

During recent years, many advanced oxidation techniques have been investigated for water and wastewater treatment to overcome the shortage of clean water. This review summarizes the background and principles of photocatalysis applied as an advanced oxidation technology. In particular, this paper focuses on modification of photocatalysts with various dopants as well as the novel photocatalytic reactors to improve the oxidation efficiency of the pollutants in water and wastewater.

* *Corresponding author.*
E-mail address: shokrollahzadeh@irost.ir

1. Introduction

Shortage and availability of clean and healthy water is one of the universal problems of our century. Enormous amount of researches have been conducted to overcome these problems with the approach of minimizing the chemicals, energy, cost and environmental damage. In both developing and industrialized countries, a growing number of contaminants are entering water supplies from human activity: from traditional compounds such as heavy metals to emerging micropollutants such as endocrine disrupters and nitrosamines [1-3]. Conventional methods of water purification and disinfection can remove contaminants from water, but these methods are often chemically, energy consuming and operationally intensive. Thus, sever chemical treatments (such as those using ammonia, chlorine compounds, ozone...) and waste from treatment (sludge, brines, toxic material) can add to the problems of contamination and polluting water sources. Advanced Oxidation Processes (AOPs) can be used for treating of water polluted by organic maters and for disinfection of pathogens. These methods are based on highly reactive chemicals formation which degrades even most refractory molecules into biodegradable compounds. The Methods that are using UV, H_2O_2/UV, O_3/UV and H_2O_2/O_3/UV combinations use photolysis of H_2O_2 and ozone to generate the OH^- radicals which are able to oxidize organic molecule. Heterogeneous photocatalysis, are based on the use of a wide band gap semiconductor and irradiation with UV-vis light [4, 5]. Among various treatment methods, chemical oxidative treatments, and especially, AOPs, are suitable for their capability of oxidizing and mineralizing almost any organic contaminant [6].

Compared to the typical chemical treatments AOP is a method that uses a range of technologies to the increase oxidation power. For effective water treatments, a range of pollutants need to be removed economically at room temperature and under atmospheric pressure [7-9]. As shown in Table 1, AOP produces OH•, which has stronger oxidation power than ordinary oxidants [10].

Technology that uses the catalytic activity of semiconducting metal oxides, such as TiO_2, ZnO, Bi_2WO_6, $Bi_{20}Ti_{20}$, Fe_2O_3, Nb_2O_5, $BiTiO_3$, $SrTiO_3$, $ZnWO_4$, CuS/ZnS, WO_3, Ag_2CO_3, ZnS, etc. [11, 12], has been researched actively with particular focus on TiO_2 photocatalytic technology [13]. The advanced technology that uses TiO_2 photocatalysis use only photo energy without additional chemicals [14]. In

Table 1.
Redox potential of major oxidizing agents that are used in water treatment [10].

Oxidizing agent	Oxidation potential (V)	Relative oxidation power*
OH radical	2.80	2.06
Ozone	2.07	1.52
Hydrogen peroxide	1.77	1.30
Perhydroxyl radical	1.70	1.25
Permanganate	1.68	1.24
Chlorine dioxide	1.57	1.15
Chlorine	1.36	1
Oxygen	1.20	0.88

*Relative oxidizing power when chlorine oxidation power is 1

semiconductor photocatalysis, electrons from the valence band of a semiconductor are excited to the conduction band by wavelength of higher energy than the respective band gap [15], resulting in the formation of e^-_{CB}/h^+_{VB} pairs. The lack of a continuum of inter-band states in semiconductors assures an adequately extended lifetime for photo generated e^-_{CB}/h^+_{VB} pairs to initiate redox reactions on the catalyst surface. Electrons in the conduction band can reduce O_2 to form superoxide radicals ($O_2^{\bullet-}$). Additional reaction of $O_2^{\bullet-}$ with holes on the valence band produce single oxygen ($1O_2$) [16, 17].

Subsequent reactions of valence band holes with surface adsorbed H_2O or HO^- result in the formation of hydroxyl radical, H_2O_2 and protonated superoxide radicals ($HOO^\bullet$) . H_2O_2 is resulting from the combination of two $HOO^\bullet$ [18, 19]. Further reaction of H_2O_2 with $HO^\bullet$ results in the formation of $HOO^\bullet$ [19, 20]. During the photocatalytic process, free electrons/holes, and reactive oxidizing species such as $HO_2^\bullet$, $HO^\bullet$ and $O_2^{\bullet-}$ react with the surface adsorbed impurities such as inorganic/organic compounds, and biological species leading to their decomposition. Photocatalytic treatment of polluted water has some advantages over other methods such as low operational and installation cost, simple pre-processing, less area and decomposition possibility of non-biodegradable matters. Limited lamp life (when UV-Lamp is used) and photocatalyst recovery (when using powder) are disadvantages of photocatalytic water treatment [19]. Photocatalytic research is basically related to the development of solar energy use [21, 22].

Conventional wastewater treatment consists of a combination of physical, chemical, and biological processes and operations to remove solids, organic matter and, sometimes, nutrients from wastewater that are almost energy and cost consuming processes [23].The water treatment process may vary slightly at different locations, depending on the technology of the plant and the water it needs to process [24]. Although many excellent review articles have been published until now discussing the importance of photocatalysts in water and wastewater treatment, but the few reviews have been published on photocatalytic reactor configuration conjugated with membrane processes. So the the purposes of the present article is to review the improving strategies of photocatalytic activity, and various types of reactor configuration and combination of photocatalysts with membranes for water and wastewater treatment. A summary of advanced oxidation processes, principles and mechanism of photocatalysis, improvement strategies of photocatalytic water treatment, and photocatalytic reactors in water and wastewater treatment have been discussed in this review.

2.Advanced Oxidation Process

The wide area of water pollution, variety and non-biodegradable problems has become a problem that cannot be solved by the conventional treatments like Coagulation / Flocculation, sedimentation, filtration, Disinfection, sludge drying, Fluoridation and pH correction [24-27]. Moreover, in the case of water treatment technology, it is very difficult to remove pollutants completely with existing biological treatment technology. Compared to the typical chemical treatments, AOP is a method that uses a range of technologies to increase oxidation power. Generated OH' radicals strongly oxidize organic compounds, toxic pesticide and manure emissions to minerals in AOP (Fig.1).

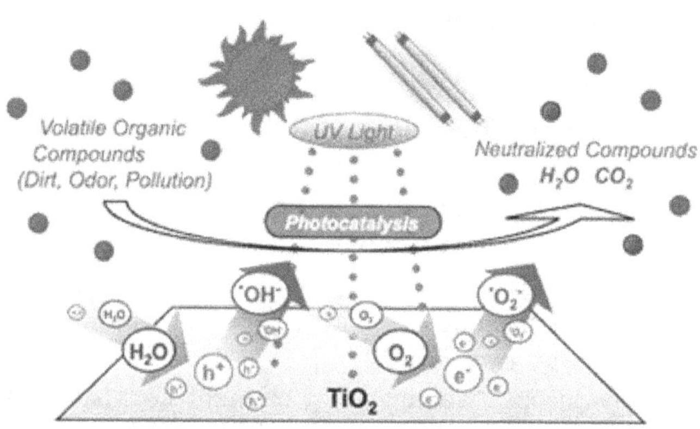

Fig. 1. Degradation of pollutants by OH radicals generated from photocatalysis process [25].

The formation of OH radicals can occur through several processes that are discussed below. AOPs are divided into two categories according to the presence or absence of light (UV) as reaction promoter [28].

2.1. Dark oxidative process

Advanced oxidation processes based on presence or absence of light, are divided into dark oxidative process and UV conjugated process. In the dark oxidative processes, the light is not effective in the reaction. These processes are consisting of using ozone, Fenton's reagent, elecctro-magnetic beam, ultrasound, microwave, and cavitation. In the following sections, these processes are discussed.

2.1.1. Ozone (O_3)

This technology injects ozone gas directly into polluted water [29]. In one mechanism the organic contaminants react directly with ozone, and in another effectively oxidation decomposition occur through an indirect reaction with $OH^\cdot$ radical, which is generated by decomposed ozone. The mechanism of ozone decomposition depends on the properties of the organic pollutants in polluted water. Hence, the organic pollutants in contaminated water can either promote or inhibit ozone decomposition and can interact with contaminants decomposition [9, 30].

2.1.2. Fenton's reagent

Fenton oxidation technology that uses the Fenton reaction is a method to generate $OH^\cdot$ in a Fe^{2+} and H_2O_2 mixture[31, 32]. This can decompose H_2O_2 by Fe^{2+} in the liquid phase without light irradiation, and generate $OH^\cdot$ radical. In the case light irradiation can promote decomposition of H_2O_2 due to ferrous or ferric ions and rapid hydroxyl radical generation is possible. The Fenton oxidation reaction is only applicable under the acidic conditions, i.e., pH< 4. Therefore, the process has disadvantages of high cost of operation due to the additional cost of the acid/base for pH adjustments [33, 34]. Fenton reaction can be express as:

$$H_2O_2 + Fe^{3+} \longrightarrow HO_2^\cdot + H^+ + Fe^{2+} \quad (1)$$

$$HO_2^\cdot + Fe^{2+} \longrightarrow HO_2^- + Fe^{3+} \quad (2)$$

$$HO_2^\cdot + Fe^{3+} \longrightarrow HO_2^\cdot + H^+ + Fe^{2+} \quad (3)$$

2.1.3. Elecctro-magneticbeam, Ultrasound, andmicrowave

$OH^{\bullet}$ radical can be generated by directly decomposing water using an external energy source, such as an electromagnetic beam, ultrasound and [35]. When water is irradiated with gamma rays (as electromagnetic beam), water molecules are decomposed to $OH^{\bullet}$, H^+, and aqueous electrons. also high-voltage discharge in water can decompose water molecules. When high voltage pulses of approximately 10 kV, are added between two electrodes located with a few centimeter gap between them, a flame is generated as water discharge is caused. This generates high concentrations of $OH^{\bullet}$ radical. $OH^{\bullet}$ can also be produced by ultrasound or microwaves [36]. When these energy sources pass through water, localized pressure differences with time generate micro bubbles, which acquire vibration energy and become unstable. These bubbles generate strong energy at high temperatures and high pressures, and the water molecule is broken down to $OH^{\bullet}$ and H^+.

2.1.4. Cavitation (sonication and hydrodynamic)

Sonication or hydrodynamic processes include the formation of cavitation micro bubbles. These bubbles collapse severely after reaching a critical resonance size and generate high temperature and highly reactive $OH^{\bullet}$ radicals. Removal of organics occurs by thermal decomposition at the bubble-water interface and by reaction with the radicals. Oxidation by cavitation is enhanced by addition of O_3 or H_2O_2 [28].

2.2. UV conjugated process

The second category of advanced oxidative processes is UV conjugated processes. These processes occur in the presence of light especially in the range of ultra-violet wavelength and are consisting of UV/O_3, UV/H_2O_2, UV/Fenton, and UV/photocatalyst processes.

2.2.1. UV/O_3 (Ozonation)

Hydroxyl radicals are generated when low pressure UV light is applied to ozonated water. Destruction of organic compounds occurs by hydroxyl radical reactions, coupled with direct photolysis and oxidation by molecular ozone [28].

$$O_3 + H_2O \xrightarrow{} O_2 + H_2O_2 \quad (4)$$
(λ<300nm) UV lamp

$$2O_3 + H_2O_2 \xrightarrow{} 2OH^{\bullet} + 3O_2 \quad (5)$$
generation of $OH^{\bullet}$

2.2.2. UV/H_2O_2

The $OH^{\bullet}$ radical route is the predominant removal mechanism in the destruction of organics [28].

$$H_2O_2 \xrightarrow{} 2OH^{\bullet} \quad (6)$$
(λ<300nm) UV lamp

2.2.3. UV/Fenton

UV/Fenton processes are like Fenton reaction. UV light can accelerate generation of $OH^{\bullet}$ radicals.

2.2.4. UV/semiconductors (photocatalysis)

Photocatalysis refers to the speed-up of the chemical reaction in the presence of substances called photocatalysts, which can absorb special wavelengths depending on the band structure [37-40]. Usually, semiconductors including TiO_2, Fe_2O_3, WO_3, ZnO, CeO_2, CdS, Fe_2O_3, ZnS, MoO_3, ZrO_2, and SnO_2 are selected as photocatalysts due to their narrow band gap and distinct electronic structure (unoccupied conduction band and occupied valence band) [20, 41-59]. In the following principles and mechanism of photocatalysis have been discussed.

3.Principles and mechanism of photocatalysis

3.1. Structural and electronic properties

The initial work of water decomposition using electrodes composed of TiO_2 was done by Fujishima and Honda [47]. The quantum efficiencies of TiO_2 in solar energy conversions are rather poor due to the usually faster recombination of electron-holes [60, 61]. Other noteworthy advances include the invention of dye sensitized solar cells by Grätzel et al., which are composed of TiO_2 anodes and the discovery of anti-fogging abilities of TiO_2 surfaces by Wang et al. [13, 62]. TiO_2 is widely used because of its non-toxicity, abundance (inexpensiveness), thermal/chemical stability, and high redox potential [47, 63, 64]. Due to superior mobility of electron–hole pairs, and improved surface hydroxyl density, anatase TiO_2 considered as the photocatalytically most active polymorph of TiO_2 [65, 66]. Factors, such as surface area, particle size, ratio of poly-morphs, type of dopants, defect concentration, synthesis method, and phase purity, affect the photocatalytic activity of semiconductors [67-69].

3.2. Principles of photocatalysts

A photocatalyst receives UV radiation from sunlight and uses the energy to decompose different substances including organic maters, organic acids, estrogens, pesticides, dyes, hydrocarbons, microbes and chlorine resistant organisms, inorganic molecules such as nitrous oxides (NO_x) [70-72]. For preventing free nanoparticles in water, photocatallysts nanoparticles are usually immobilized on a substrate [73] or integrated into thin-films and other materials. For the activation of TiO_2, UV irradiation is needed. To allow activation by visible light, TiO_2 can be modified with a second semiconductor, dyes, nitrogen, carbon or sulphur. For example, TiO_2 doped with nitrogen has better photocatalytic performance than pure TiO_2 nanoparticles [74].

Photocatalytic reactions are initiated by the absorption of illumination with energy equal to or greater than the band gap of the semiconductor. This produces electron-hole (e^-/h^+) pairs as in Equation (7), Fig. 2:

$$TiO_2 + h\upsilon = e^-_{cb}\,(TiO_2) + h^+_{vb}\,(TiO_2) \qquad (7)$$

Where cb is the conduction band and vb is the valence band. Thus, as a result of irradiation, the TiO_2 particle can behave either as an electron donor or acceptor for molecules in contact with the semiconductor. The electron and hole can recombine, releasing the absorbed light energy as heat, with no chemical reaction. On the semiconductor surface, the excited electron and the hole can participate in redox reactions with water, hydroxide ion (OH^-), organic compounds or oxygen leading to mineralization of the pollutant.

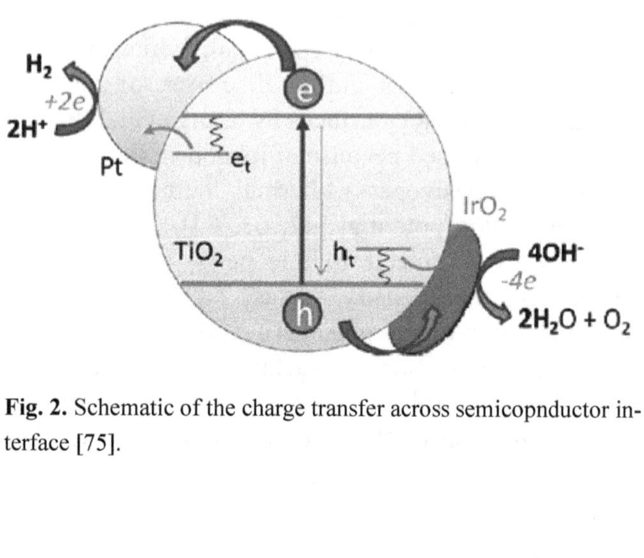

Fig. 2. Schematic of the charge transfer across semicopnductor interface [75].

Oxidation of water or OH^- by the hole generates the hydroxyl radical. OH radicals are able to rapidly attack pollutants on the semiconductor surface [76, 77].

Pervious work indicate that heterogeneous photocatalytic process involves five separate steps [78] and include (1) diffusion of reactants to the surface of photocatalyst, (2) adsorption of reactants onto the surface of semiconductor, (3) reaction on the surface of semiconductor, (4) desorption of products from the semiconductor and (5) diffusion of products from the surface of the semiconductor. There are two routes through which OH radicals can be formed the reaction of the valence-band holes with either adsorbed H_2O or with the surface OH groups on the photocatalyst particle [79].

3.3. Limitation of semiconductors

One of the main disadvantages of TiO_2 photocatalyst is the recombination of photo-generated charge carriers, which decreases the efficiency of the reaction [80]. The photo-excited electrons come back to the valence band during the recombination processes [81-83]. Methods such as hetero junction formation, doping with ions, and nano sized crystals have been demonstrated to reduce recombination [84-90]. For example, TiO_2 partially loaded with Ag and Au noble metal nanoparticles exhibited better photocatalytic activities [91], because metal nanoparticles act as electron traps during the photocatalytic reaction, and thus decreasing the rate of electron–hole recombination [92]. Notably, higher visible-light photocatalytic activities of N-doped, and S, N-codoped anatase–rutile nano hetero junctions have been reported by Etacheri et al. [93]. In conclusion, increasing the life-time of electron–hole pairs resulted in increasing the photocatalytic performance of photocatalysts. Both rutile and anatase have tetragonal crystal structure and previous reports proved that the kinetic stability of anatase is higher than that of rutile under ambient conditions. A thermodynamic phase stability calculation by Banfield and co-workers show that a critical particle size of 14 nm is required to initiate anatase to rutile transformation. Another disadvantage of the TiO_2 photocatalyst is the large band gap of TiO_2, that limits its use to UV light with wavelength lower than 390 nm. therfore only 5% of the solar radiations can be used by undopped TiO_2 photocatalysis, which significantly decreases Economic justification [94]. Studies have been done to synthesize small band gap photocatalysts that can utilize both UV and visible light. Various methods conducted for the visible light activation are explained in the following sections [94].

4.Strategies to improve photo degradation efficiency

In principle, photocatalytic activity will be enhanced if the absorption of light is increased or if the recombination losses for photogenerated charge carriers are decreased. Different strategies for enhancing the photocatalytic activity of photocatalysts under visible and UV have been conducted include sensitization with different sensitizers (spatially dyes), addition of nobel metals, doping with transient metal ions, addition of inert elements, and using compound materials [95]. In the following section, these strategies have been discussed.

4.1. Photoexcited treatment on the surface of photocatalysts

Addition of photo excited sensitizers (e.g., ruthenium-II) on the surface of photocatalysts can widen the wavelength range and enlarge the utilization range of sun light. The principle of photosensitization of photocatalyst is showed in Fig. 3. The visible light excites the sensitizer molecules adsorbed on TiO_2 and hence inject electrons to conduction band of TiO_2. While the conduction band acts as a mediator for transferring electrons from the sensitizer to substrate electron acceptors on TiO_2 surface, the valence band remains unaffected in a common photosensitization [95]. Various inorganic/organic dyes [96-103] as a sensitizer in relation to photo electrochemical cells and water splitting systems have been used. The studies of photosensitization used to degradation of pollutants using visible light are few.

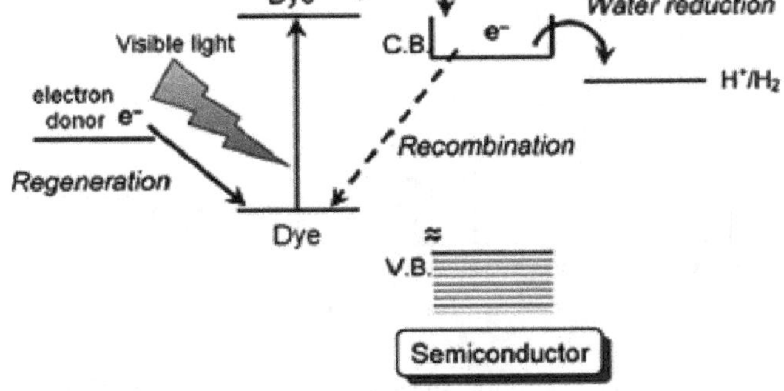

Fig. 3: Photosensitized TiO2 Particles [104].

Some dyes having redox property and visible light sensitivity can be used in photocatalytic [105, 106]. The process is similar to composite semiconductors (are discussed in section 4.5), The photo-excitation, electron injection and dye regeneration can be proposed as follows:

$$\text{dye} + h\upsilon \longrightarrow \text{dye}^* \quad (8)$$

$$\text{dye}^* \xrightarrow{TiO_2} \text{dye}^+ + e^- \quad (9)$$

$$\text{dye}^+ + e^- \longrightarrow \text{dye} \quad (10)$$

To achieve higher efficiency in oxidation reactions, fast electron injection and slow backward reaction are needed. The recombination times were found to be mostly in the order of nanoseconds to microseconds on electron/hole recombination of dyes [107], while the duration of electron injection is in the order of femtoseconds [108]. The quick electron injection and slow recombination reaction make dye sensitized semiconductors suitable. Frequently used dyes are listed in Table 2.

Sensitized photocatalysis usually leads to a rapid destruction of chromophore structure to form smaller organic species, resulting in final mineralization of the dye [110-113].

4.2. Addition of noble metal

Noble metals, such as Ag, Pt, Rh, Pd, Au have been investigated and showed to be very effective for promotion of TiO_2 photocatalyst. Photo-excited electrons can be transferred from conduction band to nobel metal particles located on the surface of TiO_2 (Fig.4). These metals reduce the recombination, resulting efficient charge separation and higher photocatalytic activity [114]. Because of high cost of Au and Pt, more research is needed to find low-cost metals with suitable improvement of photocatalytic activity [115].

The atomic radiuses of noble metal (e.g., Pt, Ag, Au) are bigger, so they are unable to enter the crystal lattice of photocatalysts. But, they can improve the surface characteristics of photocatalysts, and restrain the recombine of electron-hole [116].

It has been shown that noble metals such as platinum, palladium, silver, and gold ions allow to wide light absorption of TiO_2 to the visible light [118]. The surface of noble metal acts as visible light absorbing agent [119]. The photocatalyt-

Table 2.
Frequently used dyes [109].

Dye	Type	λ_{max}(nm)
Thionine (TH^+)	Thiazines	596
Toluidine blue (Tb^+)	Thiazines	630
Methylene blue (MB)	Thiazines	665
New methylene blue	Thiazines	650
Azure A	Thiazines	635
Azure B	Thiazines	647
Azure C	Thiazines	620
Phenosafranin (PSF)	Phenazines	520
Safranin-O (Saf-O/SO)	Phenazines	520
Safranin-T (Saf-T/ST)	Phenazines	520
Neutral red (NR)	Phenazines	534
Fluorescein	Xanthenes	490
Erythrosin	Xanthenes	530
Erythrosin B	Xanthenes	525
Rhodamin B (Rh. B)	Xanthenes	551
Rose Bengal	Xanthenes	550
Pyronine Y (PY)	Xanthenes	545
Eosin	Xanthenes	514
Rhodamin 6G	Xanthenes	524
Acridine orange (AO)	Acridines	492
Proflavin (PF)	Acridines	444
Acridine yellow (AY)	Acridines	442
Fusion	Triphenyl methane derivatives	545
Crystal violet	Triphenyl methane derivatives	578
Malachite green	Triphenyl methane derivatives	625
Methyl violet	Triphenyl methane derivatives	580

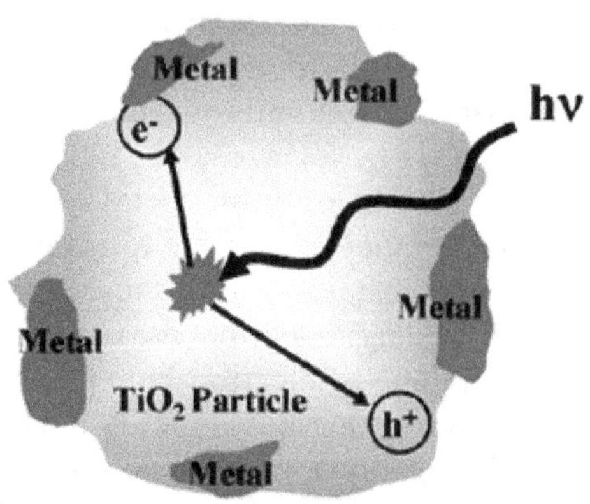

Fig. 4. Electrons capture by a metal in contact with a semiconductor surface [114].

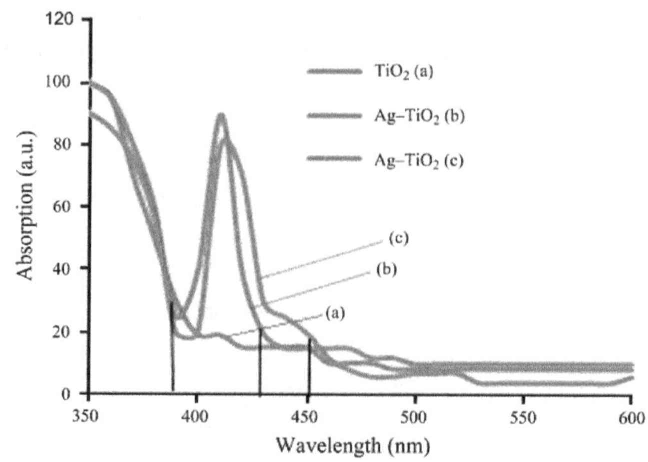

Fig. 5. UV–vis absorption spectra of the annealed (a) TiO_2 and (b) 3% and (c) 7% Ag-doped TiO_2 nanoparticles [117].

activity improvement of silver doped TiO_2 was mainly belong to the increase of surface area, which promotes more reactive sites available to take part in photoreactions. On the other hand, doping TiO_2 with Ag (Fig.5) retards the recombination rate of electron–hole pair by enhancing the charge carrier separation and hence improves the photocatalytic activity [120, 121]. Sobana [122] and Xin [123] show that Ag deposited on TiO_2 act as electron traps, enhance the electron–hole separation and lowering recombination.

4.3. Addition of transition metal ion

One of the most effective methods to create visible-light sensitive photocatalysts is adding impurity in the forbidden band through metal ion doping [124, 125].

Common metal ion dopants are V, Ni, Cr, Mo, Fe, Sn, Mn, Co, Cu. A trace quantity of transition metal ion is deposited on TiO_2, extended the available wavelength range, and increase the utilization rate of visible light. Fe^{3+}, Mo^{5+}, Ru^{2+}, V^{5+} and Rh^{2+} can improve the activity of photocatalysts. Among them the effect of Fe^{3+} is the most prominent. But Cr^{3+} is harmful to the activity of photocatalysts (Fig. 6). The doping Fe^{3+} induces the narrowing of the band gap of TiO_2. The dopant Fe^{3+} ion causes the formation of new states near the conduction band. Therefore the doping by iron ions significantly promotes the photocatalytic activity in the visible light region [126]. Also, it inhibits the recombination of the photo generated electron and hole. The Fe ions with a suitable concentration can trap the photo generated electron, which enhances the utilization efficiency of the photo generated electron and hole.

Fig.7 shows mechanism of Fe ion narrowing the band gap. According to figure various structures of iron can change the energy levels of TiO_2.

Metal ion cause creation of impurity energy levels in the band gap of TiO_2 as below:

$$Mn^{+} + h\upsilon \longrightarrow M^{(n+1)+} + e^{-}_{CB} \quad (11)$$

$$Mn^{+} + h\upsilon \longrightarrow M^{(n-1)+} + h^{+}_{VB} \quad (12)$$

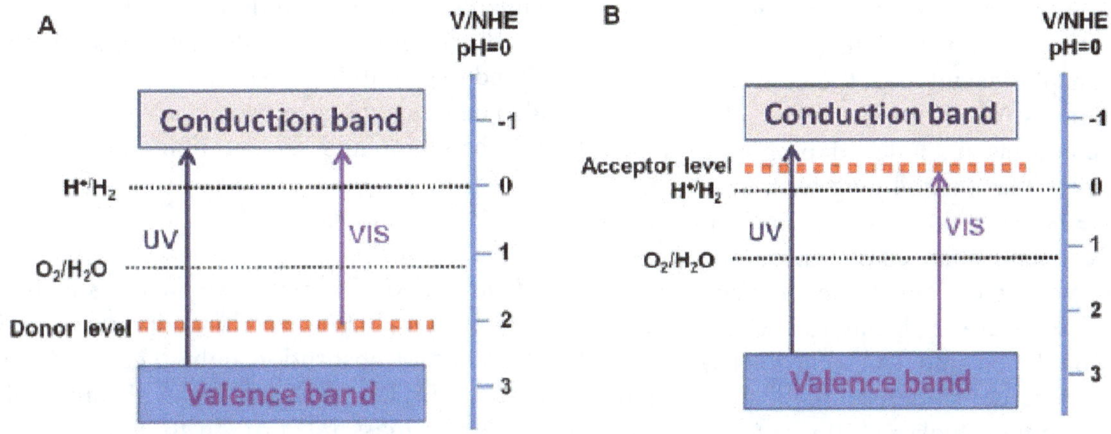

Fig. 6. Donor level (A) and acceptor level (B) formed by metal ion doping [124]

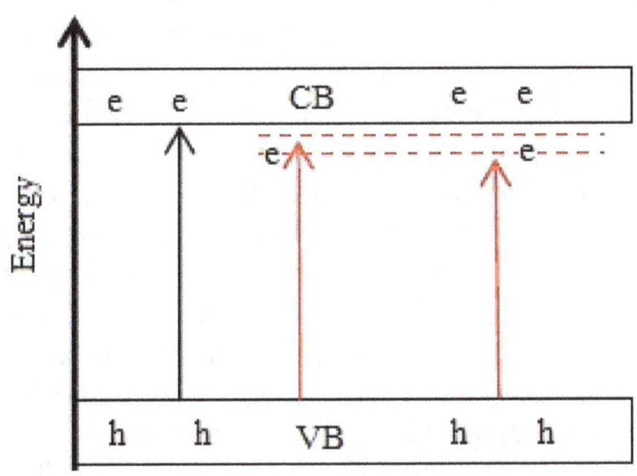

Fig. 7. Schematic diagram of the mechanism of Fe^{3+} doping on TiO_2 [125].

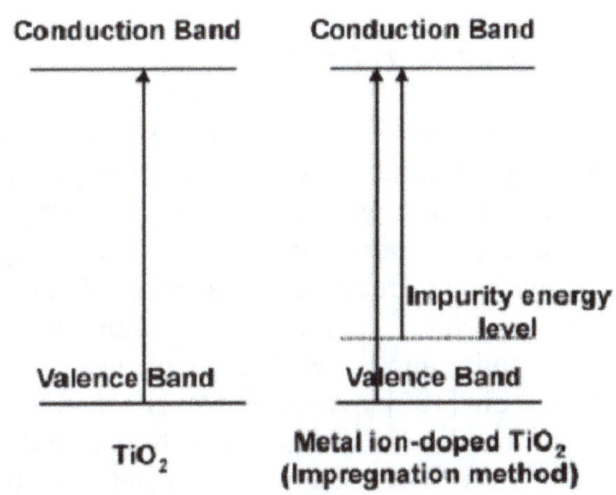

Fig. 8. Scheme of TiO_2 band structures, chemically ion-doped TiO_2 [114].

M is metal and Mn^{+} is metal ion dopant. Electron (hole) transfer between metal ions and TiO_2 can modify electron/hole recombination (Fig.8):

$$Mn^{+} + e^{-}_{CB} \longrightarrow M^{(n-1)} \quad \text{as electron trap} \quad (13)$$

$$Mn^{+} + h^{+}_{vB} \longrightarrow M^{(n+1)+} \quad \text{as hole trap} \quad (14)$$

Photocatalytic reaction only takes place when the trapped electron and hole are transferred to the surface of particle. Therefore, metal ions should be deposited near the surface of TiO_2 particles for a suitable charge transfer. There is an optimum of metal ion concentration, above which the photocatalytic activity decreases because of the increase in recombination [114].

Fe, Rh, Mo, V, Ru, Re, and Os ions can enhance photocatalytic activity, while Al and Co ions cause detrimental effects. Fe and Cu ions can trap electrons and holes, and the impurity energy levels generated are near to conduction band as well as the valence band edge of TiO_2. Therefore, doping of either Fe or Cu ions is suitable for enhancement of photocatalytic activity. In the same way, the effects of doping transition metal ions (Cr, Mn, Fe, Co, Ni and Cu) on photocatalytic efficiency of TiO_2 also have been investigated. As, Cu, Mn, and Fe ions can trap both electrons and holes, doping of these metal ions may be better than doping with Cr, Co and Ni ions, as the latter metal ions can only trap one type of charge carrier. Enhanced photocatalytic activities were observed at certain doping concentration of different rare earth metal ions (La, Ce, Er, Pr, Gd, Nd and Sm) doped into TiO_2. Gd ions were found to be most effective in increasing the photocatalytic activity due to their higher ability to transfer charge carriers to the interface of photocatalyst [125].

4.4. Addition of inert elements

The utilization of visible light could be increased significantly when the oxygen of TiO_2 was partially substituted with nitrogen. TiO_2 improved with carbon contains more pores, and its degradation capability is two times greater than that pure TiO_2. Doping of anions (N, F, C, S, etc.) in TiO_2 could increase its photoactivity into the visible wavelength.

Zhao and Liu [127] investigated N-TiO_2 and found that TiO_2 doped with substitutional N has shallow acceptor states above the valence band. In other hand, TiO_2 doped with interstitial nitrogen has isolated impurity states in the middle of the band gap

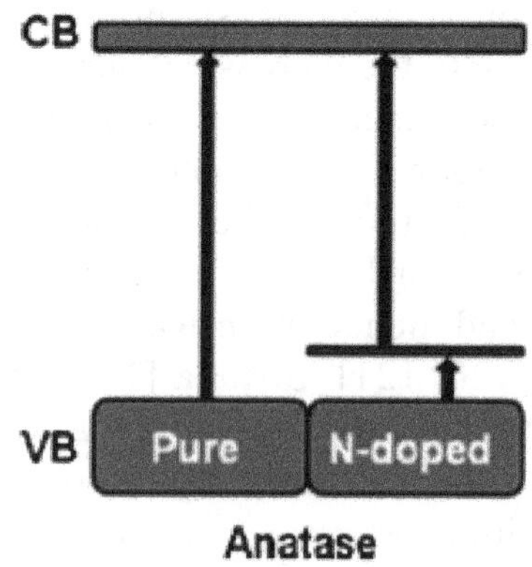

Fig. 9. Doping TiO_2 with N [114].

(Fig.9). The N-doped TiO_2 was determined to be effective for methylene blue decomposition, disinfection and for phenol degradation under visible light. S doping is leading to similar band gap narrowing. C and P are less effective as the introduced states were so deep that photo-generated charge carriers cannot be transferred to the surface of the photocatalyst. S-doped TiO_2 is more efficient than pure TiO_2 under visible light radiation. Although the valence band was shifted upwards, the oxidation ability was found to be still high. Similar to S-doping, N-doping also causes a valence band shift upward, led to narrow band gap and less oxidizing holes [128].

4.5. Compound materials

Compound materials extend the specific surface area, improve the reaction conditions, and widen the range of absorption light. The band-gap energy of $ZnFe_2O_4$ is lower, and it can absorb visible light. When the mass ratio of Zn to Ti in the compound materials of $ZnFe_2O_4/TiO_2$ is 0.05, it possesses the highest photocatalytic activity [128].

Mixing different semiconductors with different energy levels is another method, which has received great interest recently [130-134]. In photocatalytic systems, combining TiO_2 with other semiconductors was done to widen the absorption wavelength range into the visible light region and to mitigate the charge carrier recombination in individual photoelectrodes (Fig.10) [131]. Zhang [131] showed that the good matching of the conduction band and valence band of the two semiconductors could ensure an efficient transfer of the charge carriers from one to another. When the large band gap of TiO_2 is coupled with a small band gap semiconductor having more negative conduction band level, the electron can

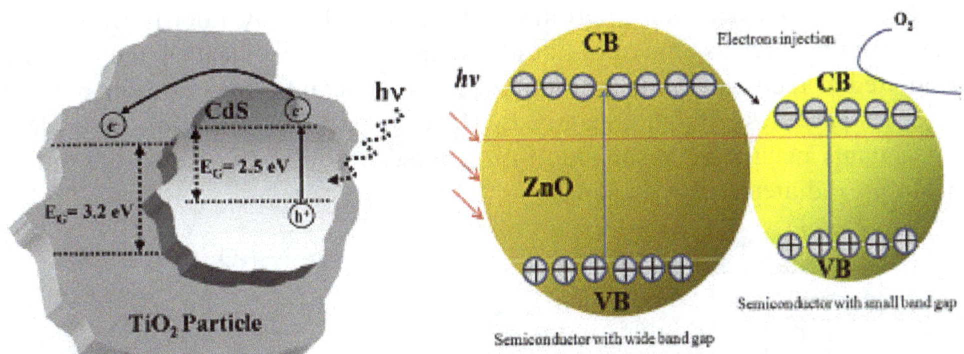

Fig. 10. The excitation process in a semiconductor–semiconductor photocatalyst [114, 129].

be injected from the small band gap of the semiconductor used as a sensitizer to the TiO_2 [138]. According to Robert [132] the efficient transfer of electron between TiO_2 and the sensitizer was occurred when the conduction band of TiO_2 is more anodic than the corresponding band of the sensitizer. Under visible irradiation, only the sensitizer is excited and the electrons photogenerated will flow into the conduction band of the adjacent TiO_2. As mentioned in certain number of studies, the photocatalytic activity of TiO_2 combined with metal oxides such as CdS,[134] SnO_2, [135]WO_3,[136] and FeO_3 [137]. Bi_2S_3, [138] among others, has been largely used for improving the process under visible irradiation. For photocatalytic application intended for water and wastewater treatment, the systems M_xO_y/TiO_2 and $M_xS_y/$ TiO_2 are the most widely used catalyst. By comparison, $SnO_2/$ TiO_2 and ZnO/TiO_2 are examples of a TiO_2/M_xO_y system, which have been used by Zhang [139] and Ge [140], respectively.

5. Photocatalytic reactors in water and wastewater treatment

Hybrid photocatalysis-membrane processes are used in the installations called "photocatalytic membrane reactors". In the literature, other names such as "membrane chemical reactor" (MCR) [9], "membrane reactor" [56], "membrane photoreactor" [56], or more specific, "submerged membrane photocatalysis reactor" [5] and "photocatalysis–ultrafiltration reactor" (PUR) [35] are used for these configurations. Because all these names mean the hybridization of photocatalysis with membrane process, it will be useful to use a general term of the "photocatalytic membrane reactor". Photocatalytic membrane reactors are divided commonly into two main groups: (I) reactors with catalyst suspended in feed solution and (II) reactors with catalyst supported in/on the membrane. In case of the reactors with photocatalyst in suspension a membrane filtration may be needed as a single step for the proper recovery of catalyst particles from the solution [56]. The main drawback of such configuration are decay of the permeate flux and membrane fouling, which are especially observed in case of pressure drive membrane methods. One solution for this problem can be application of photocatalytic membranes reactors. In this configuration, oxidation by hydroxyl radicals occurs on the outer surface and within the pores of the membrane, at a same time reactants are permeating in a one-pass flow. However, fixation of the photocatalyst often results in a loss of photoactivity of photocatalyst [82]. Moreover, selection of a proper membrane, refractory to UV irradiation and hydroxyl radicals is necessary. In both configurations, i.e. utilizing photocatalyst in suspension or fixed on/in a membrane, the membrane might act as a barrier for the molecules which are decomposed and products and by products of their decomposition. However, this role significantly depends on the separation characteristics of the membrane used and the membrane process applied. The products and by-products of photo degradation of organics are often low molecular weight compounds. Therefore, in case of pressure driven membrane processes, NF and RO only might be considered as processes able to separate these substances. In case of membrane methods in which the mechanism of separation is other than the sieve one, other properties of substances, for example volatility, should be taken into consideration. Based on these considerations, the photocatalytic reactors for water and wastewater treatment are divided into three main groups, including slurry reactors, wall and fixed bed reactors, and fluidized bed reactors.

5.1. Slurry reactors

Figures 11-14 show different configuration of slurry

photocatalytic reactors. In this reactors, micrometric particles of photocatalyst are suspended in liquid system. Light irradiated to feed tank and membrane separate micrometric particles of photocatalyst. Degradation of pollutants occurs by suspended micrometric photocatalyst radiated with UV, separation take place by microfiltration or nanofiltration. Reactions take place on the surface of photocatalysts. Reactants diffuse from bulk of liquid to surface of particles, reaction take place and then degraded products defuse to the bulk.

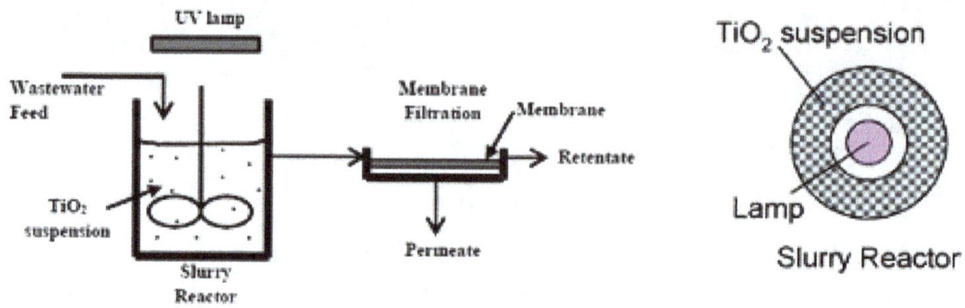

Fig.11. schematic of slurry reactors [137].

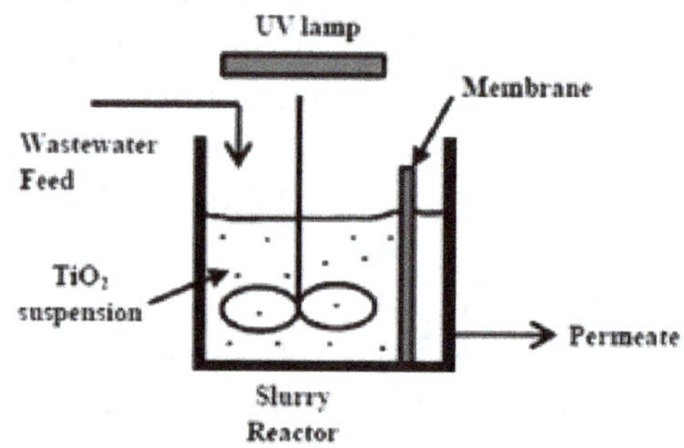

Fig. 12. schematic of slurry reactor [137].

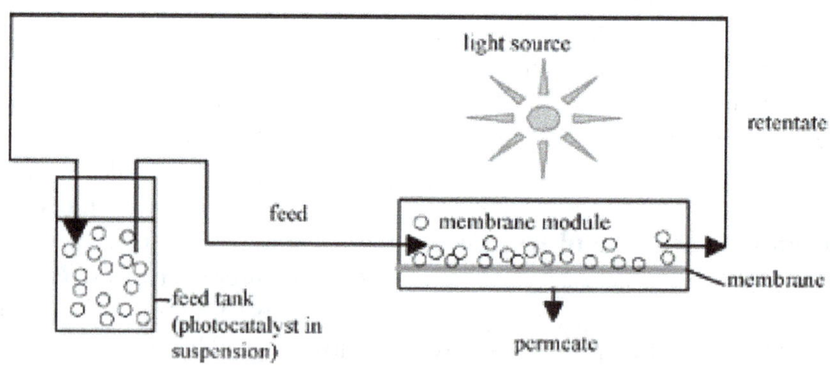

Fig. 13. Schematic of slurry reactor [138].

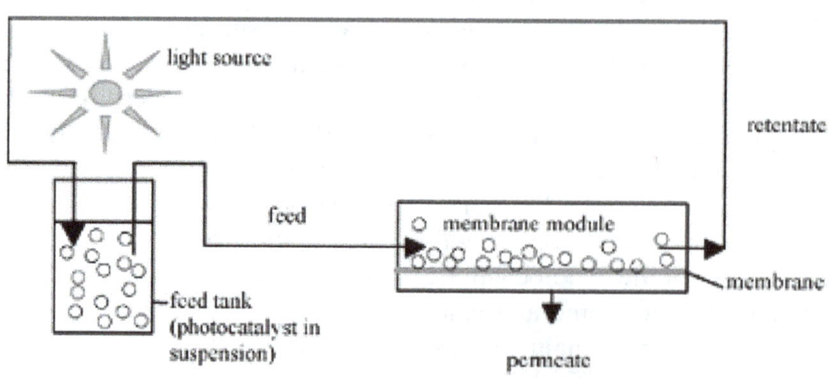

Fig. 14. Schematic of slurry reactor [138].

The fine particles of the Degussa P-25 TiO_2 have always been applied in a slurry form.

Light is irradiated to membrane vessel or feed tank, Fig. 13 and Fig. 14.

Another type of slurry photpcatalytic reactor is thin film slurry (TFS) [139]. This configuration benefits from a very large illuminated catalyst surface area per unit volume of reactor and minimal mass transfer limitations. Two feasible configurations of TFS photocatalytic reactors are shown in Table 3.

In TFSFW the water is flowing along a flat wall (FW), as a falling film or between parallel plates and the reactor is irradiated by lamps that are distributed equally. In TFSIW the liquid film is flowing along the internal wall (IW), of a cylindrical reactor as a falling film or in an annulus with the lamp that are located along the central axis. Between these two options, the TFSIW is better configuration for commercial design when using UV lamps. The advantages of this reactor geometry are:

(1) The cylindrical symmetry of the system allows all parts of the photoreactor to be uniformly illuminated with little loss of efficiency providing that the lengths of the reactor and lamp are comparable;
(2) There is improved photon utilization as back-scattered photons have a low probability of escaping from the reactor and can be recaptured by the liquid film;
(3) There is no need for a light reflector which inevitably introduces a loss in photon utilization and adds to the overall cost of the unit;
(4) Improved safety aspects such as the containment of the lamp inside the reactor in case of fire or explosions;
(5) Simple design that require the use of a single lamp and occupies minimum floor space.

The comparison of ideal flow systems has shown that falling film laminar flow and plug flow systems have high performance depending on the degree of conversion obtained in the photocatalytic reactors [139].

5.2. *Wall and fixed bed reactors*

There is another type of photocatalytic systems that photocatalyst fixed on the surface of membrane, the wall of reactor or fixed granular beads. Fig. 15-

Table 3.
TFS photocatalytic reactors [139].

Reactor configuration	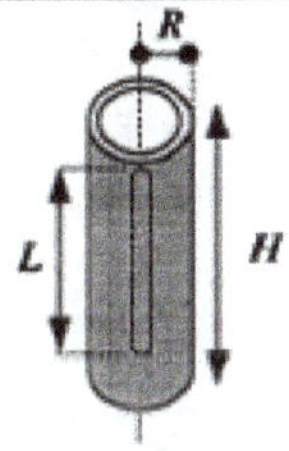	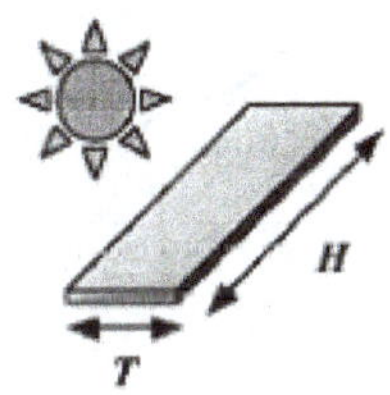
	TFSIW (UV lamps)	TFSFW (solar radiation or UV lamps)

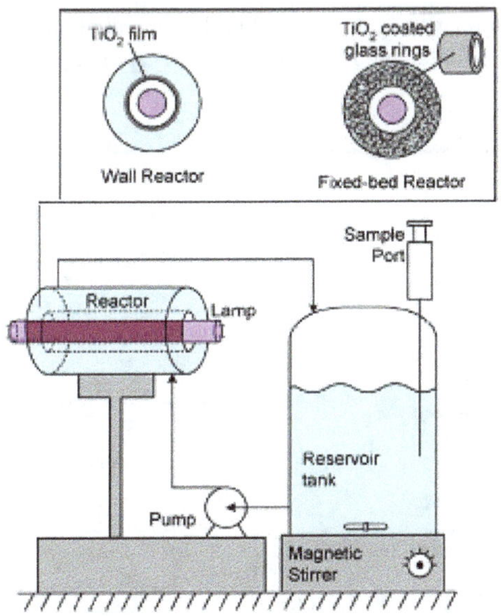

Fig. 15. Schematic representation of the fixed photocatalyst without membrane [140].

20 show various configurations of these reactors. In this configuration catalysts has been fixed on wall of the reactor or has been fixed on surface of the membrane or fixed on fixed millimetric granular beads. Reaction is only limited to the surface of wall or membrane.

Fig. 15 show a fixed photocatalytic system consists of an annular reactor 15 cm long, 3 cm inner tube diameter and 5 cm external tube diameter operating in a closed recirculation system. A centrifugal pump is used for circulating water and reactor equipped with a stirred reservoir tank of 2L volume with a device for withdrawal of samples [140]. Van Grieken et al. showed disinfection of E. coli bacteria suspensions in slurry, wall and fixed-bed reactors and concluded that immobilized systems have shown that they are stable and do not present deactivation after several cycles of reuse, being readily applicable for continuous water treatment systems.

Photocatalyst can be fixed on the wall of glassy reactor in to configuration (Fig. 16).

Fig. 17-19 show different configuration of catalyst that fixed on wall reactor.

In Fig. 20 the hybrid photo-catalytic/UF experimental set-up is reported. It consists of three concentric tubes placed in the vertical direction. The inner tube is the photocatalytic UF membrane. The intermediate and external tubes are made of Plexiglas, and define an outer flow channel the annular space between the tubes) where polluted water is fed in the lumen of the transparent Ca alginate/TiO_2 polymer fibers.

Membranes with catalyst coating (Fig. 21) have fouling problems. For long time processes, because of membrane fouling, pressure drop increases significantly at constant flux mode, while flux reduces greatly at constant pressure drop mode. In PMRs, fouling mechanisms can be mainly divided into two

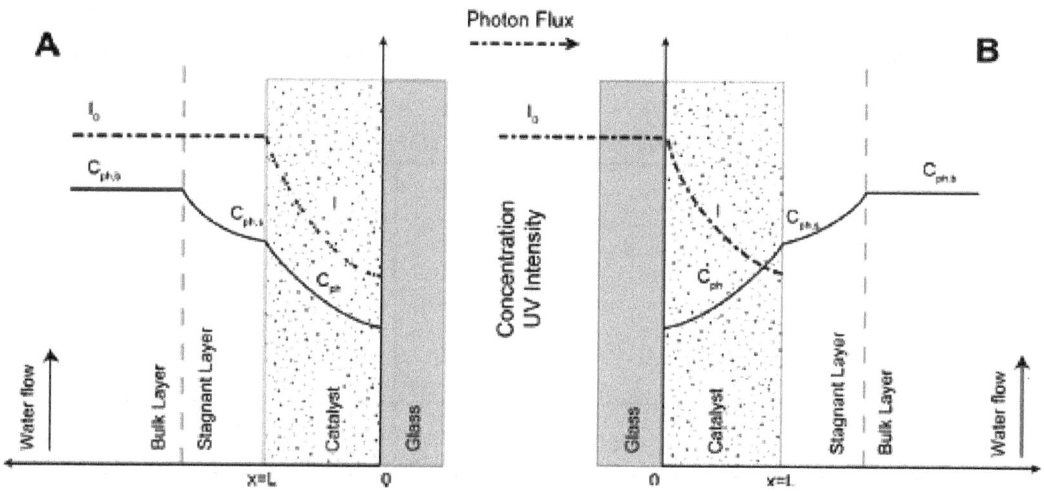

Fig. 16. (A) Schematic of the liquid side illumination system where pollutant and photon flux come from the same direction. (B) Schematic of the support side illumination where pollutant and photon flux come from opposite directions [141].

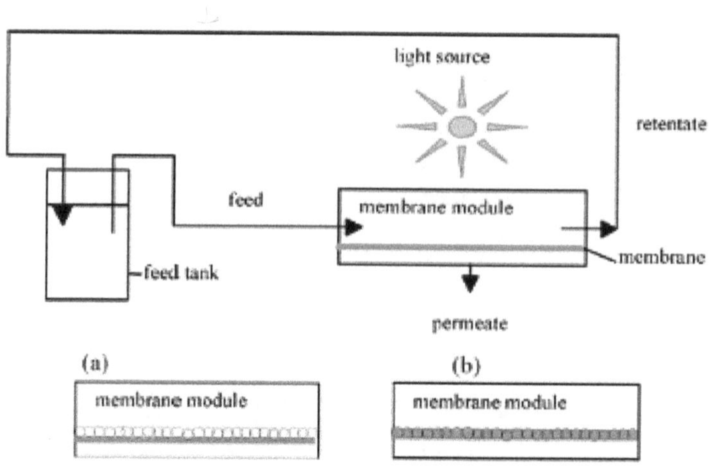

Fig. 17. Schematic of fixed on wall reactor [138].

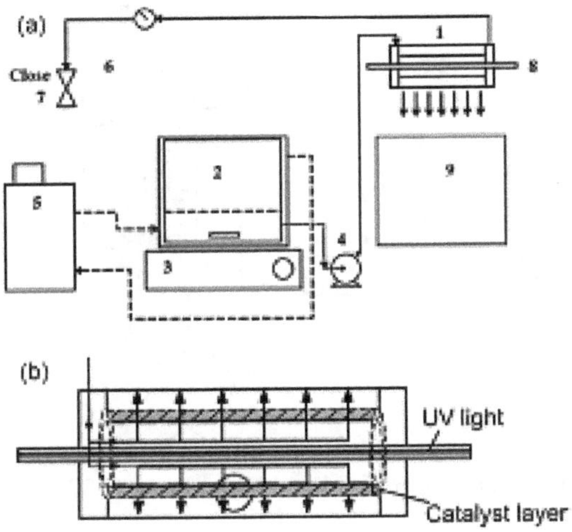

Fig. 18. Schematic of fixed on wall reactor [138].

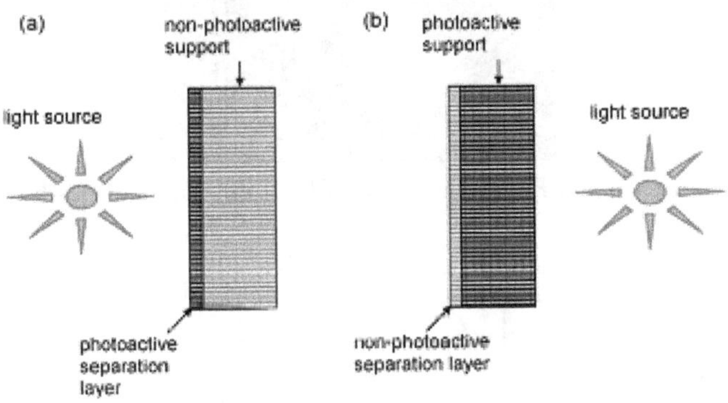

Fig. 19. Schematic of fixed on wall reactor [138].

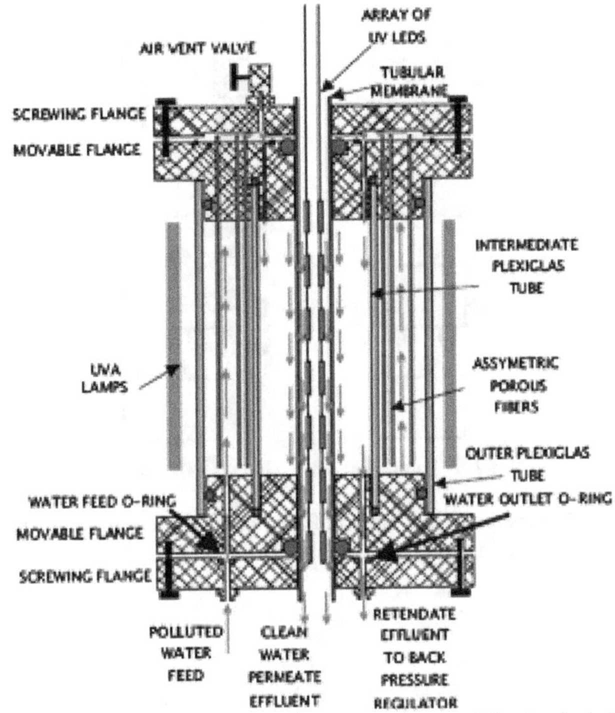

Fig. 20. Photocatalytic membrane system used in the hybrid-photocatalytic/ultrafiltration with catalyst fixed on wall of reactor [142].

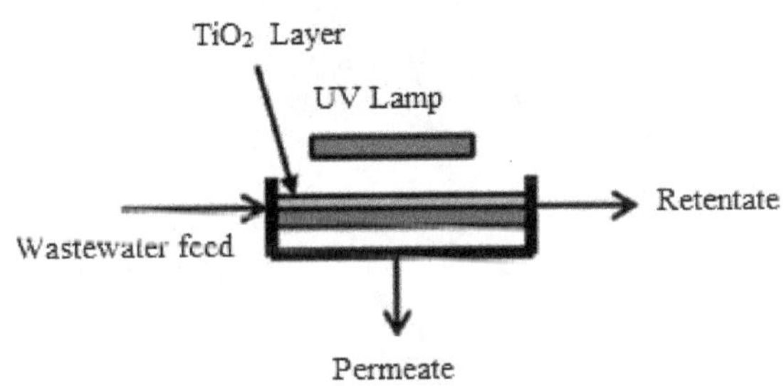

Fig. 21. Schematic of fixed on membrane wall [143].

steps: pore blocking and cake layer steps, which are shown in Fig. 22.

Both suspension and fixed catalyst reactors have their advantages and disadvantages. These properties are listed in Table 4.

Javier Marugan et.al [145] used fixed bed photocatalytic reactor for inactivation of Escherichia coli bacteria. P25 industrial titanium dioxide immobilized onto glass Raschig rings was used [Fig.23].

5.3.Fluidized bed reactors

Suspension catalytic reactors are more efficient than fixed bed reactors because mass transfer is easier and there is possibility to add desired amount of catalyst to the reactor. At the other side fixed catalyst on wall reactors do not have separation problem of photocatalyst from water. Therefore the third type of reactor configuration has been formed. This reactor is something between slurry and fixed on wall catalyst reactors. Photo catalysts fixed on milimetric balls and these balls fluidized by air force or water flow. There for by this way advantages of two previous model is accessible and overcome to their limitations. Fig. 24 show configuration that air fluidized milimetric balls and reaction take place on surface of balls.

Photocatalytic degradation of organic contaminants in wastewater by TiO_2 introduced in immobilized state on supporting material [137]. Soo et al. 2005 synthesized hollow ceramic ball (Fig. 25) with a novel method, 1.5 mm in diameter, and coated TiO_2 on wall and photocatalytic activity of fluidized bed process (Fig. 26) was investigated for RhB degradation. Results show that this process effectively photocatalyzed the degradation of RhB.

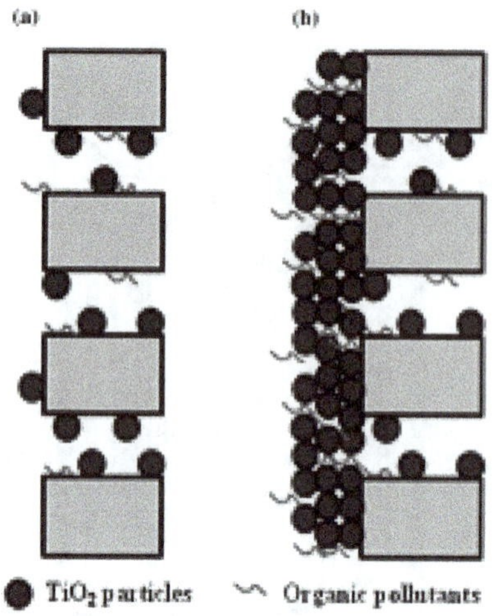

Fig. 22. Schematic illustration of membrane fouling mechanisms: (a) pore blocking and (b) cake layer [144].

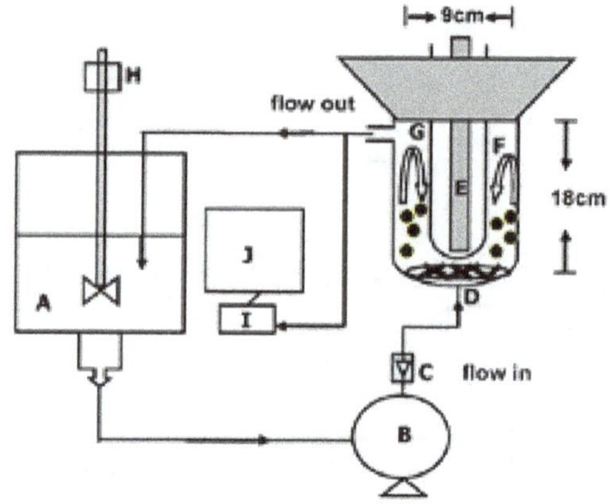

Fig. 24. Schematic of fluidized bed reactor [137].

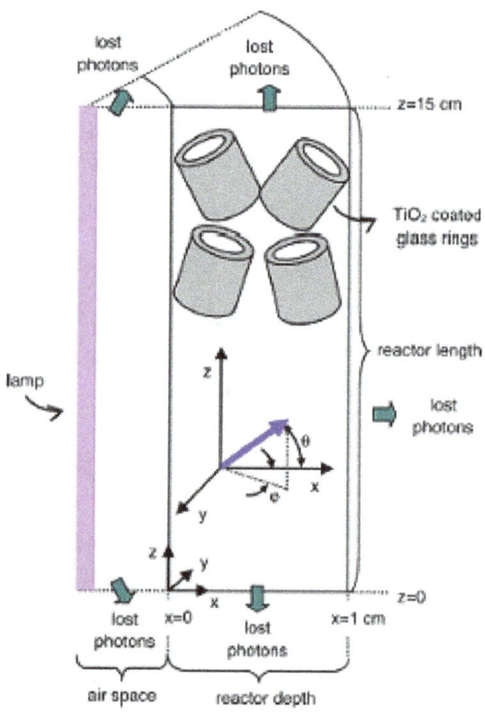

Fig. 23. schematic representation of fixed bed reactor [145].

Fig. 25. The SEM photograph of Photomedia (×40) [146].

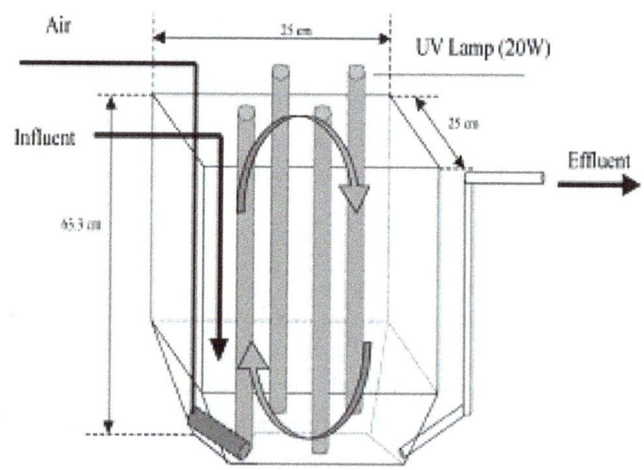

Fig. 26. The schematic diagram of photocatalytic fluidized bed reactor [146].

In another work Wang et al. used granular activated carbon as support of photocatalyst [146]. The photoatalytic oxidation of acid dye under visible light irradiation was done in a recycle fluidized bed reactor. $ZnFe_2O_4/TiO_2$-immobilized granular activated carbon ($ZnFe_2O_4/TiO_2$-GAC) was used as the photocatalyst, and the lamp (CDM-T, 150 W) was used as the light source. Granular activated carbons with an average diameter of 2.172 mm (8–9 mesh), 1.189 mm (12–16 mesh), and 0.542 mm (28–32 mesh) were hired as the supports of $ZnFe_2O_4/TiO_2$ photocatalysts. Zulfakar et al. used immobilized TiO_2 on quartz sand with 0.3 mm in diameter [147]. The performance of the supported photocatalyst was evaluated in different operating parameters such as photocatalyst loading and initial phenol concentration in a fluidized bed reactor (Fig.27).

Fluidized bed photocatalytic reactor systems have several benefits over common immobilized (coated on wall) or slurry-type photocatalytic reactors. The unique reactor configuration provides both exposure of photocatalyst to UV/sun light and good

Table 4.
Advantages and disadvantages of suspended reactors with TiO_2 immobilized in/on the membrane [137].

Advantages	Disadvantages
- No need to separate and recycle the catalyst—the membrane can be used as long as the activity of TiO_2 remains high	- The effectiveness of degradation lower than in case of catalyst in suspension
- No membrane fouling due to presence of TiO_2 particles	- Adjustment of catalyst loading to the composition of the treated solution is impossible (fixed amount of the catalyst immobilized in/on the membrane)
- Possibility of fouling mitigation due to enhanced hydrophilicity of the TiO_2-modified membrane	
- Possibility of fouling mitigation due to decomposition of organic contaminants forming the gel layer or filtration cake	- Risk of damage of polymer membranes by the UV light or hydroxyl radicals
- Contaminants could be decomposed either in feed or in permeate	- Necessity of the membrane exchange when the catalyst loses its activity

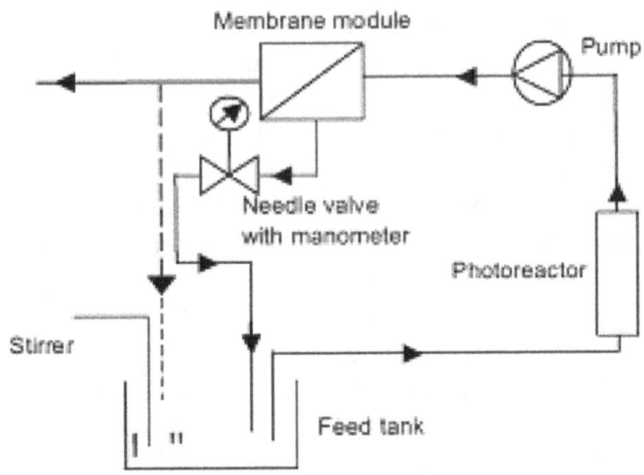

Fig. 27. Fludized bed reactor set-up [148].

penetration of the light into the photocatalyst bed that allow more contact of photocatalyst and reactant fluid. Also, UV light can be more evenly distributed within the given catalyst bed. Separation problem of catalysts from treated water is very easy in fluidized bed reactor systems.

6. Conclusions

This article reviews the principles and methods of photocatalysts preparation, promoting photocatalyst activity and improving performance, mainly for TiO_2. In addition to destruction of viruses and bacteria, heterogeneous photocatalysis has been used to decompose natural organic matters, volatile organic compounds in water, air and soil. Widening the wavelength for semiconductor application by doping is an area of increasing interest. Various configurations of photocatalytic reactors were investigated and their advantages and disadvantages were reported. The photocatalysts can be modified and used in an appropriate reactor configuration for destruction of specific pollutants.

References

[1] S. Suárez, M. Carballa, F. Omil, J.M. Lema, How are pharmaceutical and personal care products (PPCPs) removed from urban wastewaters?, Reviews in Environmental Science and Bio/Technology, 7 (2008) 125-138.

[2] C. Zwiener, S.D. Richardson, Analysis of disinfection by-products in drinking water by LC–MS and related MS techniques, TrAC Trends in Analytical Chemistry, 24 (2005) 613-621.

[3] T. Wintgens, F. Salehi, R. Hochstrat, T. Melin, Emerging contaminants and treatment options in water recycling for indirect potable use, Water Science and Technology, 57 (2008) 99-107.

[4] P.R. Gogate, A.B. Pandit, A review of imperative technologies for wastewater treatment I: oxidation technologies at ambient conditions, Advances in Environmental Research, 8 (2004) 501-551.

[5] M. Pera-Titus, V. Garcıa-Molina, M.A. Baños, J. Giménez, S. Esplugas, Degradation of chlorophenols by means of advanced oxidation processes: a general review, Applied Catalysis B: Environmental, 47 (2004) 219-256.
[6] C. Comninellis, A. Kapalka, S. Malato, S.A. Parsons, I. Poulios, D. Mantzavinos, Advanced oxidation processes for water treatment: advances and trends for R&D, Journal of Chemical Technology and Biotechnology, 83 (2008) 769-776.
[7] R. Ameta, S. Benjamin, A. Ameta, S.C. Ameta, Photocatalytic degradation of organic Pollutants: A Review, in: Materials Science Forum, Trans Tech Publ, 2013, pp. 247-272.
[8] R.C. Martins, R.M. Quinta-Ferreira, Remediation of phenolic wastewaters by advanced oxidation processes (AOPs) at ambient conditions: comparative studies, Chemical engineering science, 66 (2011) 3243-3250.
[9] S.-Y. Lee, S.-J. Park, TiO 2 photocatalyst for water treatment applications, Journal of Industrial and Engineering Chemistry, 19 (2013) 1761-1769.
[10] L.G. Devi, R. Kavitha, A review on non metal ion doped titania for the photocatalytic degradation of organic pollutants under UV/solar light: Role of photogenerated charge carrier dynamics in enhancing the activity, Applied Catalysis B: Environmental, 140 (2013) 559-587.
[11] J. Yu, X. Yu, Hydrothermal synthesis and photocatalytic activity of zinc oxide hollow spheres, Environmental science & technology, 42 (2008) 4902-4907.
[12] A. Mehdizadeh Dehkordi, An Experimental Investigation Towards Improvement of Thermoelectric Properties of Strontium Titanate Ceramics, (2014).
[13] R. Wang, K. Hashimoto, A. Fujishima, M. Chikuni, E. Kojima, A. Kitamura, M. Shimohigoshi, T. Watanabe, Light-induced amphiphilic surfaces, Nature, 388 (1997) 431-432.
[14] A. Sonune, R. Ghate, Developments in wastewater treatment methods, Desalination, 167 (2004) 55-63.
[15] T. Daimon, T. Hirakawa, M. Kitazawa, J. Suetake, Y. Nosaka, Formation of singlet molecular oxygen associated with the formation of superoxide radicals in aqueous suspensions of TiO_2 photocatalysts, Applied Catalysis A: General, 340 (2008) 169-175.
[16] Y. Nosaka, T. Daimon, A.Y. Nosaka, Y. Murakami, Singlet oxygen formation in photocatalytic TiO_2 aqueous suspension, Physical Chemistry Chemical Physics, 6 (2004) 2917-2918.
[17] H. Sakai, R. Baba, K. Hashimoto, A. Fujishima, A. Heller, Local Detection of Photoelectrochemically Produced H2O2 with a"Wired" Horseradish Peroxidase Microsensor, The Journal of Physical Chemistry, 99 (1995) 11896-11900.
[18] K. Ranjit, I. Willner, S. Bossmann, A. Braun, Lanthanide oxide-doped titanium dioxide photocatalysts: novel photocatalysts for the enhanced degradation of p-chlorophenoxyacetic acid, Environmental science & technology, 35 (2001) 1544-1549.
[19] J.R. Harbour, M.L. Hair, Radical intermediates in the photosynthetic generation of hydrogen peroxide with aqueous zinc oxide dispersions, Journal of Physical Chemistry, 83 (1979) 652-656.
[20] A. Mills, S.-K. Lee, A web-based overview of semiconductor photochemistry-based current commercial applications, Journal of Photochemistry and Photobiology A: Chemistry, 152 (2002) 233-247.
[21] N.S. Lewis, Toward cost-effective solar energy use, science, 315 (2007) 798-801.
[22] S. Bouadila, S. Kooli, M. Lazaar, S. Skouri, A. Farhat, Performance of a new solar air heater with packed-bed latent storage energy for nocturnal use, Applied Energy, 110 (2013) 267-275.
[23] http://www.lenntech.com/composition-seawater.htm, in.
[24] M. Heikkinen, H. Poutiainen, M. Liukkonen, T. Heikkinen, Y. Hiltunen, Subtraction analysis based on self-organizing maps for an industrial wastewater treatment process, Mathematics and computers in simulation, 82 (2011) 450-459.
[25] A. Biati, F. Moattar, A. Karbassi, A. Hassani, Role of saline water in removal of heavy elements from industrial wastewaters, (2010).
[26] L. Tajeddine, M. Nemmaoui, H. Mountacer, A. Dahchour, M. Sarakha, Photodegradation of fenamiphos on the surface of clays and soils, Environmental Chemistry Letters, 8 (2010) 123-128.
[27] S. Kommineni, J. Zoeckler, A. Stocking, P.S. Liang, A. Flores, R. Rodriguez, T. Browne, P.R. Roberts, A. Brown, 3.0 Advanced Oxidation Processes, ______ Treatment Technologies for removal of Methyl Tertiary Butyl Ether (MTBE) fron drinking water: air stripping, Advanced Oxidation Process, Granular Actived carbon, Sinthetic resin sorbents, 2 (2000) 109-208.
[28] J. Zhang, K.-H. Lee, L. Cui, T.-s. Jeong, Degradation of methylene blue in aqueous solution by ozone-based processes, Journal of Industrial and Engineering Chemistry, 15 (2009) 185-189.
[29] S.C. Kwon, J.Y. Kim, S.M. Yoon, W. Bae, K.S. Kang, Y.W. Rhee, Treatment characteristic of 1, 4-dioxane by ozone-based advanced oxidation processes, Journal of Industrial and Engineering Chemistry, 18 (2012) 1951-1955.
[30] Z. Li, S. Yuan, C. Qiu, Y. Wang, X. Pan, J. Wang, C. Wang, J. Zuo, Effective degradation of refractory organic pollutants in landfill leachate by electro-peroxone treatment, Electrochimica Acta, 102 (2013)

174-182.

[31] S. Sabhi, J. Kiwi, Degradation of 2, 4-dichlorophenol by immobilized iron catalysts, Water Research, 35 (2001) 1994-2002.

[32] D. Tabet, M. Saidi, M. Houari, P. Pichat, H. Khalaf, Fe-pillared clay as a Fenton-type heterogeneous catalyst for cinnamic acid degradation, Journal of environmental management, 80 (2006) 342-346.

[33] H.-J. Jung, J.-S. Hong, J.-K. Suh, A comparison of fenton oxidation and photocatalyst reaction efficiency for humic acid degradation, Journal of Industrial and Engineering Chemistry, 19 (2013) 1325-1330.

[34] S. Şahinkaya, COD and color removal from synthetic textile wastewater by ultrasound assisted electro-Fenton oxidation process, Journal of Industrial and Engineering Chemistry, 19 (2013) 601-605.

[35] G. Muthuraman, I.-S. Moon, A review on an electrochemically assisted-scrubbing process for environmental harmful pollutant›s destruction, Journal of Industrial and Engineering Chemistry, 18 (2012) 1540-1550.

[36] S. Banerjee, S.C. Pillai, P. Falaras, K.E. O'shea, J.A. Byrne, D.D. Dionysiou, New insights into the mechanism of visible light photocatalysis, The journal of physical chemistry letters, 5 (2014) 2543-2554.

[37] M. Pelaez, N.T. Nolan, S.C. Pillai, M.K. Seery, P. Falaras, A.G. Kontos, P.S. Dunlop, J.W. Hamilton, J.A. Byrne, K. O. shea, A review on the visible light active titanium dioxide photocatalysts for environmental applications, Applied Catalysis B: Environmental, 125 (2012) 331-349.

[38] J. Schneider, M. Matsuoka, M. Takeuchi, J. Zhang, Y. Horiuchi, M. Anpo, D.W. Bahnemann, Understanding TiO_2 photocatalysis: mechanisms and materials, Chemical reviews, 114 (2014) 9919-9986.

[39] S. Banerjee, D.D. Dionysiou, S.C. Pillai, Self-cleaning applications of TiO_2 by photo-induced hydrophilicity and photocatalysis, Applied Catalysis B: Environmental, 176 (2015) 396-428.

[40] A. Fujishima, Electrochemical photolysis of water at a semiconductor electrode, nature, 238 (1972) 37-38.

[41] P.V. Kamat, K. Tvrdy, D.R. Baker, J.G. Radich, Beyond photovoltaics: semiconductor nanoarchitectures for liquid-junction solar cells, Chemical reviews, 110 (2010) 6664-6688.

[42] V. Etacheri, M.K. Seery, S.J. Hinder, S.C. Pillai, Oxygen Rich Titania: A Dopant Free, High Temperature Stable, and Visible-Light Active Anatase Photocatalyst, Advanced Functional Materials, 21 (2011) 3744-3752.

[43] D.A. Keane, K.G. McGuigan, P.F. Ibáñez, M.I. Polo-López, J.A. Byrne, P.S. Dunlop, K. O›Shea, D.D. Dionysiou, S.C. Pillai, Solar photocatalysis for water disinfection: materials and reactor design, Catalysis Science & Technology, 4 (2014) 1211-1226.

[44] S.C. Pillai, P. Periyat, R. George, D.E. McCormack, M.K. Seery, H. Hayden, J. Colreavy, D. Corr, S.J. Hinder, Synthesis of high-temperature stable anatase TiO_2 photocatalyst, The Journal of Physical Chemistry C, 111 (2007) 1605-1611.

[45] V. Etacheri, M.K. Seery, S.J. Hinder, S.C. Pillai, Highly Visible Light Active TiO_2− x N x Heterojunction Photocatalysts†, Chemistry of Materials, 22 (2010) 3843-3853.

[46] L. Vayssieres, Growth of arrayed nanorods and nanowires of ZnO from aqueous solutions, Advanced Materials, 15 (2003) 464-466.

[47] V. Etacheri, R. Roshan, V. Kumar, Mg-doped ZnO nanoparticles for efficient sunlight-driven photocatalysis, ACS applied materials & interfaces, 4 (2012) 2717-2725.

[48] G.-S. Li, D.-Q. Zhang, J.C. Yu, A new visible-light photocatalyst: CdS quantum dots embedded mesoporous TiO2, Environmental science & technology, 43 (2009) 7079-7085.

[49] A. Kudo, M. Sekizawa, Photocatalytic H_2 evolution under visible light irradiation on Ni-doped ZnS photocatalyst, Chemical Communications, (2000) 1371-1372.

[50] F. Kong, L. Huang, L. Luo, S. Chu, Y. Wang, Z. Zou, Synthesis and characterization of visible light driven mesoporous nano-photocatalyst MoO_3/TiO_2, Journal of nanoscience and nanotechnology, 12 (2012) 1931-1937.

[51] V. Etacheri, M.K. Seery, S.J. Hinder, S.C. Pillai, Nanostructured Ti1-x S x O2-y N y Heterojunctions for Efficient Visible-Light-Induced Photocatalysis, Inorganic chemistry, 51 (2012) 7164-7173.

[52] M.R. Hoffmann, S.T. Martin, W. Choi, D.W. Bahnemann, Environmental applications of semiconductor photocatalysis, Chemical reviews, 95 (1995) 69-96.

[53] V. Etacheri, G. Michlits, M.K. Seery, S.J. Hinder, S.C. Pillai, A Highly Efficient TiO_2–x C x Nano-heterojunction Photocatalyst for Visible Light Induced Antibacterial Applications, ACS applied materials & interfaces, 5 (2013) 1663-1672.

[54] N. Wetchakun, S. Chaiwichain, B. Inceesungvorn, K. Pingmuang, S. Phanichphant, A.I. Minett, J. Chen, $BiVO_4/CeO_2$ nanocomposites with high visible-light-induced photocatalytic activity, ACS applied materials & interfaces, 4 (2012) 3718-3723.

[55] S. Polisetti, P.A. Deshpande, G. Madras, Photocatalytic activity of combustion synthesized ZrO_2 and ZrO_2–TiO_2 mixed oxides, Industrial & Engineering Chemistry Research, 50 (2011) 12915-12924.

[56] L. Li, M. Krissanasaeranee, S.W. Pattinson, M.

Stefik, U. Wiesner, U. Steiner, D. Eder, Enhanced photocatalytic properties in well-ordered mesoporous WO_3, Chemical Communications, 46 (2010) 7620-7622.

[57] Y. Liu, L. Yu, Y. Hu, C. Guo, F. Zhang, X.W.D. Lou, A magnetically separable photocatalyst based on nest-like γ-Fe_2 O_3/ZnO double-shelled hollow structures with enhanced photocatalytic activity, Nanoscale, 4 (2012) 183-187.

[58] D. Chu, J. Mo, Q. Peng, Y. Zhang, Y. Wei, Z. Zhuang, Y. Li, Enhanced Photocatalytic Properties of SnO_2 Nanocrystals with Decreased Size for ppb-level Acetaldehyde Decomposition, ChemCatChem, 3 (2011) 371-377.

[59] M. Grätzel, Heterogeneous photochemical electron transfer, CRC Press Boca Raton, FL, 1989.

[60] S. Chakrabarti, B.K. Dutta, Photocatalytic degradation of model textile dyes in wastewater using ZnO as semiconductor catalyst, Journal of hazardous materials, 112 (2004) 269-278.

[61] S.N. Frank, A.J. Bard, Heterogeneous photocatalytic oxidation of cyanide and sulfite in aqueous solutions at semiconductor powders, The journal of physical chemistry, 81 (1977) 1484-1488.

[62] Y. Wang, Y. Huang, W. Ho, L. Zhang, Z. Zou, S. Lee, Biomolecule-controlled hydrothermal synthesis of C–N–S-tridoped TiO_2 nanocrystalline photocatalysts for NO removal under simulated solar light irradiation, Journal of Hazardous materials, 169 (2009) 77-87.

[63] C. Su, C.-M. Tseng, L.-F. Chen, B.-H. You, B.-C. Hsu, S.-S. Chen, Sol–hydrothermal preparation and photocatalysis of titanium dioxide, Thin Solid Films, 498 (2006) 259-265.

[64] R. Asahi, Y. Taga, W. Mannstadt, A.J. Freeman, Electronic and optical properties of anatase TiO_2, Physical Review B, 61 (2000) 7459.

[65] M. Muruganandham, M. Swaminathan, Solar photocatalytic degradation of a reactive azo dye in TiO_2-suspension, Solar Energy Materials and Solar Cells, 81 (2004) 439-457.

[66] T. Miyagi, M. Kamei, T. Mitsuhashi, T. Ishigaki, A. Yamazaki, Charge separation at the rutile/anatase interface: a dominant factor of photocatalytic activity, Chemical Physics Letters, 390 (2004) 399-402.

[67] M. Toyoda, Y. Nanbu, Y. Nakazawa, M. Hirano, M. Inagaki, Effect of crystallinity of anatase on photoactivity for methyleneblue decomposition in water, Applied Catalysis B: Environmental, 49 (2004) 227-232.

[68] E. Beyers, P. Cool, E.F. Vansant, Anatase formation during the synthesis of mesoporous titania and its photocatalytic effect, The Journal of Physical Chemistry B, 109 (2005) 10081-10086.

[69] A.O. Ibhadon, P. Fitzpatrick, Heterogeneous photocatalysis: recent advances and applications, Catalysts, 3 (2013) 189-218.

[70] P. Schmitt-Kopplin, N. Hertkorn, H.-R. Schulten, A. Kettrup, Structural changes in a dissolved soil humic acid during photochemical degradation processes under O_2 and N_2 atmosphere, Environmental science & technology, 32 (1998) 2531-2541.

[71] A. Paleologou, H. Marakas, N.P. Xekoukoulotakis, A. Moya, Y. Vergara, N. Kalogerakis, P. Gikas, D. Mantzavinos, Disinfection of water and wastewater by TiO_2 photocatalysis, sonolysis and UV-C irradiation, Catalysis Today, 129 (2007) 136-142.

[72] K. Sunada, T. Watanabe, K. Hashimoto, Studies on photokilling of bacteria on TiO_2 thin film, Journal of Photochemistry and Photobiology A: Chemistry, 156 (2003) 227-233.

[73] N. Daneshvar, A. Niaei, S. Akbari, S. Aber, N. Kazemian, Photocatalytic disinfection of water polluted by Pseudomonas aeruginosa, Global Nest J, 9 (2007) 1-5.

[74] P.V. Kamat, Graphene-based nanoarchitectures. Anchoring semiconductor and metal nanoparticles on a two-dimensional carbon support, The Journal of Physical Chemistry Letters, 1 (2009) 520-527.

[75] D.S. Bhatkhande, V.G. Pangarkar, A.A. Beenackers, Photocatalytic degradation for environmental applications–a review, Journal of Chemical Technology and Biotechnology, 77 (2002) 102-116.

[76] K. Pirkanniemi, M. Sillanpää, Heterogeneous water phase catalysis as an environmental application: a review, Chemosphere, 48 (2002) 1047-1060.

[77] A. Huang, L. Cao, J. Chen, F.-J. Spiess, S.L. Suib, T.N. Obee, S.O. Hay, J.D. Freihaut, Photocatalytic degradation of triethylamine on titanium oxide thin films, Journal of catalysis, 188 (1999) 40-47.

[78] I. Arabatzis, S. Antonaraki, T. Stergiopoulos, A. Hiskia, E. Papaconstantinou, M. Bernard, P. Falaras, Preparation, characterization and photocatalytic activity of nanocrystalline thin film TiO_2 catalysts towards 3, 5-dichlorophenol degradation, Journal of Photochemistry and Photobiology A: Chemistry, 149 (2002) 237-245.

[79] W. Choi, A. Termin, M.R. Hoffmann, The role of metal ion dopants in quantum-sized TiO_2: correlation between photoreactivity and charge carrier recombination dynamics, The Journal of Physical Chemistry, 98 (1994) 13669-13679.

[80] X. Chen, S.S. Mao, Titanium dioxide nanomaterials: synthesis, properties, modifications, and applications, Chem. Rev, 107 (2007) 2891-2959.

[81] A. Sclafani, J. Herrmann, Comparison of the photoelectronic and photocatalytic activities of various anatase and rutile forms of titania in pure liquid organic phases and in aqueous solutions, The Journal of Physical Chemistry, 100 (1996) 13655-13661.

[82] J. Liqiang, Q. Yichun, W. Baiqi, L. Shudan, J. Baojiang, Y. Libin, F. Wei, F. Honggang, S. Jiazhong, Review of photoluminescence performance of nano-sized semiconductor materials and its relationships with photocatalytic activity, Solar Energy Materials and Solar Cells, 90 (2006) 1773-1787.

[83] N. Serpone, Relative photonic efficiencies and quantum yields in heterogeneous photocatalysis, Journal of Photochemistry and Photobiology A: Chemistry, 104 (1997) 1-12.

[84] J.C. Yu, L. Zhang, J. Yu, Direct sonochemical preparation and characterization of highly active mesoporous TiO_2 with a bicrystalline framework, Chemistry of Materials, 14 (2002) 4647-4653.

[85] Y. Do, W. Lee, K. Dwight, A. Wold, The effect of WO_3 on the photocatalytic activity of TiO_2, Journal of Solid State Chemistry, 108 (1994) 198-201.

[86] J. Engweiler, J. Harf, A. Baiker, WO_x/TiO_2 catalysts prepared by grafting of tungsten alkoxides: morphological properties and catalytic behavior in the selective reduction of NO by NH_3, Journal of Catalysis, 159 (1996) 259-269.

[87] K. Vinodgopal, P.V. Kamat, Enhanced rates of photocatalytic degradation of an azo dye using SnO_2/TiO_2 coupled semiconductor thin films, Environmental science & technology, 29 (1995) 841-845.

[88] A. Maira, K.L. Yeung, C. Lee, P.L. Yue, C.K. Chan, Size effects in gas-phase photo-oxidation of trichloroethylene using nanometer-sized TiO_2 catalysts, Journal of Catalysis, 192 (2000) 185-196.

[89] Y. Li, D.-S. Hwang, N.H. Lee, S.-J. Kim, Synthesis and characterization of carbon-doped titania as an artificial solar light sensitive photocatalyst, Chemical Physics Letters, 404 (2005) 25-29.

[90] R. Georgekutty, M.K. Seery, S.C. Pillai, A highly efficient Ag-ZnO photocatalyst: synthesis, properties, and mechanism, The Journal of Physical Chemistry C, 112 (2008) 13563-13570.

[91] G. Balasubramanian, D.D. Dionysiou, M.T. Suidan, I. Baudin, J.-M. Laıné, Evaluating the activities of immobilized TiO_2 powder films for the photocatalytic degradation of organic contaminants in water, Applied Catalysis B: Environmental, 47 (2004) 73-84.

[92] P. Periyat, S.C. Pillai, D.E. McCormack, J. Colreavy, S.J. Hinder, Improved high-temperature stability and sun-light-driven photocatalytic activity of sulfur-doped anatase TiO2, The Journal of Physical Chemistry C, 112 (2008) 7644-7652.

[93] J.A. Gamboa, D.M. Pasquevich, Effect of Chlorine Atmosphere on the Anatase-Rutile Transformation, Journal of the American Ceramic Society, 75 (1992) 2934-2938.

[94] A. Hagfeldt, M. Graetzel, Light-induced redox reactions in nanocrystalline systems, Chemical Reviews, 95 (1995) 49-68.

[95] N.J. Cherepy, G.P. Smestad, M. Grätzel, J.Z. Zhang, Ultrafast electron injection: implications for a photoelectrochemical cell utilizing an anthocyanin dye-sensitized TiO_2 nanocrystalline electrode, The Journal of Physical Chemistry B, 101 (1997) 9342-9351.

[96] A. Kay, R. Humphry-Baker, M. Graetzel, Artificial photosynthesis. 2. Investigations on the mechanism of photosensitization of nanocrystalline TiO_2 solar cells by chlorophyll derivatives, The Journal of Physical Chemistry, 98 (1994) 952-959.

[97] B. Patrick, P. Kamat, Photosensitization of large-bandgap semiconductors. Charge injection from triplet excited thionine into ZnO colloids, Journal of Physical Chemistry, 96 (1992).

[98] P.V. Kamat, Picosecond charge-transfer events in the photosensitization of colloidal titania, Langmuir, 6 (1990) 512-513.

[99] P.V. Kamat, K. Gopidas, D. Weir, Photoelectrochemistry in particulate systems. Photosensitized charge injection into opaque TiO_2 semiconductor powder as probed by time-resolved diffuse reflectance laser flash photolysis, Chemical physics letters, 149 (1988) 491-496.

[100] K. Kalyanasundaram, N. Vlachopoulos, V. Krishnan, A. Monnier, M. Grätzel, Sensitization of titanium dioxide in the visible light region using zinc porphyrins, Journal of Physical Chemistry, 91 (1987) 2342-2347.

[101] R. Eichberger, F. Willig, Ultrafast electron injection from excited dye molecules into semiconductor electrodes, Chemical physics, 141 (1990) 159-173.

[102] Y. Cho, W. Choi, C.-H. Lee, T. Hyeon, H.-I. Lee, Visible light-induced degradation of carbon tetrachloride on dye-sensitized TiO_2, Environmental science & technology, 35 (2001) 966-970.

[103] S. Malato, P. Fernández-Ibáñez, M. Maldonado, J. Blanco, W. Gernjak, Decontamination and disinfection of water by solar photocatalysis: recent overview and trends, Catalysis Today, 147 (2009) 1-59.

[104] X. Zhang, T. Peng, S. Song, Recent advances in dye-sensitized semiconductor systems for photocatalytic hydrogen production, Journal of Materials Chemistry A, 4 (2016) 2365-2402.

[105] R. Argazzi, N.Y.M. Iha, H. Zabri, F. Odobel, C.A. Bignozzi, Design of molecular dyes for application in photoelectrochemical and electrochromic devices based on nanocrystalline metal oxide semiconductors, Coordination Chemistry Reviews, 248 (2004) 1299-1316.

[106] A.S. Polo, M.K. Itokazu, N.Y.M. Iha, Metal complex sensitizers in dye-sensitized solar cells, Coordination Chemistry Reviews, 248 (2004) 1343-1361.

[107] I. Martini, J.H. Hodak, G.V. Hartland, Effect of

water on the electron transfer dynamics of 9-anthracenecarboxylic acid bound to TiO_2 nanoparticles: demonstration of the Marcus inverted region, The Journal of Physical Chemistry B, 102 (1998) 607-614.

[108] J.M. Rehm, G.L. McLendon, Y. Nagasawa, K. Yoshihara, J. Moser, M. Grätzel, Femtosecond electron-transfer dynamics at a sensitizing dye-semiconductor (TiO_2) interface, The Journal of Physical Chemistry, 100 (1996) 9577-9578.

[109] A.K. Jana, Solar cells based on dyes, Journal of Photochemistry and Photobiology A: Chemistry, 132 (2000) 1-17.

[110] G. Liu, J. Zhao, Photocatalytic degradation of dye sulforhodamine B: a comparative study of photocatalysis with photosensitization, New Journal of Chemistry, 24 (2000) 411-417.

[111] F. Zhang, J. Zhao, T. Shen, H. Hidaka, E. Pelizzetti, N. Serpone, TiO_2-assisted photodegradation of dye pollutants II. Adsorption and degradation kinetics of eosin in TiO_2 dispersions under visible light irradiation, Applied Catalysis B: Environmental, 15 (1998) 147-156.

[112] F. Zhang, J. Zhao, L. Zang, T. Shen, H. Hidaka, E. Pelizzetti, N. Serpone, Photoassisted degradation of dye pollutants in aqueous TiO_2 dispersions under irradiation by visible light, Journal of Molecular Catalysis A: Chemical, 120 (1997) 173-178.

[113] J. Zhao, T. Wu, K. Wu, K. Oikawa, H. Hidaka, N. Serpone, Photoassisted degradation of dye pollutants. 3. Degradation of the cationic dye rhodamine B in aqueous anionic surfactant/TiO_2 dispersions under visible light irradiation: evidence for the need of substrate adsorption on TiO_2 particles, Environmental science & technology, 32 (1998) 2394-2400.

[114] M. Ni, M.K. Leung, D.Y. Leung, K. Sumathy, A review and recent developments in photocatalytic water-splitting using TiO_2 for hydrogen production, Renewable and Sustainable Energy Reviews, 11 (2007) 401-425.

[115] K. Gupta, R. Singh, A. Pandey, A. Pandey, Photocatalytic antibacterial performance of TiO_2 and Ag-doped TiO2 against S. aureus. P. aeruginosa and E. coli, Beilstein journal of nanotechnology, 4 (2013) 345-351.

[116] V.N. Nguyen, N.K.T. Nguyen, P.H. Nguyen, Hydrothermal synthesis of Fe-doped TiO_2 nanostructure photocatalyst, Advances in Natural Sciences: Nanoscience and Nanotechnology, 2 (2011) 035014.

[117] E. Kowalska, S. Rau, Photoreactors for wastewater treatment: a review, Recent Patents on Engineering, 4 (2010) 242-266.

[118] M.V. Dozzi, A. Saccomanni, E. Selli, Cr (VI) photocatalytic reduction: effects of simultaneous organics oxidation and of gold nanoparticles photodeposition on TiO_2, Journal of hazardous materials, 211 (2012) 188-195.

[119] J.-W. Yoon, T. Sasaki, N. Koshizaki, Dispersion of nanosized noble metals in TiO_2 matrix and their photoelectrode properties, Thin Solid Films, 483 (2005) 276-282.

[120] M.K. Seery, R. George, P. Floris, S.C. Pillai, Silver doped titanium dioxide nanomaterials for enhanced visible light photocatalysis, Journal of Photochemistry and Photobiology A: Chemistry, 189 (2007) 258-263.

[121] S. Tomás, A. Luna-Resendis, L. Cortés-Cuautli, D. Jacinto, Optical and morphological characterization of photocatalytic TiO_2 thin films doped with silver, Thin Solid Films, 518 (2009) 1337-1340.

[122] N. Sobana, M. Muruganadham, M. Swaminathan, Nano-Ag particles doped TiO_2 for efficient photodegradation of direct azo dyes, Journal of Molecular Catalysis A: Chemical, 258 (2006) 124-132.

[123] B. Xin, Z. Ren, H. Hu, X. Zhang, C. Dong, K. Shi, L. Jing, H. Fu, Photocatalytic activity and interfacial carrier transfer of Ag–TiO_2 nanoparticle films, Applied Surface Science, 252 (2005) 2050-2055.

[124] A.J. Nozik, M.C. Beard, J.M. Luther, M. Law, R.J. Ellingson, J.C. Johnson, Semiconductor quantum dots and quantum dot arrays and applications of multiple exciton generation to third-generation photovoltaic solar cells, Chemical reviews, 110 (2010) 6873-6890.

[125] B. Yu, Z. Hu, M. Liu, H. Yang, Q. Kong, Y. Liu, Review of research on air-conditioning systems and indoor air quality control for human health, International journal of refrigeration, 32 (2009) 3-20.

[126] S. Kment, H. Kmentova, P. Kluson, J. Krysa, Z. Hubicka, V. Cirkva, I. Gregora, O. Solcova, L. Jastrabik, Notes on the photo-induced characteristics of transition metal-doped and undoped titanium dioxide thin films, Journal of colloid and interface science, 348 (2010) 198-205.

[127] D. Robert, Photosensitization of TiO_2 by M_xO_y and M_xS_y nanoparticles for heterogeneous photocatalysis applications, Catalysis Today, 122 (2007) 20-26.

[128] H. Zhang, G. Chen, D.W. Bahnemann, Photoelectrocatalytic materials for environmental applications, Journal of Materials Chemistry, 19 (2009) 5089-5121.

[129] K.M. Lee, C.W. Lai, K.S. Ngai, J.C. Juan, Recent developments of zinc oxide based photocatalyst in water treatment technology: a review, Water research, 88 (2016) 428-448.

[130] M.D. Hernández-Alonso, F. Fresno, S. Suárez, J.M. Coronado, Development of alternative

photocatalysts to TiO_2: challenges and opportunities, Energy & Environmental Science, 2 (2009) 1231-1257.

[131] M. Ilieva, A. Nakova, V. Tsakova, TiO_2/WO_3 hybrid structures produced through a sacrificial polymer layer technique for pollutant photo-and photo-electrooxidation under ultraviolet and visible light illumination, Journal of Applied Electrochemistry, 42 (2012) 121-129.

[132] Y. Bessekhouad, D. Robert, J. Weber, Bi_2S_3/TiO_2 and CdS/TiO_2 heterojunctions as an available configuration for photocatalytic degradation of organic pollutant, Journal of Photochemistry and Photobiology A: Chemistry, 163 (2004) 569-580.

[133] G. Ren, Y. Gao, J. Yin, A. Xing, H. Liu, Synthesis of High Activity TiO_2/WO_3 Photocatalyst via Environmentally Friendly and Microwave Assisted Hydrothermal Process, Journal of Chemical Society Pakistan, 33 (2011) 666-670.

[134] S. Kuang, L. Yang, S. Luo, Q. Cai, Fabrication, characterization and photoelectrochemical properties of Fe_2O_3 modified TiO_2 nanotube arrays, Applied Surface Science, 255 (2009) 7385-7388.

[135] A.H. Zyoud, N. Zaatar, I. Saadeddin, C. Ali, D. Park, G. Campet, H.S. Hilal, CdS-sensitized TiO_2 in phenazopyridine photo-degradation: Catalyst efficiency, stability and feasibility assessment, Journal of Hazardous Materials, 173 (2010) 318-325.

[136] M. Ge, C. Guo, X. Zhu, L. Ma, Z. Han, W. Hu, Y. Wang, Photocatalytic degradation of methyl orange using ZnO/TiO_2 composites, Frontiers of Environmental Science & Engineering in China, 3 (2009) 271-280.

[137] S. Leong, A. Razmjou, K. Wang, K. Hapgood, X. Zhang, H. Wang, TiO_2 based photocatalytic membranes: a review, Journal of Membrane Science, 472 (2014) 167-184.

[138] S. Mozia, Photocatalytic membrane reactors (PMRs) in water and wastewater treatment. A review, Separation and Purification Technology, 73 (2010) 71-91.

[139] G.L. Puma, P.L. Yue, Modelling and design of thin-film slurry photocatalytic reactors for water purification, Chemical engineering science, 58 (2003) 2269-2281.

[140] R. van Grieken, J. Marugán, C. Sordo, C. Pablos, Comparison of the photocatalytic disinfection of E. coli suspensions in slurry, wall and fixed-bed reactors, Catalysis Today, 144 (2009) 48-54.

[141] M. Vezzoli, T. Farrell, A. Baker, S. Psaltis, W.N. Martens, J.M. Bell, Optimal catalyst thickness in titanium dioxide fixed film reactors: Mathematical modelling and experimental validation, Chemical engineering journal, 234 (2013) 57-65.

[142] R. Molinari, C. Lavorato, P. Argurio, Recent progress of photocatalytic membrane reactors in water treatment and in synthesis of organic compounds. A review, Catalysis Today, (2016).

[143] S. Mozia, D. Darowna, K. Szymański, S. Grondzewska, K. Borchert, R. Wróbel, A.W. Morawski, Performance of two photocatalytic membrane reactors for treatment of primary and secondary effluents, Catalysis Today, 236 (2014) 135-145.

[144] W. Zhang, L. Ding, J. Luo, M.Y. Jaffrin, B. Tang, Membrane fouling in photocatalytic membrane reactors (PMRs) for water and wastewater treatment: A critical review, Chemical Engineering Journal, 302 (2016) 446-458.

[145] J. Marugán, R. van Grieken, C. Pablos, M.L. Satuf, A.E. Cassano, O.M. Alfano, Photocatalytic inactivation of Escherichia coli aqueous suspensions in a fixed-bed reactor, Catalysis Today, 252 (2015) 143-149.

[146] Y.-S. Na, D.-H. Kim, C.-H. Lee, S.-W. Lee, Y.-S. Park, Y.-K. Oh, S.-H. Park, S.-K. Song, Photocatalytic decolorization of rhodamine B by fluidized bed reactor with hollow ceramic ball photocatalyst, Korean Journal of Chemical Engineering, 21 (2004) 430-435.

[147] R.-C. Wang, K.-S. Fan, J.-S. Chang, Removal of acid dye by $ZnFe_2O_4/TiO_2$-immobilized granular activated carbon under visible light irradiation in a recycle liquid–solid fluidized bed, Journal of the Taiwan Institute of Chemical Engineers, 40 (2009) 533-540.

[148] M. Zulfakar, A.H. Hairul Nazirah, A. Hadi, A. Rahman, Photocatalytic degradation of phenol in a fluidized bed reactor utilizing immobilized TiO_2 photocatalyst: Characterization and process studies, (2011).

10

Experimental investigation of metal powder compaction without using lubricant

Seyed Mohammad Zahraee*

Department of Advanced Materials and Renewable Energy, Iranian Research Organization for Science and Technology (IROST), Tehran, Iran

HIGHLIGHTS

- It allows to compacted metal powders without admix or die wall lubrication.
- The force required to eject the compact from the die (the ejection force) was almost zero.
- It was deduced that 0.076 mm contraction of the die diameter was indeed a realistic estimate
- This value provided sufficient shrinkage to cover both, elastic deformation during compaction and elastic spring back of green compact as well as a sufficient clearance between die wall and compact.

GRAPHICAL ABSTRACT

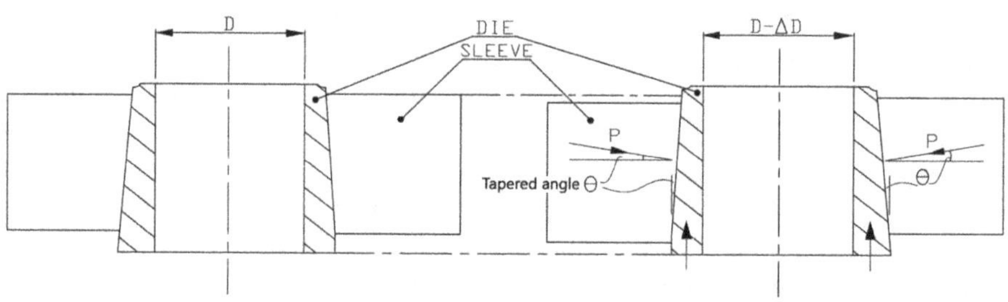

a. - Engagement of die-sleeeve before pressing together.

b. - Engagement of die-sleeeve when pressed together (Die penetrated 2mm into the sleeve) .

ARTICLE INFO

Keywords:
Powder Compaction
P/M Lubricants
Die Wall Lubrication
Elastic Die Deformation
Elastic Spring Back

ABSTRACT

The main objective of this work was to design a novel device for compaction of metal powders so that the green parts could be ejected with applying a negligible force and without the need for any lubricant in either an admixed form or applied to the die wall. For this purpose a 40 mm diameter one-piece die was envisaged which would elastically contracted 0.076 mm before compaction and after completion of powder compacting operation, it would be allowed to expand, thus releasing the green compact and so it could be ejected with a force near to weight of the compacts. The experiment indicated that this shrinkage value of 0.076 mm was indeed a realistic estimate which provided sufficient shrinkage to cover both:

I– The "elastic die deformation of 40 mm diameter during compaction" which shows 0.0433mm elastic deformation.

II–The "elastic spring back of the specimen of 40 mm diameter" which was 0.0227 mm after completion of compaction and releasing the compaction force.

The design also provided sufficient clearance of 0.010mm between the compact and the die wall on release of compacting pressure to allow ejection of compact with a force near to weight of compact while no lubricant was used on the die wall nor admixed with powder.

* *Corresponding author.*
E-mail address:smzahraee@irost.org

1. Introduction

In powder metallurgy technique, when metal powders are pressed in rigid dies, there are several types of friction (1,2) such as:
(1) Friction between moving punch (es) and die wall.
(2) Friction between powder particles.
(3) Friction between powder particles and die wall.
(4) Friction between the compact and die wall during ejection of part after compaction.

These frictional effects can be reduced by use of suitable lubricant. Lubricant can either be mixed with the metal powder or applied on to the die wall. The effects of the lubricant on compaction and ejection behavior have been extensively reported by many researchers (3-5). The practice of admixing the lubricant with metal powder and its advantage has been discussed in detail (6-8). In other hand, die wall lubrication rather than admixing and disadvantages of admixing method have been reported by some researcher (9-14).

It seems that one of the main functions of using lubricant is to reduce ejection force, this is associated with reducing die wall friction and consequently tool wear. Since the friction between green part and die wall cause cracks on the compacted part as it emerges from the die. So, this study was intended to explore the possibility of a novel die design such that subsequent reduction in ejection forces could be achieved. For this purpose a novel die system was designed so that elastically contract just before powder compaction then the powder compact without using any lubricant in either an admixed formed or applied to the die wall and then the die elastically expands when the powder is compacted and the green part ejected with an almost zero ejection force.

2. Principle of tooling design

Based on previous experience (15) with forging solid bodies, in this design it was estimated that a realistic upper bound value of 0.076 mm lateral shrinkage on 40 mm diameter of the die could be used as a guide.

So, it was therefore decided to attempt to design a solid die rather than a segmental construction. A one-piece die would have obvious advantages over a split die (segmental) which would present the possibility of powder particles jamming between the segments. This one-piece die which would elastically contract radially before powder compaction and after compaction of powder would be allowed to expand radially, this resulted in releasing the green compact, while the die is held within a sleeve (supporting ring). The outer surface of the die and the inner surface of the sleeve are tapered 1.5° as shown in Fig 1.

Both tapered surfaces are carefully polished and coated with suitable lubricants. The object of this was to keep the two metal surfaces a part and to avoid metal to metal contact.

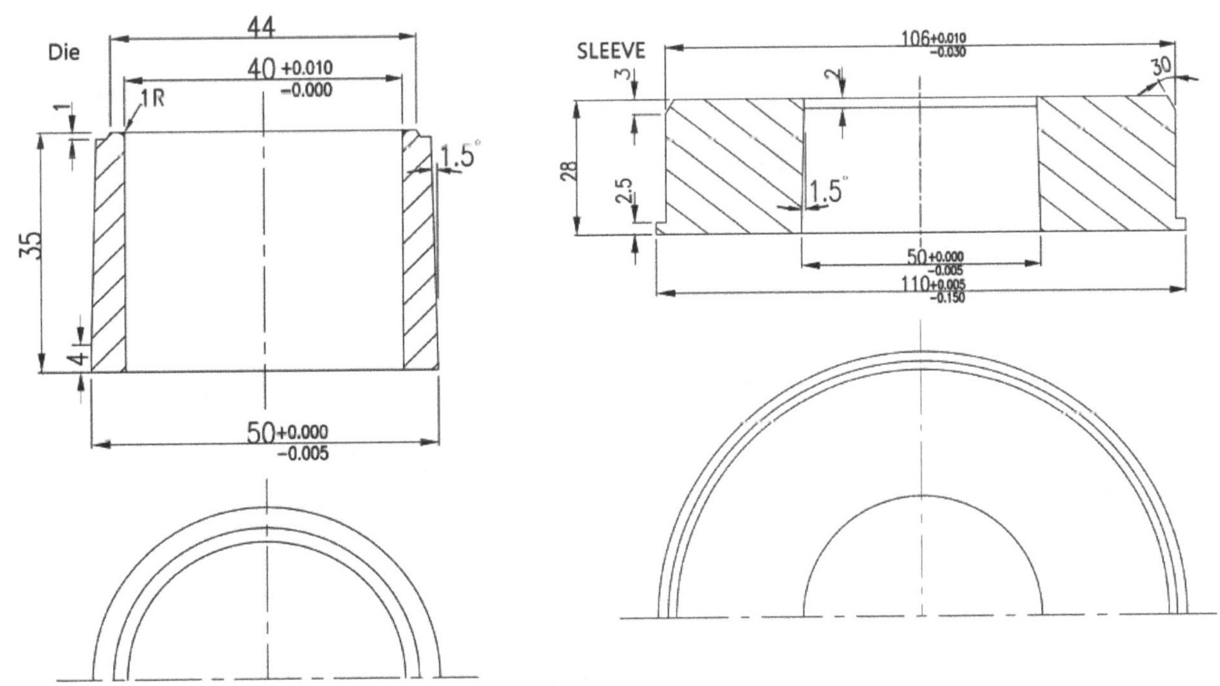

(a) Die (b) Sleeve

Fig. 1. Die and sleeve shown separately

The combined effects of a 1.5° radial taper on both parts (i.e. outer surface of the die and inner surface of the sleeve) and an axial movement of 2 mm of the sleeve with respect to the die by application of an axial force on the sleeve causes a pressure at the interface. Since the outer supporting ring (sleeve) is designed to be substantially thicker than the die, so this interface pressure results a radial displacement of 0.076 mm on the inner 40 mm diameter of the die. The radial displacement (i.e. shrinkage) is mainly inward and tensile hoop stresses in the sleeve are acceptably low. So, it would contract elastically putting the die in hoop compression which resulting contraction of 0.076 mm (i.e. just before starting the compaction). After powder compaction when the axial force is removed, they will disengaged (self- release) and the die expanded (release state) as shown in Fig 2.

3. Elastic shrinkage requires:

To design such a die it is necessary to determine the amount of elastic shrinkage required. In determining the amount of elastic contraction (shrinkage) to be imposed on the die, consideration must be given to:

(I) The elastic die deformation during compaction

(II) The elastic spring back of the pressed part.

Since the elastic shrinkage of the die must cover these two values.

3.1. Theoretical estimates of elastic die deformation during compaction

The stresses and deformations produced in powder metallurgy dies depends on several factors, such as the axial compaction pressure, the radial (lateral pressure), the type of material being compacted and the physical characteristics and dimension of the die.

It is well known that a powder mass undergoing compaction in a conventional die exerts a lateral pressure on the die walls at right angles to the axis of compaction. Since the lateral pressure (P1) acts over only part of the die length so local deformation occurs, as shown in Fig 3.

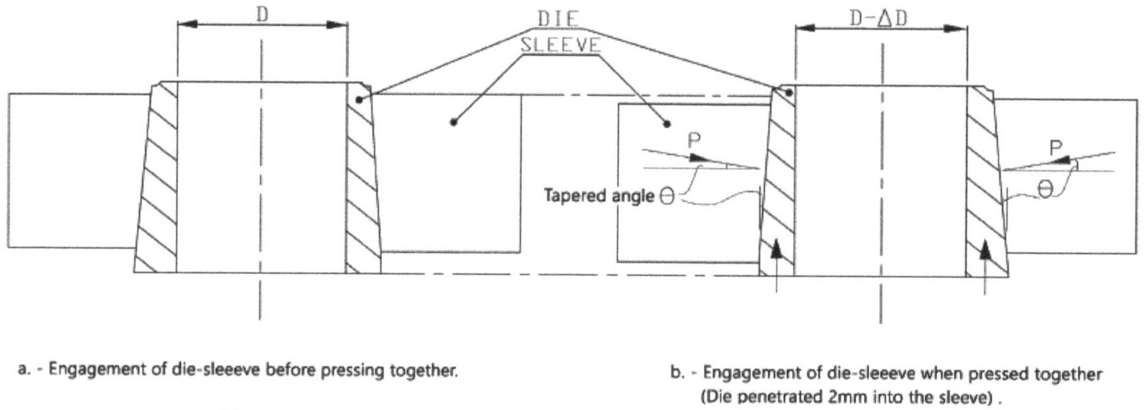

(a) (b)

Fig. 2. Assembled die and sleeve (a) release state i.e. before and after compaction, (b) Just before and during compaction

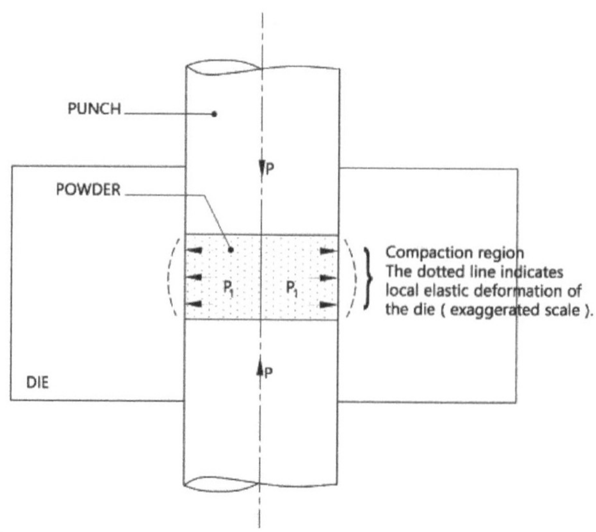

Fig. 3. Shows exaggerated local elastic deformation of a die by dotted lines

Removal of axial force allows the die wall to partially recover. Full recovery is resisted by the compacted part, therefore the amount of elastic deformation remain in the die wall is depend on the yield strength of the compact.

Bustamante,S.j etal (16) studied hoop stresses induced in a cylindrical steel die for compacting metal powder, they concluded that maximum deformation is obtained for the same powder height and applied load with finer powder particles, more ductile powders, addition of lubricant and also with powder height in the die cavity. Aren et al (17) in their research, they conclude that the lateral die wall pressure due to the powder is expected to have an exponential distribution. However, details of the distributions for specific powders for actual pressure levels are not given.

To find the amount of elastic deformation of the die, thick walled cylinder theory is used to calculate hoop and radial stresses. However, it must be noted that the die length is partially pressurized as shown in fig 3 where as in thick walled cylinder theory the total length of the bore is considered to be pressurized.

The hoop and radial stresses at any point in the wall cross- section of a thick cylinder at radius "r" are given by the lam´e equations:

$$\sigma_\theta = A + \frac{B}{r^2} \qquad (1)$$

$$\sigma_r = A + \frac{B}{r^2} \qquad (2)$$

Where σ_θ is hoop stress, θ_r is radial stress and r is radius of the die.

As it can be seen in Fig 4, at $r = r_o$ $\sigma_r = 0$

By substitute in equation (2) gives:

$$A = \frac{B}{r_o^2} \qquad (3)$$

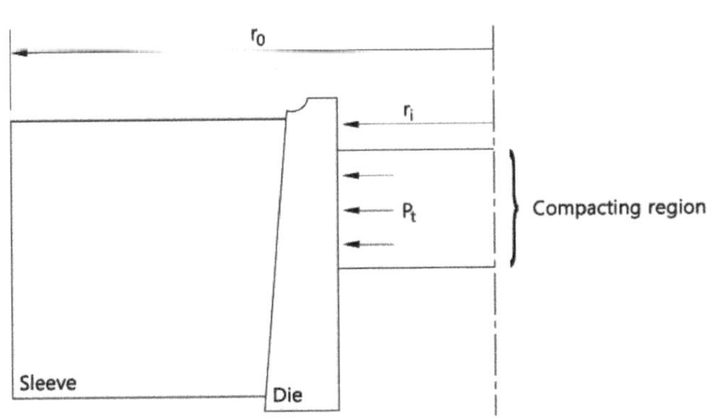

Fig.4. Shows boundary conditions on the die & sleeve

And at $r = r_i$ $\sigma_r = - Y_{compact}$

Where $Y_{compact}$ is yield strength of compacted part.

By substitute in equation (2) gives:

$$\sigma_{r_i} = -Y = A - \frac{B}{r_i^2} \qquad (4)$$

And substitute in (3) gives:

$$B = \frac{-Y}{\frac{1}{r_o^2} - \frac{1}{r_i^2}} \qquad (5)$$

Where r_i is inside radius of die = 0.020 m
r_o is outside radius of sleeve = 0.053m
Y is yield strength of compacted part.

Since three types of powders (Fe, Cu & Al) have been used, the extreme case is when iron powder is compacted with Y_{Fe}= 165 MN/m^2

Substitute in (5) gives:

B= 77×10^3 N

Substitute in (4) gives:

A= 27.4 × 10^6 N/m^2

By substituting A & B in equation (1) at $r = r_i$ gives:

σ_θ=220 MN/m^2

To find the corresponding radial strain which is related to the stresses, by using equation:

$$\varepsilon_r = \frac{1}{E}\sigma_r - \nu(\sigma_\theta + \sigma_z) \qquad (6)$$

As Den Hartog,J.P.(18) assumed: $\sigma_z = 0$

By this assumption and substituting in equation (6) gives:

ε_r– 1.083× 10^{-3} m/m

Therefore the calculated elastic deformation on the die with 40 mm inside diameter when iron powder is compacted would be 0.0433 mm.

3.2. Elastic Spring Back of green part

It is known that when compaction has completed and upper punch and die base removed the compact does not fall out of the die. The compact remains in its die under a residual radial pressure of appreciable magnitude and has to be forced out. When the compact ejected from the die expands radially. This behavior is usually called "elastic spring back".

However the elastic behavior of green compact has received little attention, but Morgan, V.T (19) mentions that the elastic spring back amounting to 0.2% on the diameter is quite common in practice. He does not give any evidence in support of this value. Anyway, the elastic spring back of the compact on release of applied pressure depends on several factors such as the magnitude of the compacting pressure, the type of powder being consolidated and the size and shape of the product. But the most effective parameters are the magnitude of compaction pressure and the type of powder which has been used for compaction.

3.2.1. Theoretical estimates of elastic spring back of green compact

The phenomenon of elastic spring back of a specimen appears when the powder is fully compacted, and can be treated as a solid plug, according to the following analysis.

A solid plug is placed in a die in which it is a perfect fit, and a punch exerts pressure on the plug. Assuming that the material of the plug is isotropic, and friction between the plug and the die is negligible. When the axial pressure is increased so that yield takes place in the material of the plug, by removal of axial pressure the principal stresses in the plug are σθ and σr (assuming $\sigma_z = 0$). The radial pressure (P_1) will then be determined by conditions of yield i.e. when $(P_1)_{max} = Y_{\text{compacted material}}$ as shown in Fig 5.

Now, by considering Mohr´s theory (circle) when $\sigma_r = P_1$, $\sigma_\theta = \sigma_r$, $\sigma_z = 0$ and $(P_1)_{max} = Y$, where Y is yield strength of compacted material. The stress system and corresponding Mohr´s circle are shown in Fig 6.

The radial strain in the plug which is related to the stresses is given by stated equation (6):

$$\varepsilon_r = \frac{1}{E}\sigma_r - \nu\left(\sigma_\theta + \sigma_z\right) \tag{6}$$

Substitute mentioned value in (6) gives:

$$\varepsilon_r = \frac{Y}{E}\left(1-\nu\right) \tag{7}$$

Now, by substitute the values of Y, E and ν of the material which has been used the radial strain of the green compact is determined. Since three types of

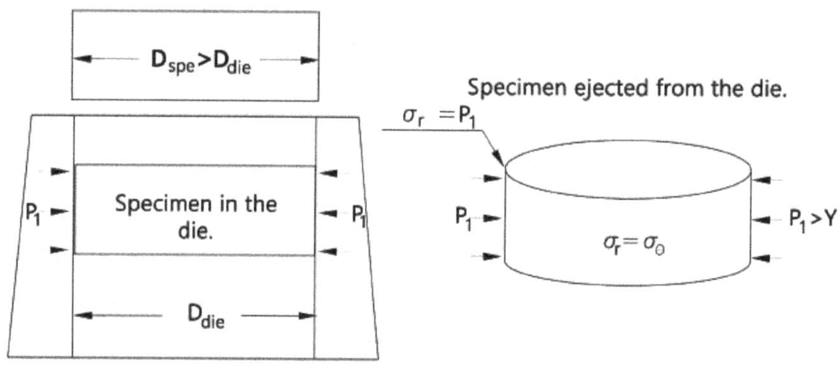

Fig.5. Shows green compact in the die & ejected from the die

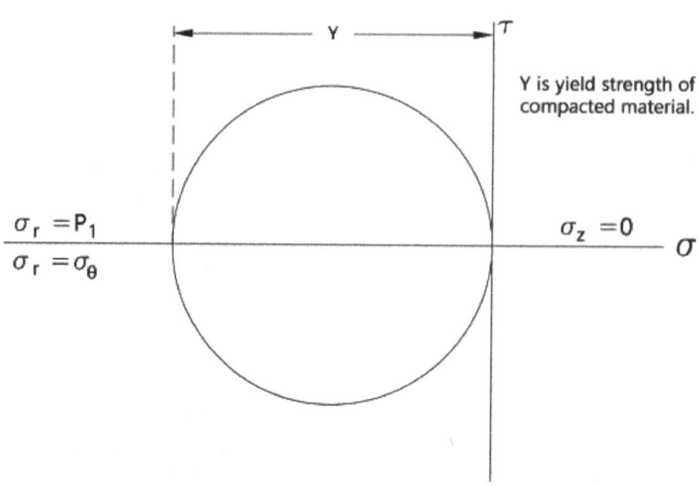

Fig.6. Mohr´s circle of stress system on the die

powder have been used the results are shown in table 1.

Table 1.
Radial strain for 3 types of compact

Material	Y MN/m^2	E GN/m^2	v	ε_r m/m (calculated Value)
AL	50	71	0.34	4.65×10^{-4}
Cu	75	117	0.35	4.17×10^{-4}
Fe	165	206	0.29	5.69×10^{-4}

So, as it can be seen the largest elastic spring back of the green compact of 40 mm diameter is when the Iron powder has been used and it is 0.0227 mm.

4. Estimation of external pressure on the die for the shrinkage

Calculations of the contraction and expansion of the die were based on the mechanics of a thin-walled cylinder which is subjected to an external uniform pressure along its length. This theory is used for simplicity since the ratio of the mean wall thickness to the internal diameter of the die is $\frac{4.65}{40}=0.11$ however the die wall thickness is not uniform along its length so the mean wall thickness has been used.
From elastic theory, the hoop stress in a thin walled cylinder subjected to an external pressure is:

$$\sigma_\theta = \frac{P.D_i}{2t} \tag{8}$$

Where 'P' is External pressure on die, produced at interface of die and sleeve as shown in Fig 2, 'D_i' is inside diameter of die and 't' is mean wall thickness of the die.
Knowing that

$$\sigma_\theta = E.\varepsilon_\theta \tag{9}$$

$$\varepsilon_\theta = \frac{\Delta D_i}{D_i} \tag{10}$$

gives

$$\sigma_\theta = E\frac{\Delta D_i}{D_i} \tag{11}$$

By substitute in (8) gives:

$$P = \frac{2t.E.\Delta D_i}{D_i^2} \tag{12}$$

Where $E=210\ GN/m^2$, $\Delta D_i=0.076\times10^{-3}$ m (i.e. inward radial shrinkage), $D_i = 40\times10^{-3}$ m, $t = 4.65\times10^{-3}$ m
By substitute these values in (12) gives: $P=92.8\ MN/m^2$

This is the external pressure on the die which produces 0.076 mm shrinkage on the die diameter of 40mm.

5.Preparation of the tooling

Initially, the two matching tapered surfaces i.e. outer surface of the die and inner surface of the sleeve which have important functions in the operation are carefully polished using a sequence of polishing media (Diamond paste) ranging from 6μm to 1/4μm diamond particles. Then the surfaces were cleaned using acetone. The surface roughness of the tapered surfaces was found to be 0.08μm for the inner surface of the sleeve and 0.10μm for the outer surface of the die. Finally both tapered surfaces were coated, the sleeve with molybdenum disulphide spray and the other (die) with P.T.F.E. (Polytetrafluorethylene) and left to dry. Also, before assembly of the die unit both coated surfaces were greased using molybdenum disulphide grease. This combination prevent the problem of die seizure inside the sleeve, so that they disengaged easily (self-releasing). The object of this operation was to keep the two metal surfaces a part and to avoid metal to metal contact which would have two inherent advantages:
(a) To obtain a low coefficient of friction which would facilitate the self- releasing action i.e. disengagement of the die from the sleeve.
(b) To minimize damage due to sliding metal to metal contact i.e. on engagement-disengagement cycles.

5.1.Performance testing of the tooling

Cyclic loading tests were carried out on an Instron machine to determine the required axial load to push the die a distance of 2mm into the sleeve as shown in Fig7.
It was found that about 16KN force was required to push the die 2mm into the sleeve.
To determine the die shrinkage corresponding to axial movement of the die with respect to the sleeve, the die and its associated parts were put together. The die penetration into the sleeve was measured by a depth micrometer while the diameter shrinkage of the die was measured simultaneously by a bore gauge. The extents of axial movements in relation to die shrinkages are shown on Fig 8.

6. Materials Studied

Compacts were made from three types of materials:

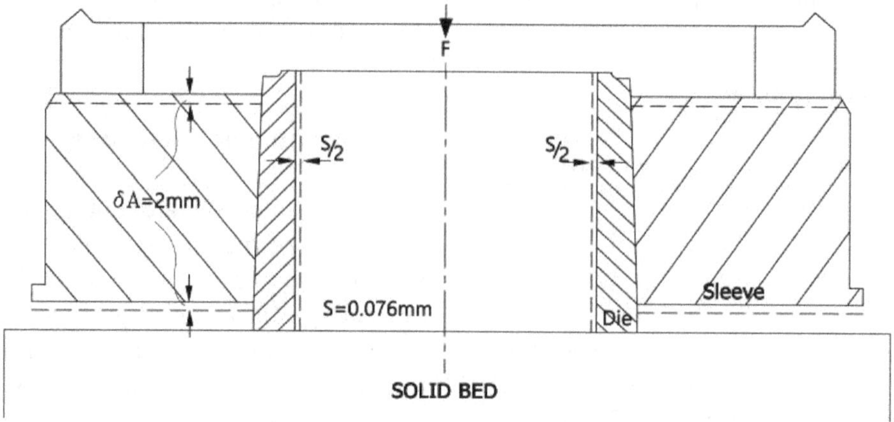

Fig.7. Die & sleeve assembly under cyclic loading

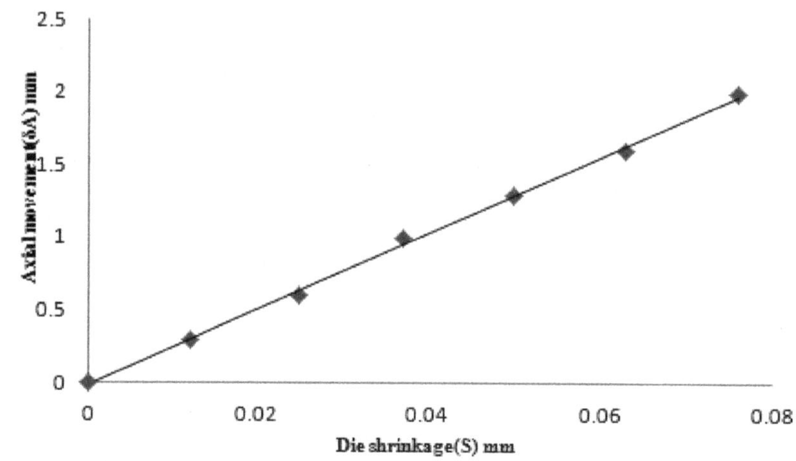

Fig.8. Axial movement of the die/its shrinkage

(a) Irregular atomized pure copper powder.
(b) Irregular atomized pure Iron powder.
(c) Irregular atomized pure Aluminum powder.

Two particle size ranges of each powder were selected for compaction with mean particles size range, 150μmm>d>75μmm and 75μmm>d>50μmm.

Since the clearance between the die base (lower punch) and the die is 0.076mm on diameter i.e. the circumferential gap is 0.038mm on release state i.e. before compaction, the powder particle ranges were chosen so that no powder particles could pass through the gap or trapped in the clearance gap, this could have interfered with the design of the die.

These powder were used for producing simple flat cylindrical disc, no admix lubricant or die wall lubrication was used during compaction.

7. Experimental Work

The die and its associated parts which have been designed and manufactured were put together (assembled) and fixed on a platform of a double action press for powder compaction. The standard experimental sequence consisted of weighting powder (50 gr) and filling the die cavity by hopper feeding and then leveling the powder in the die cavity before compaction. As mentioned compacts were made from three types of powders: copper, Iron and aluminum powders. By removing the hopper powder compacted. From each kind of powder particles, 20 compacts of same weight were produced for different destructive and non-destructive tests as green and sintered state.

8. Results and discussion

By application of almost 16KN axial force on the sleeve, an axial movement of 2mm of the sleeve with respect to the die caused a pressure of about 93 MN/m^2 at the interfaces. This interface pressure on the outer surface of the die resulted a radial shrinkage of 0.076 mm on the inner diameter of the die just before compacting of the powder.

Thus, powder compacted in the die cavity is 0.076 mm smaller in diameter than its normal state (i.e.40mm). By completion of compaction the upper punch was removed, it was found that all compacts could be ejected from the die with a force near to weight of compacts i.e. no extra force needed for die wall friction between compact and the die. All green compacts diameter was measured and are given in table 2 as a sample.

These values are in agreement with the predicted behavior of the die i.e. specimen diameters were smaller than the die diameter on normal position (the diameter of the die was 40.018mm) and larger when the die 0.076mm shrank (i.e. the diameter was 39.942mm). So, the experiment indicated that this shrinkage value of 0.076 mm was indeed a realistic estimate which provided sufficient shrinkage to cover both the local elastic die deformation of 0.043 mm and the elastic spring back of the green compact of 0.0227mm. Moreover, this value also provided sufficient clearance of 0.010mm in diameter between the compact and the die wall so that on release of compacting pressure to allow ejection of compacted part at a force near to weight of the compact while no lubricant was used on die wall nor admixed with powder.

Durability tests on the die were carried out after completion of powder compactions. These tests were devised to obtain some indications of the:

(a) potential life (cycles)

(b) wear characteristics

(c) General performance of the die assembly.

The outer surface of the die was coated with P.T.F.E. lubricant and inner surface of the sleeve with molybdenum disulphide lubricant spray. A simulated test was conducted in which the die was cycled relative to the sleeve. Displacement was equal to the movement of the die during compaction of powders (2mm).The die and sleeve set up on an Instron machine and cycled at 6 cycles per minute, approximately to expected industrial rate.

9.Conclusions

This novel die design shows that:

1. It allows to compacted metal powders (Fe, Cu and Al) without admix or die wall lubrication.
2. The force required to eject the compact from the die (the ejection force) was near to compacts weight (i.e. no extra force needed for die wall friction between compact and the die).
3. It was deduced that o.o76mm contraction of 40 mm die diameter was indeed a realistic estimate which provided sufficient shrinkage to cover both, the elastic deformation during compaction (0.043mm) and the elastic spring back of the green compact (0.0227mm) as well as a sufficient clearance (0.010mm) between the die wall and the compact.
4. It was shown that the 1.5° tapered on outer surface of the die and inner surface of the sleeve was an excellent taper angle for self releasing the die and sleeve, so that by completion of compacting cycle and removing the applied load which was on the sleeve, the die and sleeve easily disengaged i.e. self- releasing.

Acknowledgment

The author expresses his gratitude to the Niavand Shargh Technolgy Group (Mashhad) for financial support.

References

[1] F.V. Lenel, Powder Metallurgy - Principles and Applications, 1st ed., Metal Powder Industries, New Jersey, 1980.

[2] R.M. German, Powder Metallurgy Science, 1st ed., M.P.I.F. 1984.

Table 2.
Diametral details of compacted powders

Samples	Fe 150>d>75 μmm	Fe 75>d>50 μmm	Cu 150>d>75mmμ	Cu 75>d>50μmm	Al 150>d>75 μmm	Al 75>d>50 μmm
1	39.987	39.992	39.992	39.990	39.997	39.992
2	39.992	39.975	39.990	39.988	39.995	39.990
3	40.005	39.992	39.997	39.992	39.996	39.989
4	39.992	39.993	39.993	39.985	39.997	39.989
5	39.991	39.975	39.989	39.999	39.997	39.987
6	39.990	39.985	40.007	39.987	39.995	39.995

[3] R.k.Ennti, A. Lusin, S.Kumar, R..M. German, S. V. Atre, Effect of lubricant on green strength, compressibility and ejection of parts in die compaction process, Powder Technology, 233(2013) 22-29.

[4] MA.Siddiqui, M.Hamiuddin, Effect of lubricants and compacting pressure on green properties of iron powder premixes, Powder Metallurgy International, 24,2, (1992) 79-84.

[5] GF. Bochini, HI. Srl, R. Genoa, International Conference on Powder Metallurgy and Particular Materials, 1 (1995) 3.

[6] M.M.Rahman, S.S.M.Nor, A.K.Ariffin, Effect of lubricant content to the properties of Fe based component, Procedia Engineering, 68(2013) 425-430.

[7] H.S.Huang, Y.C.Lin, K.S.Hwang, Effect of lubricant addition on powder properties and compacting performance of spray-dried molybdenum powders, Inter. J. of refractory metals and hard materials, 20, 3, (2002) 175-180

[8] Ann.I. Lawrence, Sydney H. Luk and Jack A. Hamill, A performance comparison of current P/M lubricants and routes to improvement, International Conference on Powder Metallurgy & Particulate Materials, Chicago, IL USA, (1997),1-22.

[9] J.M.Capus, Die wall lubrication aids higher density, Metal Powder Report, 53, 9 (1998) 28-34

[10] S.Rastogi, S.Srinivasan, S.Ashok, G.K.Murali, Die wall lubrication studies at SFL. Proceedings of the PM World Congress, Powder Compaction, Granada, (1998)171–176.

[11] G.Jiang, , G.S.Daehn, J.J.Lannutti, Y.Fu, R.H.Wagoner, Effects of lubrication and aspect ratio on the consolidation of metal matrix composites under cyclic pressure, Acta Materialia, 49,8 (2001)1471 –1477.

[12] B.Wikman, N.Solimannezhad, R.Larsson, Wall friction coefficient estimation through modeling of powder die pressing experiment, Powder Metallurgy, 43,2 (2000)132 –138.

[13] Gasbarre, Die wall lubrication system manual, Gasbarre Products Inc., USA, DuBois, PA.15801.

[14] A.Rawlings, S. Luk, F. Hanejko, Engineered approach to high density forming using internal and external lubricants, Cited in Hoeganaes homepage (www.hoeganaes.comyhome.htm)

[15] F.R.Navaratne, An Investigation into Gear Rolling, M.S.C Dissertation (1973), Mech. Eng. Dept., UMIST, U.K.

[16] S.J. Bustamante, H.Sheinberg, Determination ofhoop stresses induced in a cylindrical steel die by compacting metal powders, Powder Metallurgy,3, 6 (1960) 36-51.

[17] B.Aren, A.Nilsson, Die tool dimensioning by finite element method, Powder Metallurgy, 30, 2 (1987) 87-96.

[18] J.P.Den Hartog, Advanced Strength of Materials, 1st ed., Dover publications INC, New York, 1987.

[19] V.T.Morgan, Techniques for the Evaluation of Powders.

III - Observations on Compacts, Powder Metallurgy, 4, 7, (1961) 44-65.

CFD simulations on natural convection heat transfer of alumina-water nanofluid with Brownian motion effect in 3-D enclosure

Mohammad Eftekhari Yazdi[1,*], Alireza Kalani Nejad[1], Saeed Dinarvand [1], Hossein Tamim[2]

[1] *Mechanical Engineering Department, Islamic Azad University, Central Tehran Branch, Tehran, Iran*

[2] *Mechanical Engineering Department, Amirkabir University of Technology, Tehran, Iran*

HIGHLIGHTS

- Natural convection in an enclosure, filled with a nanofluid, has been studied numerically.
- The numerical simulation has been carried out using the CFD approach.
- The effect of Brownian motion of nanoparticles on the heat transfer was examined.
- Heat transfer decreases with increase in nanoparticles considering Brownian motion effect.
- Nusselt number increases with Rayleigh number.

GRAPHICAL ABSTRACT

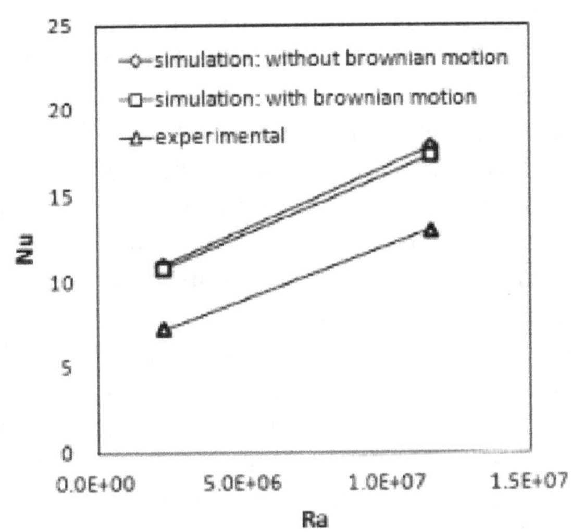

ARTICLE INFO

Keywords:
CFD simulation
Natural convection
3-D enclosure
Nanofluid
Brownian motion

ABSTRACT

The CFD simulation has been undertaken concerning natural convection heat transfer of a nanofluid in vertical square enclosure, whose dimension, width×height ×length (mm), is 40×40×90, respectively. The nanofluid used in the present study is Al_2O_3-water with various volumetric fractions of the alumina nanoparticles ranging from 0-3%. The Rayleigh number is 10^6, 10^7. Fluent v6.3 is used to simulate nanofluid considering it as a single phase. The effect of Brownian motion on the heat transfer is examined. A comparison between the two studies of with and without the Brownian motion, shows that when the Brownian motion is considered, the solid volume fraction of nanoparticles has dissimilar effects on the heat transfer. The numerical results show a decrease in heat transfer with increase in particle volume fraction considering Brownian motion effects. Moreover, computed result demonstrates an increase of Nusselt number with Rayleigh number as depicted by experimental results.

* *Corresponding author.*
E-mail address: moh.eftekhari_yazdi@iauctb.ac.ir

1. Introduction

Nanofluids are defined as the new solid-liquid composite materials with nanometer-sized solid particles, typically 1-100 nm, suspended in the base fluid. Nanofluid, a colloid composed of nanoparticles or carbon nanotubes in a base fluid, has been proposed as a highly-effective heat transfer medium in view of its abnormally higher thermal conductivity. The term of nanofluid was first used by Choi [1]. Study various aspects of nanofluids can be examined [2–4]. One of these is natural convection. Natural convection heat transfer is an important phenomenon in engineering systems due to its wide applications in electronic cooling, cooling systems for nuclear reactors, solar energy collectors, heat exchangers, and various thermal systems [5-7]. Enhancement of heat transfer in these systems is an essential topic from an energy saving perspective. The low thermal conductivity of convectional heat transfer fluids such as water and oils is a primary limitation in enhancing the performance and the compactness of such systems. An innovative technique to improve heat transfer is by using nano-scale particles in the base fluid.

Studies on natural convection using nanofluids are very limited and they are related with differentially heated enclosures. The first study concerning natural convection of a nanofluid confined in a differentially heated enclosure seems to be due to Khanafer et al. [8]. They concluded that the heat transfer rate increases with the particle fraction at any given Grashof number. Natural convection heat transfer of nanofluid in a square cavity, heated isothermally from the vertical sides, has been investigated numerically by Ho et al. [9] and santra et al. [10]. Review of pertinent literatures, two different formulas can be found for the effective viscosity and thermal conductivity of the nanofluids which have been considered by [9] while the ostwald-de Waele model for a non-Newtonian shear thinning fluid has been used by [10] to calculate the shear stress. It was found that the uncertainties associated with different models adopted to modelized the nanofluids have a great influence on the natural convection heat transfer characteristics in the enclosure. Putra et al. [11] conducted an experiment on natural convection of alumina-water and Cuo-water nanofluids inside a horizontal cylinder heated from one end and cooled from the other. They found a systematic and definite deterioration in natural convection heat transfer. The deterioration increase with particle fraction and appears more significant for Cuo-water nanofluid. Hwang et al. [12] investigated the buoyancy-driven heat transfer of Al_2O_3-water nanofluid in a rectangular cavity. They showed that the ratio of heat transfer coefficient of nanofluids to that of base fluid is decrease as the size of nanoparticles increases, or the average temperature of nanofluids is decreased. Jang and Choi [13] studied free convection in a rectangular cavity. They reported an increase in heat transfer rate with increase in particle volume fraction. Dinarvand et al. [14] studied mixed convection of a nanofluid over a vertical circular cylinder. Wang et al. [15] conducted a study on natural convection in nanofluid filled vertical and horizontal enclosures. Also, a recent study by Polidori et al. [16] analyzed the heat transfer enhancement in natural convection using nanofluids. References [17-20] present some other interesting researches in this field.

With regard to the CFD simulation using FLUENT software, the effect of gravity on sedimentation and clustering of nano ferro-fluids on natural convection heat transfer was studied by Jafari et al. [21]. They used two models in their study: the single phase approach and mixture model. Buoyancy driven heat transfer of nanofluids was studied by Ismail et al. [22] using FLUENT. They studied the effect of volume fraction and Raileigh number in their work using single phase approach.

Motivated by the above-mentioned investigations, the main objective of the present study is to numerically simulate natural convection heat transfer of nanofluid and validate them with the previous experimental results of Ho et al. [23]. In this work, FLUENT v6.3 is employed to study natural convection heat transfer of nanofluid. Simulations have been carried out in a vertical square enclosure, as used in the experimental setup of [23]. The numerical results are presented in terms of non-dimensional parameters such as Nusselt number and Rayleigh number.

2. Thermophysical properties of alumina-water nanofluid

The effective properties of the nanofluid are defined as follows:

Density:

$$\rho_{nf} = (1-\phi)\rho_f + \phi\rho_{np} \quad (1)$$

Heat capacity:

$$C_{nf} = \frac{(1-\phi)(\rho C)_f + \phi(\rho C)_{np}}{\rho_{nf}} \quad (2)$$

The Equations (1) and (2) were introduced by Buongiorno [20].

Thermal expansion coefficient:

$$\beta_{nf} = \frac{(1-\phi)(\rho\beta)_f + \phi(\rho\beta)_{np}}{\rho_{nf}} \quad (3)$$

For the thermal conductivity and viscosity two scenarios are considered: the first scenario neglects Brownian motion and uses $\mu_{eff}=\mu_{static}$ and $k_{eff}=k_{static}$. The second scenario includes Brownian motion and uses $\mu_{eff}=\mu_{static}+\mu_{Brownian}$ and $k_{eff}=k_{static}+k_{Brownian}$.

Thermal conductivity:

$$k_{static} = k_f \left[\frac{(k_{np}+k_f) - 2\phi(k_f - k_{np})}{(k_{np}+k_f) + \phi(k_f - k_{np})} \right] \quad (4)$$

This was introduced by Maxwell and Garnett [25]. For the Brownian motion, we have [26]:

$$k_{Brownian} = 5\times10^4 \beta\phi\rho_f C_{p,f} \sqrt{\frac{\kappa T}{2\rho_{np} R_{np}}} f(T,\phi) \quad (5)$$

Viscosity:

$$\mu_{static} = \frac{\mu_f}{(1-\phi)^{2.5}} \quad (6)$$

This was introduced by Brinkman [27]. For the Brownian motion, we have [26]:

$$\mu_{Brownian} = 5\times10^4 \beta\phi\rho_f \sqrt{\frac{\kappa T}{2\rho_{np} R_{np}}} f(T,\phi) \quad (7)$$

In equations 5 and 7:

$$\begin{aligned} &\kappa = 1.3807\times10^{-23}\ J/K \\ &\beta = 0.0011(100\phi)^{-0.7272} \\ &f(T,\phi) = (-6.04\phi + 0.4705)T + (1722.3\phi - 134.63) \end{aligned} \quad (8)$$

In equation 8, κ is Boltzmann constant and β, $f(T,\phi)$ are modeling functions. The radius of the Al_2O_3 nanoparticles (R_{np}) is 33 nm was used [23]. Thermophysical properties of the pure water and nanoparticles tabulated in table 1, also thermophysical properties of the nanofluid used in this study are tabulated in table 2.

3. Numerical simulation

3.1 Geometry creation and grid arrangement

The geometry and the grid were generated using GAMBIT the preprocessing module of the FLUENT v6.3. GAMBIT is an integrated preprocessor for CFD analysis. The geometry is shown in Fig. 1. The geometry is a vertical square enclosure. This enclosure was differentially heated across two vertical walls; while the remaining side walls of

Table 1. Thermophysical properties of pure water and nanoparticles [23]

Fluid	Density (kg/m³)	Specific heat (J/kg K)	Thermal conductivity (W/mK)	Viscosity (Pa s)	Thermal expansion (1/K)
Pure water					
At Tm=296 K	997	4179	0.51	0.000930	0.000210
At Tm=299 K	995	4179	0.50	0.000875	0.000242
Al_2O_3	3600	765	40	-	

Table 2. Thermophysical properties of nanofluid

Fluid	Density (kg/m³)	Specific heat (J/kg K)	Thermal conductivity (W/mK)	Viscosity (Pa s)	Thermal expansion (1/K)
Nanofluid properties					
Nanofluid 1%					
At Tm=296 K	1023.03	4058.86	0.5248	0.0009536	0.0002029
At Tm=299 K	1021.05	4058.63	0.5146	0.0008972	0.0002337
Nanofluid 2%					
At Tm=296 K	1049.06	3944.68	0.54	0.0009781	0.0001961
At Tm=299 K	1047.10	3944.24	0.5294	0.0009203	0.0002259
Nanofluid 3%					
At Tm=296 K	1075.09	3836.04	0.5555	0.001003	0.0001897
At Tm=299 K	1073.15	3835.42	0.5446	0.0009442	0.0002185

enclosure were thermally insulated. The physical boundary conditions for the geometry are defined as hot wall (left), cold wall (right) and adiabatic walls.

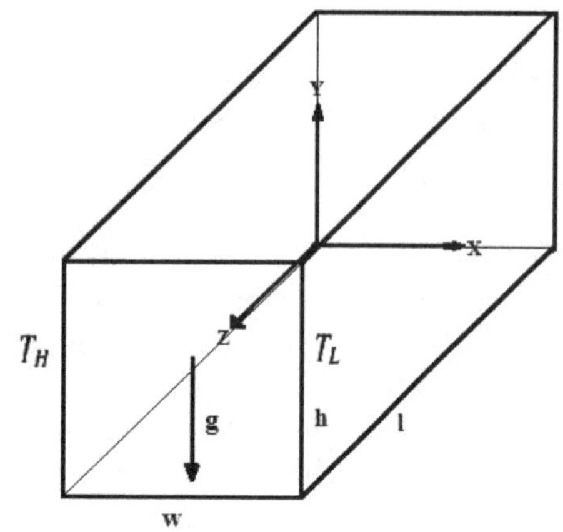

Fig. 1. schematic of 3-D enclosure

4. Computational model

4.1. Assumption

The nanofluid is incompressible and the flow is laminar. Also it is assumed that the liquid and solid are in thermal equilibrium with each other and thus no slip occurs between them. The nanoparticles in the base fluid may be easily fluidized and consequently the effective mixture behaves like a single phase fluid. The resultant mixture may be considered as a conventional single phase fluid. The effective thermophysical properties of nanofluid are dependent upon the temperature and volume fraction. The problem investigated is a three dimensional steady state laminar flow in vertical square enclosure and are governed by the following model equations.

4.2. Governing equations

4.2.1. Continuity equation

The principle of conversation of mass given by:

$$\frac{\partial \rho}{\partial t}+\nabla.(\rho U)=0 \tag{9}$$

4.2.2. Momentum equation

The principle of conversation of momentum is in fact an application of Newtonian's second law of motion to an element of fluid:

$$\frac{\partial}{\partial t}(\rho U)+\nabla.(\rho UU)=-\nabla P+\nabla \tau+B \tag{10}$$

4.2.3. Energy equation

The principle of conversation of energy given by:

$$\frac{\partial}{\partial t}(\rho h)+\nabla.(\rho U C_p T)=\nabla.(k\nabla T) \tag{11}$$

5.2. Numerical procedure

The governing equations were numerically solved using segregated solver. Laminar model was used to simulate the natural convection flow using SIMPLE scheme for pressure-velocity coupling and PRESTO was used for pressure. Steady state solver was used for all simulations. Second order upwind discretization scheme were employed for all simulations. Iteration is terminated when the maximum of the residues a value less than 10^{-6}. All simulations carried out in FLUENT software. FLUENT solver uses control volume approach to solve heat transfer problem.

The total surface heat flux (q") was computed from hot wall in each case using surface integrals. With the q", the surface-averaged heat transfer coefficient, $\bar{h}_{nf}$ was then calculated based on the measured temperature difference between the hot and cold walls, as

$$\overline{h}_{nf}=\frac{\overline{q''}}{T_h-T_c} \tag{12}$$

Further, the heat transfer coefficient was presented as the average Nusselt number, which may be defined based on the measured thermal conductivity of the nanofluid as

$$\overline{Nu_{nf}}=\frac{\overline{h}_{nf}w}{k_{nf}} \text{ or } \overline{Nu_{nf}}=\frac{\overline{q''}w}{k_{nf}(T_h-T_C)} \tag{13}$$

Moreover, based on the temperature difference across and the characteristics length, w, of the enclosure, the Rayleigh number was determined as

$$Ra_{nf}=\frac{g\beta_{nf}w^3(T_h-T_c)}{\alpha_{nf}\upsilon_{nf}} \tag{14}$$

Thermophysical properties of the nanofluid were evaluated based on the mean temperature between the hot and cold walls,

$$T_m=\frac{T_h+T_c}{2} \tag{15}$$

5. Grid independence study

The grid independence study has been conducted

in enclosure for pure water with three different grid sizes such as 41^3, 61^3 and 81^3. The hot and cold walls are maintained isothermally at temperature of 297 K and 295 K, respectively. Velocity magnitude profiles are plotted at the mid-section of the enclosure as shown in Fig. 2. From Fig. 2, it is very clear that grid size 61^3 and 81^3 gave same results for velocity magnitude. Grid size of 61^3 is shown in Fig. 3. Thus, to save time and speed up the convergence of computing, the grid size of 61^3 was used in all simulations.

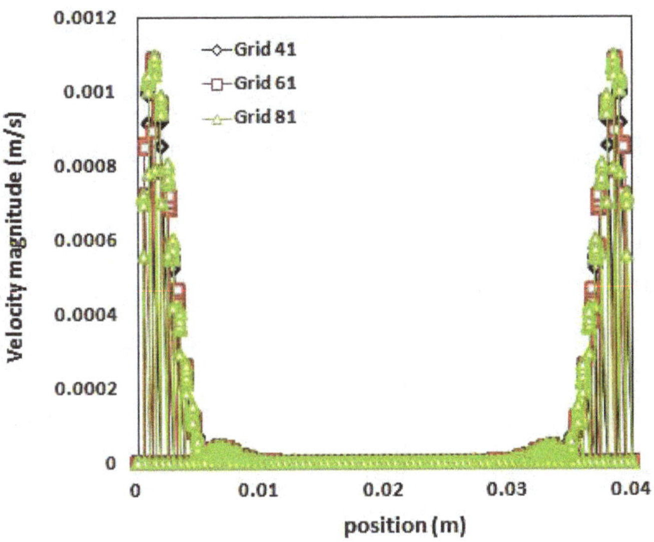

Fig. 2. Velocity magnitude at middle enclosure for showing grid independent

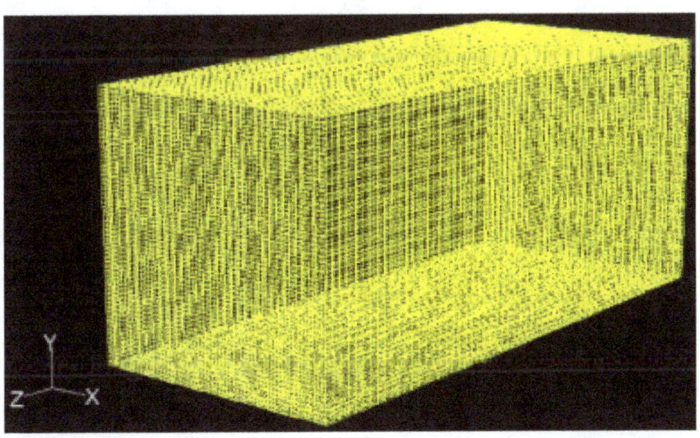

Fig. 3. Grid for vertical square enclosure

6. Results and discussion

In this section, we will discuss about the results obtained from numerical simulations. Nusselt number versus Rayleigh number is displayed.

6.1. Pure water (Ø=0%)

Numerical simulation conducted for pure water (0% volume fraction). Nusselt number calculated for two different Rayleigh numbers corresponding to two different temperatures of 296 and 303 K at hot wall and 295 K at cold wall. Fig. 4 shows the numerical results validated against experimental results for pure water. As expected, both the numerical and experimental results show increase in Nusselt number with increase in Rayleigh number. The computed results are seen to be in good agreement with the experimental results. Also, computed results show that the grid size of 61^3 is suitable to the simulation of problem.

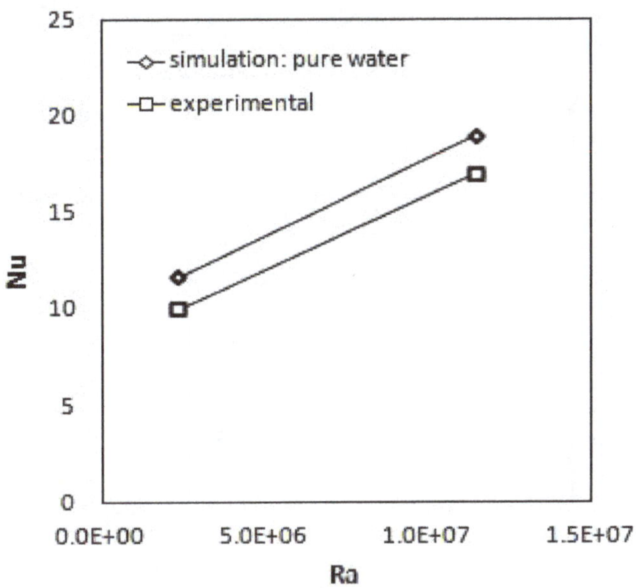

Fig. 4. Variation of Nusselt number with Rayleigh number for pure water

6.2. Nanofluid (Ø=1-3%)

Fig. 5 shows the variation of the Nusselt number versus the Rayleigh number for 1% volume fraction of nanoparticles. When the nanoparticles are added to pure water, the Nusselt number decreases compared to that of pure water. Also, Figs. 6 and 7 represent the variation of the Nusselt number with the Rayleigh number for 2 and 3% volume fraction of nanoparticles, respectively. In these cases, again, Nusselt number decreases with increase in volume fraction of nanoparticles. For all cases, the numerical results predict the similar trend as that of the experimental results. This phenomenon is paradoxical which is pointed by Jafari et al [21]. Typically, when the nanoparticles are added to the base fluid, the thermal

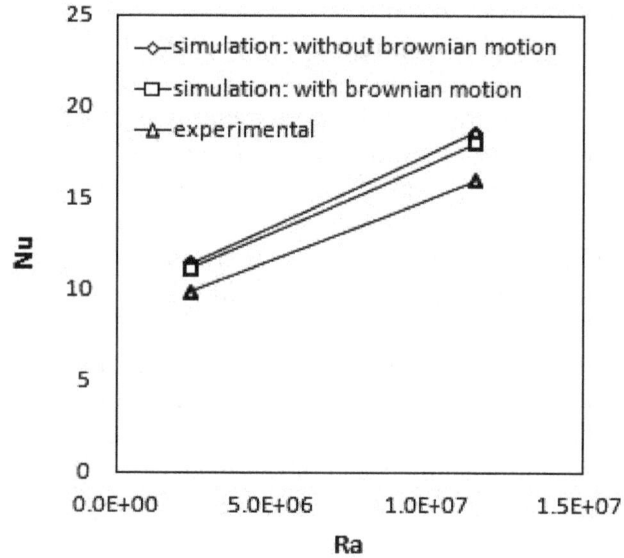

Fig. 5. Variation of Nusselt number with Rayleigh number for Ø=1%

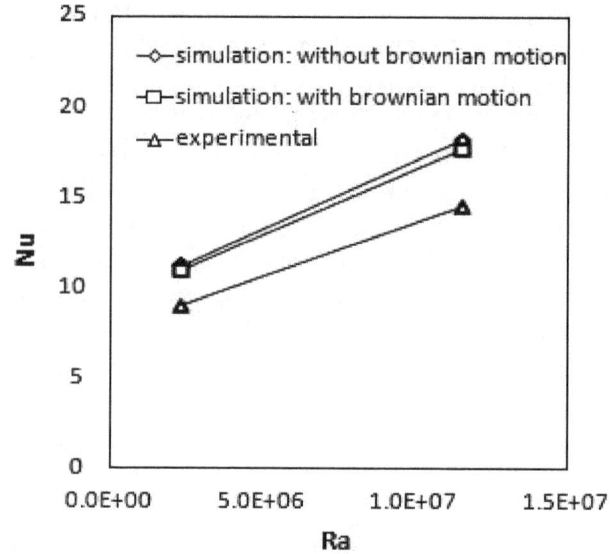

Fig. 6. Variation of Nusselt number with Rayleigh number for Ø=2%

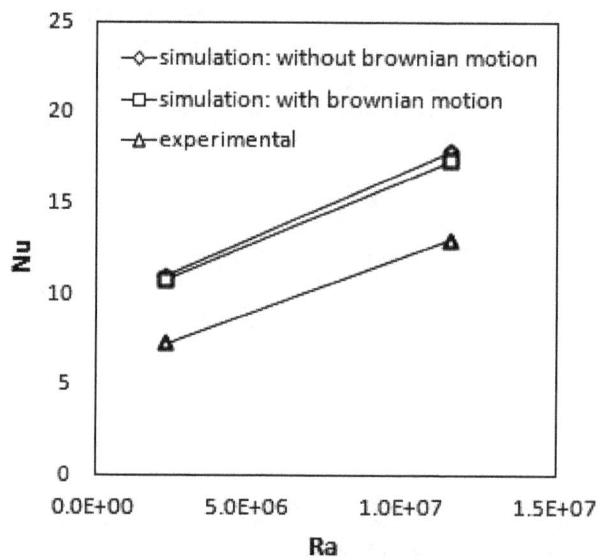

Fig. 7. Variation of Nusselt number with Rayleigh number for Ø=3%

Comparison of numerical results and experimental results shows that, with increase in particle volume fraction, a considerable difference in experimental and numerical results are observed. There are several reasons for this problem. This may be due to the selection of single phase approach which takes into account the effective mixture properties but not the particles into consideration. Thermal conductivity models have also played a decisive role in the numerical results. Theoretical models for predicting the thermal conductivity of nanofluids are different experimental measurements, thus leading to different numerical result and the experimental. It should be noted that the Rayleigh's number of the nanofluids used in computations are different from experimental value due to difference in the values of material properties.

6.3. Brownian motion effect

According to Figures 5–7, it can be seen that the values of the Nusselt number are generally lower when the Brownian motion is considered. Also, these values are closer to the experimental results.

6.4. Temperature contours

Figures 8–11 show isotherms for pure water and the nanofluid with 1–3% volume fraction of nanoparticlrs, respectively. It is observed that with

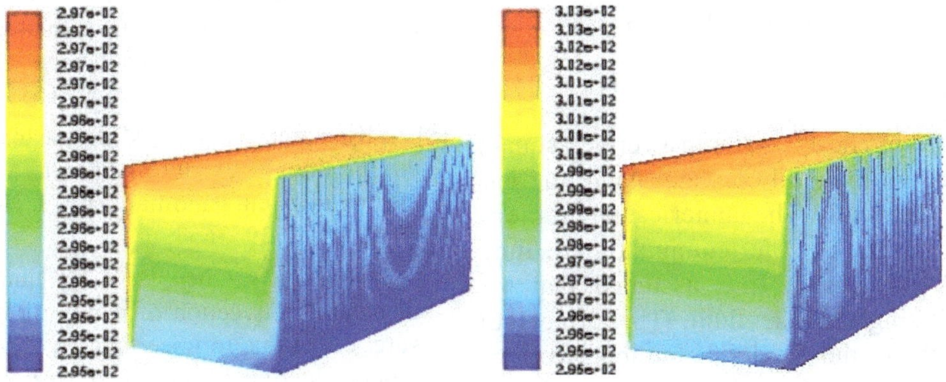

Fig. 8. Temperature contour for pure water

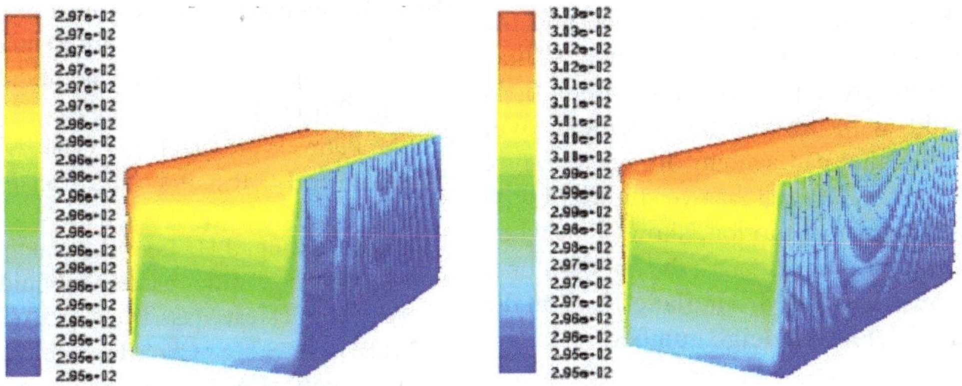

Fig. 9. Temperature contour for Ø=1%

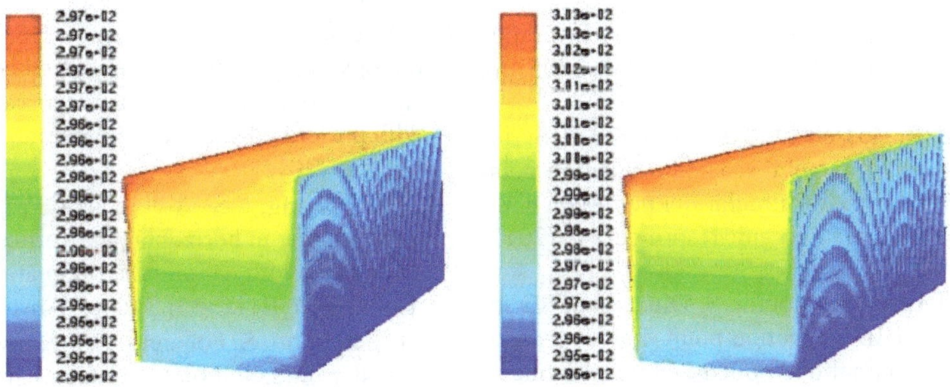

Fig. 10. Temperature contour for Ø=2%

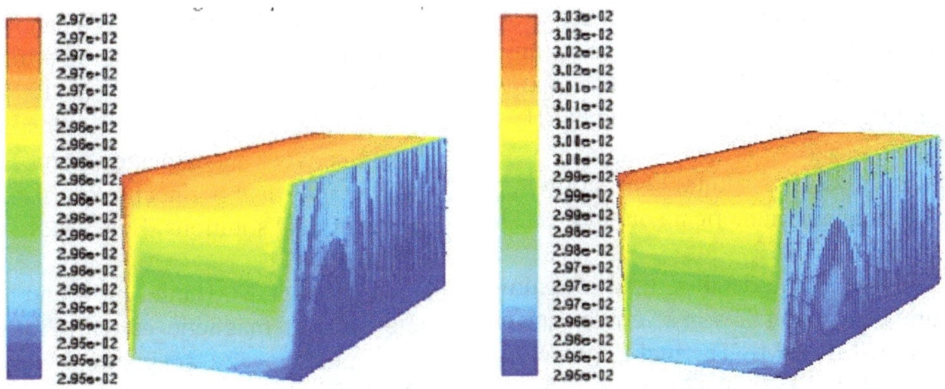

Fig. 11. Temperature contour for Ø=3%

increase in hot wall temperature, the temperature drops very gradually from top to bottom.

7. Conclusions

Natural convection heat transfer in a vertical square enclosure, filled with an Al_2O_3-water nanofluid has been studied numerically. The numerical simulations were carried out using the computational fluid dynamic (CFD) approach. It is assumed that the nanofluid is a single phase fluid. The effects of parameters such as the Rayleigh number, volume fraction of nanoparticles and Brownian motion on the heat transfer are examined. Comparisons with previously published work were performed and the results were found to be in good agreement. Thus, CFD can be effectively implemented for simulations of the nanofluid with further improvement over theoretical models that can account for temperature effects. A comparison between the two studies of with and without the Brownian motion, shows that when the Brownian motion is considered, the solid volume fraction of nanoparticles has dissimilar effects on the heat transfer at different Rayleigh number.

References

[1] S.U.S. Choi, Enhancing thermal conductivity of fluids with nanoparticles, In: ASME FED, Vol. 231, 1995, pp. 99–105.

[2] S. Dinarvand, R. Hosseini, E. Damangir, I. Pop, Series solution for steady three-dimensional stagnation point flow of a nanofluid past a circular cylinder with sinusoidal radius variation, Meccanica (springer) 48 (2013) 643–652.

[3] M. Eftekhari Yazdi, A. Moradi, S. Dinarvand, Radiation effects on MHD stagnation point flow in a nanofluid, Research Journal of Applied Science, Engineering and Technology 5 (22) (2013) 5201–5208.

[4] S. Khalili, S. Dinarvand, R. Hosseini, I. Roohi Dehkordi, H. Tamim, Stagnation-point flow and heat transfer of a nanofluid adjacent to linearly stretching/shrinking sheet: a numerical study, Research Journal of Applied Science, Engineering and Technology (2013), In press (2013).

[5] G. De Vahl Davis, I.P. Jones, Natural convection in a square cavity: a benchmark numerical solution, Int. J. Numer. Meth. Fluid3 (1983) 227–248.

[6] R.J. Krane, J. Jessee, Some detailed field measurements for a natural convection flow in a vertical square enclosure,, In: 1st ASME-JSME Thermal Engineering Joint Conference, Vol. 1 (1983) pp 323–329.

[7] T. Fusegi, J.M. Hyun, K. Kuwahara, B. Farouk, A Numerical study of three dimensional natural convection in a differentially heated cubical enclosure, Int. J. Heat Mass Transfer 34 (1991) 1543–1557.

[8] K. Khanafer, K. Vafai, M. Lightstone, Buoyancydriven heat transfer enhancement in a two dimensional enclosure utilizing nanofluids, Int. J. Heat Mass Transfer 46 (2003) 3639–3653.

[9] C.J. Ho, M.W. Chen, Z.W. Li, Numerical simulation of natural convection of nanofluid in a square enclosure: effect due to uncertainties of viscosity and thermal conductivity, Int. J. Heat Mass Transfer 51 (2008) 4506–4516.

[10] A.K. Santra, S. Sen, N. Chakraborty, Study of heat transfer augmentation in a differentially heated square cavity using copper-water nanofluid, Int. J. Therm. Sci. 47 (2008) 1113–1122.

[11] N. Putra, W. Roetzel, S.K. Das, Natural convection of nanofluids, Heat and Mass Transfer 39 (2003) 775–784.

[12] K.S. Hwang, J.H. Lee, S.P. Jang, Buoyancydriven heat transfer of water based nanofluids in a rectangular cavity, Int. J. Heat Mass Transfer 50 (2007) 4003–4010.

[13] S.P. Jang, S.U.S. Choi, Free convection in a rectangular cavity (Benard convection) with nanofluids, In: Proceedings of the IMECE, Anaheim, California, USA (2004).

[14] S. Dinarvand, A. Abbassi, R. Hosseini, I. Pop, Homotopy analysis method for mixed convection boundary-layer of a nanofluid over a vertical circular cylinder with prescribed surface temperature, Thermal Science, In Press (2015).

[15] X.Q. Wang, A.S. Mujumdar, C. Yap, Free convection heat transfer in horizontal and vertical rectangular cavities filled with nanofluids, International Heat Transfer Conference IHTC-13, Sydnes, Australia (2006).

[16] G. Polidori, S. Fohanno, C.T. Nguyen, A note on heat transfer modeling of Newtonian nanofluids in laminar free convection, Int. J. Therm. Sci. 46 (2007) 739–744.

[17] S. Dinarvand, R. Hosseini, I. Pop, Unsteady convective heat and mass transfer of a nanofluid in Howarth's stagnation point by Buongiorno's model, Int. J. Num. Meth. Heat & Fluid Flow, 25 (5) (2015) 1176–1197.

[18] H. Tamim, S. Dinarvand, R. Hosseini, I. Pop, MHD mixed convection stagnation-point flow of a nanofluid over a vertical permeable surface: a comprehensive report of dual solutions, Heat Mass Transfer 50 (2014) 639–650.

[19] M. Bahiraei, M. Hangi, Flow and heat transfer characteristics of magnetic nanofluids: A review, Journal of Magnetism and Magnetic Materials, 374 (2015) 125–138.

[20] H. Tamim, S. Dinarvand, R. Hosseini, S. Khalili, I. Pop, Unsteady mixed convection flow of a nanofluid near orthogonal stagnation-point on a vertical permeable surface, J. Process Mech. Eng. 228 (3) (2014) 226–237.

[21] A. Jafari, S.M. Mousavi, T. Tynjala, P. Sarkomaa, CFD simulation of gravitational sedimentation and clustering effects on heat transfer of a nanoferrofluid, In: PIERS Proceedings, Beijing, China, March 23–27 (2009).

[22] A.F. Ismail, W. Rashmi, M. Khalid, Numerical study on buoyancy driven heat transfer utilizing nanofluids in a rectangular enclosure, In: Proceedings of the UK-Malaysia engineering conference 2008, London (2008) 118–123.

[23] C.J. Ho, W.K. Liu, Y.S. Chang, C.C. Lin, Natural convection heat transfer of alumina-water nanofluid in vertical square enclosures: An experimental study, International Journal of Thermal Science 49 (2010) 1345–1353.

[24] J. Buongiorno, Convective transport in nanofluids, ASME Journal for Heat Transfer 128 (3) (2006) 240–250.

[25] Maxwell, J.C. Garnett, Colours in metal glasses and in metallic films, Philos Trans. Roy. Soc. A. 203 (1904) 385–420.

[26] B. Ghasemi, S.M. Aminossadati, Brownian motion nanoparticles in a triangular enclosure with natural convection, Int. J. of Thermal Science 49 (2010) 931–940.

[27] H.C. Brinkman, The viscosity of concentrated suspensions and solution, Journal of chemical and physics 20 (1952) 571–581.

12

Removal of crystal violet from aquatic environment by surfactant-modified dolomite

Maryam Arabpour, Ahmad Rahbar-Kelishami*, Reza Norouzbeigi

Department of Chemical, Petroleum and Gas Engineering, Iran University of Science and Technology, Tehran, Iran

HIGHLIGHTS

- Evaluation of crystal violet removal using dolomite for the first time.
- dolomite modified with SDS shows good adsorption capacity to remove CV ions.
- Adsorption kinetic, isotherm, and thermodynamic were studied in detail.

GRAPHICAL ABSTRACT

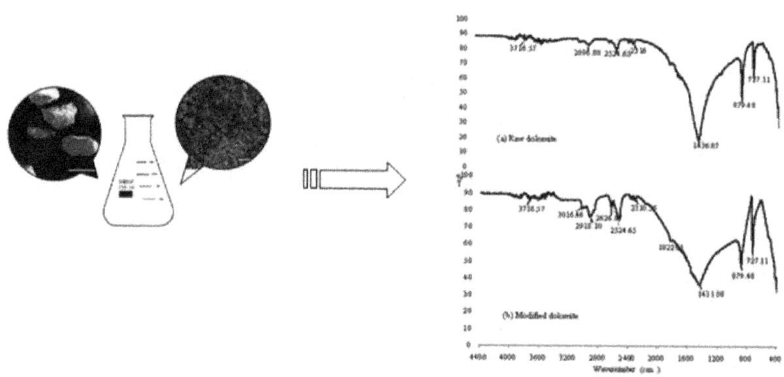

ARTICLE INFO

Keywords:
Adsorption
Crystal violet (CV^+)
Kinetics
Surfactant-modified dolomite
Taguchi method

ABSTRACT

This paper presents the adsorption of crystal violet (CV^+) from aqueous solution on surfactant-modified dolomite. Liquid phase batch operations were carried out to observe the effect of various experimental parameters such as initial dye concentration, adsorbent dosage, pH and NaCl concentration. The optimum conditions for these parameters were calculated by Taguchi method. Equilibrium isotherm data were analyzed according to Langmuir, Freundlich and Temkin equations. The experimental data fitted well to Freundlich adsorption model. The maximum adsorption capacities by Langmuir analysis were determined to be 49.261mg/g at 22°C that indicates suitable performance of this adsorbent. Pseudo-first-order and pseudo-second-order kinetics and the intraparticle diffusion models were also evaluated for the adsorption of CV^+ onto surfactant-modified dolomite. The positive value of free energy change confirmed the nature of physisorption of dye onto adsorbent.

* *Corresponding author.*
E-mail address: ahmadrahbar@iust.ac.ir

1. Introduction

Many industries use cationic dyes to color their products [1]. Crystal violet (CV^+) as a cationic dye is belongs to the class of triarylmethane, used for dyeing in varying industries [2]. When these dyes are left in the industrial effluent, could be a hazardous to the environmental. Thus, the removal of dyes from wastewater is important [3]. There are several chemical or physical processes to treat dye wastewaters. Adsorption process is one of the efficient physical methods to remove dyes from wastewater [4].

In recent years, low-cost clays have been applied as one of the adsorbents for the removal of dyes from dye effluent [5]. Dolomite as a low cost adsorbent is belong to calcite group and general formula of dolomite is $AB(CO_3)_2$ that A can be calcium, barium and or strontium and the B can be iron, magnesium or manganese [6]. Dolomite was applied in adsorption of many elements including the removal of Cu^{2+}, Pb^{2+}, strontium (II) and barium(II) [7,8], phosphate and ortho-phosphate [9–10], oil [11], fluoride [12] and arsenate [13].

Taguchi method is a powerful technique which can be used as an optimization technique for minimizing experiments number and expense. Taguchi method is a suitable method for analyzing interaction effects when ranking and screening various controllable factors [14].

In the present work, the batch study was undertaken for the removal of CV^+ onto Sodium Dodecyl Sulfate (SDS)-modified dolomite by Taguchi method. The effect of pH, adsorbent mass, NaCl concentration and initial dye concentration has been studied and the obtained experimental data were analyzed using adsorption isotherm models. Kinetic experiments have been also conducted to determine the rate of crystal violet adsorption onto modified dolomite and thermodynamic parameters were determined.

2. Experimental and analytical methods

2.1. Adsorbent

The dolomite used was obtained from Hamedan, Iran. The sample identified to be suitable for the chemical and mineralogical analysis, washed in deionized water and dried at 60°C. Dolomite was composed of 31.77% CaO, 20.19% MgO, 0.57% SiO_2, 0.23% Al_2O_3, 0.06% Na_2O, 0.043% P_2O_5, 0.03% K_2O, Fe_2O_3 0.01% (X-ray fluorescence spectroscopy (XRF) instrument (Philips, Netherlands). The ignition loss of the dolomite was also found to be 46.96%.

2.2. Preparation of surfactant-modified dolomite

200g of raw dolomite and 5g NaCl were immersed in 2L water while 40g SDS was used as a surfactant. Temperature of the reaction mixture was kept at the room condition (25°C) and the pH value was adjusted at 4.4±0.1. The mixture was stirred for 24h and then kept standing overnight. After the clear liquid was decanted off the top, the modified samples were washed several times using tap water and then washed with deionized water. The resulting products were finally dried at 60°C for 24h.

2.3. Characterization methods

In order to demonstrate the dolomite structure, SEM (VEGA-TESCAN Model instrument coating with gold) micrographs of the grain cross-sections were reported. Fourier transform infrared spectrometer spectra were recorded (KBr) on a Shimdzu 8400s FTIR to observe surface modification.

A surface analyzer (PHS-1020, PHSCHINA) was used to measure the nitrogen adsorption isotherm at 77.3 K in the range of relative pressure (P/P_o) 0.0268 to 0.957. The surface areas were calculated by the BET (Brunauer–Emmett–Teller) method. The total pore volume value was 0.472cm^3/g, the average pore diameter was 0.698 nm and the BET surface area value was 8.907 (m^2/g).

2.4. Batch adsorption experiments

Adsorption of CV^+ (obtained from Merck) was carried out by a batch technique to obtain equilibrium data. For optimizing the different factors which affect the amount of adsorption, Taguchi method was utilized. To reduce the error rate tests, were performed randomly and each adsorption sample was carried out in twice to avoid any discrepancy between experimental results and for determine the optimal conditions, three levels and four factors with three interactions (A*B, A*C and A*D) were considered (Table 1).

Designed experiments by Taguchi method was shown in Table 1. Adsorption experiments were carried out by adding 0.05, 0.1 and 0.5g of the dolomite sample to varying concentration (5, 20 and

Table 1. Designed experiments by Taguchi method

Num	A) adsorbent dosage (g)	B) initial dye concentration (mg/L)	C) pH	D) NaCl concentration (mg/L)	S/N ratio
1	0.05	5	6	0.1	-0.1225
2	0.05	5	9	10	1.3711
3	0.05	5	4	100	-3.8358
4	0.05	20	6	10	15.0563
5	0.05	20	9	100	19.0462
6	0.05	20	4	0.1	13.4420
7	0.05	50	6	100	24.8259
8	0.05	50	9	0.1	26.5308
9	0.05	50	4	10	15.1388
10	0.1	5	6	0.1	1.3265
11	0.1	5	9	10	1.3711
12	0.1	5	4	100	-5.0829
13	0.1	20	6	10	13.6501
14	0.1	20	9	100	14.3916
15	0.1	20	4	0.1	8.5335
16	0.1	50	6	100	20.0924
17	0.1	50	9	0.1	24.9614
18	0.1	50	4	10	12.7897
19	0.5	5	6	0.1	-11.4050
20	0.5	5	9	10	-10.2572
21	0.5	5	4	100	-18.8619
22	0.5	20	6	10	-2.0475
23	0.5	20	9	100	-0.8192
24	0.5	20	4	0.1	-7.3509
25	0.5	50	6	100	1.6844
26	0.5	50	9	0.1	6.4856
27	0.5	50	4	10	-0.1313

50 mg/L) of CV^+ solution in a series of 100 mL flasks. NaCl concentration was varied in 0.1, 10 and 100 mg/L. Each flask was filled with 50mL aqueous solution of CV^+ and adjusted to desired pH 6.0, 9.0 and 12.0. Temperature was kept in 22±1°C.

The pH values were adjusted by adding a few drops of dilute NaOH or HCl, and were measured by a digital pH meter (PL-500, EZODO). The pH-meter was calibrated using buffer solutions of pH 4.0 and 9.0 before use. The suspensions were stirred in shaker at 170 rpm for 1h, Then suspensions were centrifuged at 4000 rpm for 15 min at the end of the adsorption process. The concentration of the dye in the solution was analyzed using a (T80 +UV/VIS) Spectrophotometer. The measurements were made at the wavelength λ_{max}= 583nm, which corresponds to maximum absorbance. The amount of CV^+ adsorbed (mg/g), (q_e), onto modified dolomite was calculated from the mass balance equation (1).

$$q_e = (C_0 - C_e) \times V/W \quad (1)$$

where, C_0 the initial concentration of the dye solution (mg/L), C_e the equilibrium concentration of the dye solution (mg/L), V the volume of dye solution (L), and W the mass of adsorbent sample used (g).

3. Results and discussion

3.1. SEM analysis

The scanning electron microscopy results of the raw dolomite obtained by SEM (VEGA-TESCAN Model instrument coating with gold) were given in Fig. 1. The sample a (Fig. 1a) shows the presence of discrete grains with some grains having sharp edges and the images illustrate that the raw dolomite has significant amounts of calcite impurities. The sample b (Fig. 1b)

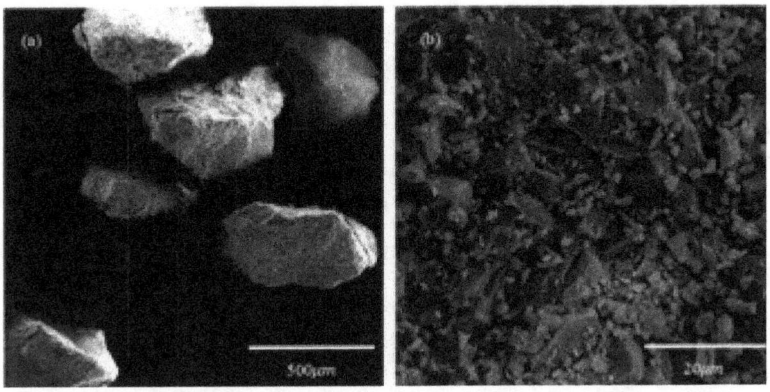

Fig. 1. SEM image of a) dolomite grains, b) raw dolomite surface.

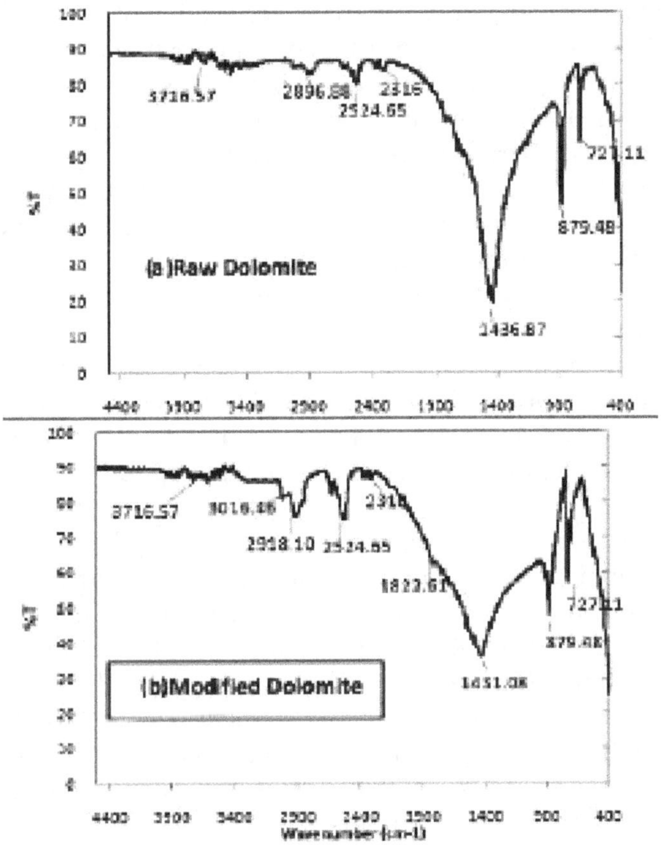

Fig. 2. FTIR spectrum of a) raw dolomite, b) modified dolomite.

shows the dolomite structure consists of rough and uneven surface with suitable sites for adsorption of dye molecule.

3.2. FTIR analysis

FTIR spectra of the raw dolomite and surfactant modified dolomite were recorded in the region 4400–400cm^{-1}. Dolomite FTIR analysis (Fig. 2a) shows main absorption bands of raw dolomite at 2524, 1436, 879 and 727cm^{-1}. The results are in good agreement with the absorption frequencies, indicated by previous researchers [15, 16]. The weak bands at 1040, 799, 525cm^{-1} shows the presence of Si-O vibration of silicate phase [17].

The band at 3716cm^{-1} that corresponds to (OH) vibrations of hydroxyl groups attached to Mg ions (belong to Mg_3OH) located in the natural dolomite and surfactant modified dolomite. Fig. 2b shows the band at 1436cm^{-1} was decreased and shifted to 1431, can be appointed to the intercalation of surfactant molecules between the silica layers [18]. The band intensity at 1822cm^{-1} could be demonstrated characteristics of adsorbed carbonate on the oxide surfaces [19]. The peak at 2316cm^{-1} was decreased and shifted to 2310 which could belong to presence of surfactant molecules and two peaks at 2626 cm^{-1} are illustrated the bending vibration of HCO^{3-} ions both of samples. The strong bands at 2918 and 3016 cm^{-1} was considered with surfactant-modified dolomite can be based the symmetric and asymmetric vibrations of the methylene group (CH_2) [20].

3.3. Taguchi method results

S/N ratio as a summary statistic analysis of recovery is used to find the optimum level of S/N ratio related to recovery (Table 1). By taking the excremental data, the average recovery for the optimum level of the four factors can be received. In optimum conditions (initial concentration of 50 mg/L, initial pH of 9.0, NaCl concentration of 0.1 mg/L, adsorbent dosage of 0.05 g) amount of dye adsorbed (mg/g), (q_e), for CV^+ were calculated 21.21mg/g.

The ANOVA result is reported in Table 2. According to Table 2, the (f) refers to the degrees of freedom of each factor and (S) refers to the factor variation and error variation. The factor variances and the error variance (V), are shown in the fourth column and also (F) in fifth column refers to the variance ratio of each factor variance to the error variance. The data in the sixth column (P) refers to the contribution ratio.

According to the results the optimal conditions are indeed those with most significant contribution to the quality characteristic, in the last column of Table 2 shown by their contribution ratio. Referring to the contribution ratio, the amount of adsorbent and pH has the greatest impact. Uncontrolled or unwanted factors make about 2.26% of significance to the quality characteristic. This pointes show that the factors effect on the quality characteristic have a great influence of about 97.74% on the quality characteristic.

3.4. Adsorption Isotherm

According to the optimal condition Langmuir, Freundlich and Temkin isotherm models were fitted to the experimental data. Adsorption isotherms help in describing how molecules of adsorbate interact with adsorbent surface.

$$C_e/q_e = C_e/q_m + 1/bq_m \qquad (2)$$

The Langmuir equation is commonly expressed as in the linear form [21]:

In this equation, C_e and q_e are the equilibrium concentration of dye (mg/L) and the equilibrium dye concentration on the adsorbent (mg/g), respectively. Where q_m is the amount of the dye adsorbed (mg) per unit of adsorbent (g) and b is the Langmuir adsorption constant (L/mg).

The adsorption equilibrium data was also fitted with the Freundlich model given below [22]:

$$\log q_e = \log K_f + 1/n \log C_e \qquad (3)$$

where K_f and n are Freundlich constants related to adsorption capacity, n is an indicator of how is favorable the adsorption process and K_f (dm^3/g) is the adsorption capacity of the adsorbent. In this equation, C_e is equilibrium concentration of dye (mg/L) and q_e is the amount of the dye adsorbed (mg) per unit of adsorbent (g).

In the Temkin isotherm model, the heat of adsorption of all the molecules in a layer decreases linearly with coverage due to the effects of some indirect sorbate/adsorbate interactions. Adsorption is specified by a similar distribution of the bonding energies, up to some maximum binding energy. The Temkin isotherm has been usually used in the following form [23]:

$$q_e = B \ln A + B \ln C_e \qquad (4)$$

Table 2. The ANOVA results in CV^+ removal efficiency with modified dolomite.

Factor	f	S	V	F	P (%)
A	2	1566.48	783.241	393.71	0
B	2	1826.47	913.236	459.05	0
C	2	275.16	137.581	69.16	0
D	2	14.05	7.026	3.53	0.097
AB	4	46.04	11.511	5.79	0.030
AC	4	1.86	0.466	0.23	0.909
AD	4	17.64	4.409	2.22	0.183
Error	6	11.94	1.989	-	-
Total	26	375965	783.241	-	-

where A (1/g) is the Temkin model constant, corresponding to the maximum binding energy and B is the Temkin constant related to heat of sorption. Therefore, the linear plots of q_e versus lnC_e, enables one to determine the constants A and B.

The calculated constants for Langmuir, Freundlich and Temkin isotherms based on the data collected as shown in Table 3. The results indicate the Freundlich isotherm fits quite well with the experimental data. Based on the results, n value is 1.349 which indicates that effective adsorption intensity.

The isotherm results of the present investigation are comparable with the reported values by previous researchers [24–30]. The value of the maximum adsorption capacity (q_m) for modified-dolomite calculated from the Langmuir isotherm in this study has a higher ability to remove CV from aqueous solution compared to other adsorbents (Table 4).

3.5. Adsorption kinetic considerations

Several kinetic models are available to investigate the behavior of the adsorption process and to test the mechanism of the dye adsorption and also to examine the experimental data. In this study, the adsorption data were analyzed using the pseudo-first-order and pseudo-second-order kinetics and the intraparticle diffusion models.

The pscudo-first-order rate expression of Langergren is given as [31]:

$$\ln\left(q_e - q_t\right) = lnq_e - k_1 t \quad (5)$$

The pseudo-second-order kinetic model is expressed as [32]:

$$t/q_t = 1/k_2 q_2^{\,2} + 1/q_2\, t \quad (6)$$

where qt is the amount of dye adsorbed (mg/g) at various times (t), q_e is the maximum adsorption capacity (mg/g) for pseudo-first-order adsorption, k_1 is the pseudo-first-order rate constant for the adsorption process (min^{-1}), q_2 is the maximum adsorption capacity (mg/g) for the pseudo-second-order adsorption and k_2 is the rate constant of pseudo-second-order adsorption (g/mg min). Table.5 shows the calculated constants for pseudo-first-order and pseudo-second-order, there is an agreement between q_e experimental and q_2 calculated values for the pseudo-second-order model. The results indicate the pseudo-second-order model better represents the adsorption kinetics. For the pseudo-second-order model with the increase of the temperature the rate constants for surfactant-modified dolomite decrease. The pseudo-second-order plots at different temperature to the adsorption of CV^+ onto modified dolomite are shown in Fig. 3.

The intraparticle diffusion model later is given by the Weber–Morris equation [33]:

$$q_t = k_p t^{1/2} + C \quad (7)$$

where the parameter C is the intercept, k_p is the intraparticle diffusion rate constant (mg/g $min^{1/2}$) and q_t is the amount of dye adsorbed (mg/g) at various times (t).

Table 3. Langmuir, Freundlich and Temkin isotherm constants for the adsorption of CV^+.

T (°C)	Freundlich isotherm parameters			Langmuir isotherm parameters			Temkin isotherm parameters		
	K_f	n	r^2	q_m	b	r^2	A	B	r^2
22	1.795	1.349	0.9988	49.261	0.0261	0.90	0.376	8.5237	0.9463

Table 4. Adsorption results of CV+ from the literature by various adsorbents.

Adsorbent	q_m (mg/g)	Ref. nos.
polymer	12.9	24
jute fiber carbon (JFC)	27.99	25
Activated carbon prepared from waste apricot	32.89	26
NaOH-modified rice husk	44.876	27
Sawdust	37.83	28
Semi-interpenetrated networks hydrogels	35.09	29
Neem Sawdust	4.44	30
Modified dolomite	49.261	In this study

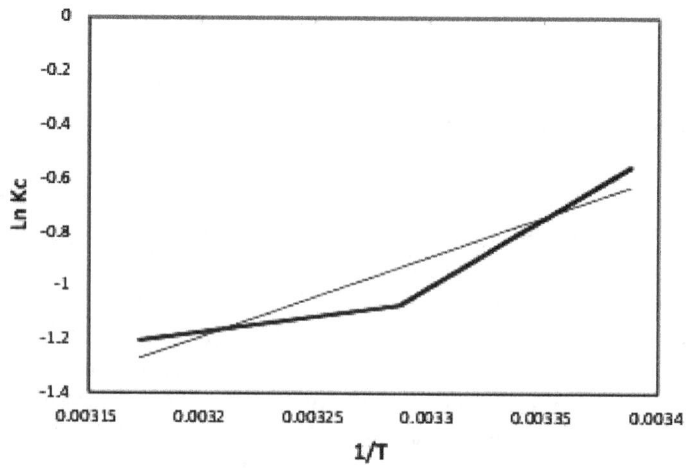

Fig. 3. Pseudo-second-order plots for the adsorption of CV^+ onto modified dolomite.

Table 5. Kinetic parameters for the adsorption of CV^+ onto surfactant-modified dolomite at various temperatures.

T (°C)	k_1 (min^{-1})	q_e(mg/g)	r^2	k_2 (g/mg min)	q_2 (mg/g)	r^2	k_p (mg/g min$^{1/2}$)	C (mg/g)	r^2
22	$1.77*10^{-2}$	9.606	0.809	$3.461*10^{-3}$	23.036	0.94	2.0472	3.78	0.908
31	$1.81*10^{-2}$	5.929	0.918	$2.09*10^{-3}$	22.222	0.932	1.9214	1.98	0.909
42	$1.55*10^{-2}$	2.630	0.850	$1.728*10^{-3}$	21.692	0.941	1.8773	0.87	0.896

Based on the evidence, the adsorption of dye onto surfactant-modified dolomite is a complex process, so the whole process cannot be only described by a single kinetic model. Table.5 shows the calculated intraparticle diffusion constant. The regression of q_t versus $t^{1/2}$ is linear (not shown), but it was not the main rate characterizing step during the whole process.

3.6. Thermodynamic parameters

To study the influence of the temperature on the adsorption capacities of the sample towards CV^+ in optimum conditions, experiments were performed using 50 mg/L dye solutions, 0.1 mg/L NaCl concentration, 0.05 g modified dolomite and pH value 9.0.

The Arrhenius type relationship parameters were determined by the pseudo-second-order rate constant as a function of temperature [34].

$$\ln k_2 = \ln A_1 - E_a/RT \quad (8)$$

where E_a is the Arrhenius activation energy and A_1 is the Arrhenius factor, which can be evaluated from the slope of the linear plot of $\ln k_2$ versus 1/T. In this equation R is the gas universal constant which is equal to 8.314 J mol^{-1} K^{-1}, k_2 is the pseudo-second-order rate constant and T is the operating temperature. The other thermodynamic parameters of dye adsorption are expressed by the following equations:

$$lnKc = \Delta S/R - \Delta H/RT \quad (9)$$

$$\Delta G = -RT\, lnKc \quad (10)$$

$$\text{Where}: Kc = C_A/C_e \quad (11)$$

when lnKc is plotted versus 1/T, ΔS°,ΔH° and ΔG° are the changes of entropy, enthalpy and the Gibbs energy were determined (Fig.4). Where, K_C is the equilibrium constant, C_A is the amount of dye adsorbed on the adsorbent at equilibrium (mg/L) and C_e is the equilibrium concentration of the dye in the solution (mg/L).

The enthalpy change ΔH° for the adsorption process was −24.763 kJ/mol, which did not suggested very strong chemical forces between the CV^+ molecules and surfactant-modified dolomite. On the other hand, the obtained value shows that adsorption process is an exothermic process (Table 6).

Low values of activation energies between 5 to 40 kJ/mol is a specification for physisorption [35] and the result obtained is +26.597 kJ/mol for the adsorption of CV^+ onto surfactant-modified dolomite,

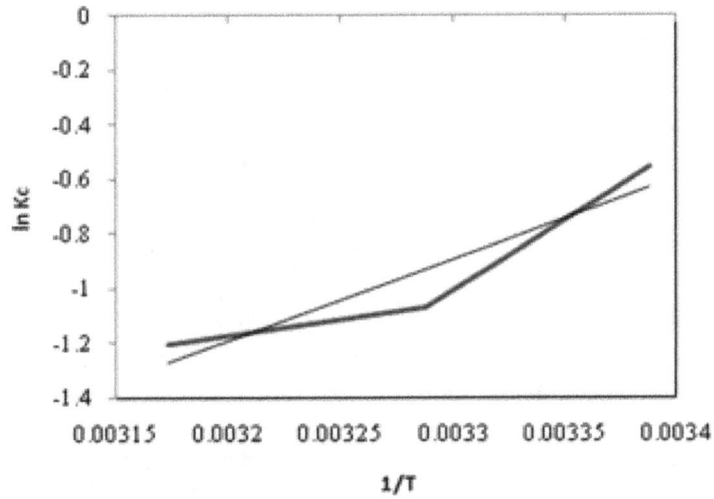

Fig. 4. Plot of lnKc vs. 1/T for estimation of thermodynamic parameter.

Table 6. Thermodynamic parameters calculated with the pseudo-second-order rate constant.

T (K)	1/T	ln k_2	E_a (kJ/mol)	K_c	ln K_c	ΔG° (kJ/mol)	ΔH° (kJ/mol)	ΔS° (J/K mol)
295.15	0.003388	-5.666		0.5744	-0.5544	1.36		
304.15	0.003287	-6.171	26.597	0.3427	-1.071	2.708	-24.763	-84.137
315.15	0.003173	-6.361		0.3002	-1.2033	3.153		

indicating low potential barrier and corresponding to a physisorption for adsorption process.

The ΔS° obtained is -84.137 J/K mol therefore suggested that no considerable change occurs in the structure of modified dolomite and also there is decreased randomness at the interface of modified dolomite and solution during adsorption process. On the other hand, negative value of ΔS° indicated that reduction of degree of freedom for adsorption species. The ΔG° values were positive therefore reflects that the presence of an energy barrier in the adsorption of CV^+ during adsorption system and need additional energy to complete.

4. Conclusions

This study investigated the adsorption equilibrium and dynamics parameters of crystal violet (CV^+), onto surfactant-modified dolomite. For optimizing the different factors which affect the amount of adsorption, Taguchi method is utilized. In optimum conditions, initial concentration of 50 mg/L, initial pH of 9.0, NaCl concentration of 0.1 mg/L and adsorbent dosage of 0.05 g were chosen. According to the Taguchi results the amount of NaCl concentration has the lowest impact. The most effective parameters for maximizing the capacity of modified dolomite to adsorb CV^+ was found pH of solution and the amount of adsorbent.

The equilibrium data indicate the best fit obtained with Freundlich isotherm model. The maximum adsorption capacities by Langmuir analysis were determined to be 49.261 mg/g. The pseudo-first-order, pseudo-second-order and the intraparticle diffusion models were used to fit the experimental data and the adsorption results indicated that adsorption system was best described by the pseudo-second-order model. The nature of physisorption of CV^+ onto modified dolomite confirms by the positive low value of E_a (+26.597 kJ/mol). The Gibbs energy (ΔG°) values for the adsorption process were positive, which indicates the system achieved energy from an external source and the adsorption was not spontaneous.

The surfactant modified dolomite was shown to be effective as low-cost material in removing CV^+ dye from aqueous solutions.

Symbols

A [1/g]
Temkin adsorption constant
A_1 [-]
Arrhenius factor
b [L/mg]
Langmuir adsorption constant
B [-]
Temkin constant related to heat of sorption

C_e [mg/L]
dye concentration at equilibrium
C_0 [mg/L]
initial dye concentration
C_A [mg/L]
amount of dye adsorbed
E_a [KJ/mol]
Arrhenius energy
k_1 [min^{-1}]
pseudo-first-order adsorption rate constant
k_2 [g /mg min]
pseudo-second-order adsorption rate constant
K_f [L/g]
Freundlich adsorption constant
k_p [mg /g (min0.5)]
intraparticle diffusion rate constant
K_C [-]
equilibrium constant
n [-]
Freundlich adsorption constant
q_m [mg/g]
monolayer capacity of the adsorbent
q_t [mg/g]
adsorbed quantity of dye per g of adsorbent at any time
q_2 [mg/g]
maximum adsorption of pseudo second order
q_e [mg/g]
adsorbed quantity of dye per g of adsorbent at equilibrium
r^2 [-]
correlation coefficient
t [min]
time
V [L]
volume of dye solution
W [g]
weight of adsorbent used

Acknowledgements

The authors have declared no conflict of interest.

References

[1] C.C. Wang, L.C. Juang, T.C. Hsu, C. K. Lee, J.F. Lee, F.C. Huang, Adsorption of basic dyes onto montmorillonite, J. Colloid Interface Sci. 273 (2004) 80–86.

[2] A. Mittal, J. Mittal, A. Malviya, D. Kaur, V.K. Gupta, Adsorption of hazardous dye crystal violet from wastewater by waste materials, J. Colloid Interface Sci. 343 (2010) 463–473.

[3] G. Crini, Kinetic and equilibrium studies on the removal of cationic dyes from aqueous solution by adsorption onto a cyclodextrin polymer, Dyes Pigm. 77 (2008) 415-426.

[4] E. Eren, Removal of basic dye by modified Unye bentonite, Turkey, J. Hazard. Mater. 162 (2009) 1355–1363.

[5] G. Crini, Non-conventional low-cost adsorbents for dye removal: A review, Bioresour. Technol. 97 (2006) 1061-1085.

[6] G.M. Walker, L. Hansen, J.-A. Hanna, S.J. Allen, Kinetics of a reactive dye adsorption onto dolomitic sorbents, Water Res. 37 (2003) 2081–2089.

[7] E. Pehlivan, A.M. Ozkan, S. Dinc¸ S. Parlayici, Adsorption of Cu^{2+} and Pb^{2+} ion on dolomite powder, J.Hazard. Mater. 167 (2009) 1044–1049.

[8] A. Ghaemi, M. Torab-Mostaedi, M. Ghannadi- Maragheh, Characterizations of strontium (II) and barium (II) adsorption from aqueous solutions using dolomite powder, J.Hazard. Mater. 190 (2011) 916– 921.

[9] S. Karaca, A. G¨urses, M. Ejder, M. Acıkyıldız, Adsorptive removal of phosphate from aqueous solutions using raw and calcinated dolomite, J.Hazard. Mater. 128 (2006) 273–279.

[10] C. Mangwandi, A. B. Albadarin, Y. Glocheux , G. M. Walker, Removal of ortho-phosphate from aqueous solution by adsorption onto dolomite, J. Environ. Chem. Eng. 2 (2014) 1123–1130.

[11] C. Ngamcharussrivichai, W. Wiwatnimit, S. Wangnoi, Modified dolomites as catalysts for palm kernel oil transesterification, J. Mol. Catal. A: Chemical. 276 (2007) 24–33.

[12] K. Sasaki, M. Yoshida, B. Ahmmad, N. Fukumoto, T. Hirajima, Sorption of fluoride on partially calcined dolomite, Colloids Surf. 435 (2013) 56– 62.

[13] G.M. Ayoub, M. Mehawej, Adsorption of arsenate on untreated dolomite powder, J. Hazard. Mater. 435 (2007) 259–266.

[14] Y. Cao, A. Pawlowski, J. Zhang, Preparation of activated carbons with enhanced adsorption of cationic and anionic dyes from Chinese Hickory husk using the Taguchi method, Environ. Protect. Eng.36 (2010) 68.

[15] B.K. Shahraki, B. Mehrabi, R. Dabiri, Thermal behaviour of zefreh dolomite mine (Central Iran), J. Min. Metall. 45 (2009) 35–44.

[16] V.Ramasamy,V. Ponnusamy, S. Sabari, S.R.Anishia, S.S. Gomathi, Effect of grinding on the crystal structure of recently excavated dolomite, Indian journal of Pure & Applied Physics. 47 (2009) 586–591.

[17] S. Gunasekaran, G. Anbalagan, Thermal decomposition of natural dolomite, J. Indian Academy Sci. 30 (2007) 339–344.

[18] J. Madejov´a, FTIR techniques in clay mineral studies, Vib. Spectrosc.31 (2003) 1–10.

[19] D.P. Das, J. Das, K. Parida, Physicochemical characterization and adsorption behavior of calcined Zn/Al

hydrotalcite-like compound (HTlc) towards removal of fluoride from aqueous solution, J. Colloid Interface Sci. 261 (2003) 213–220.

[20] A. O zcan, A.S. O zcan, Adsorption of Acid Red 57 from aqueous solutions onto surfactant-modified sepiolite, J.Hazard. Mater. 125 (2005) 252–259.

[21] I. Langmuir, The adsorption of gases on plane surfaces of glass, mica and platinum, J. Am. Chem. Soc. 40 (1918) 1361–1403.

[22] ZH.M.F. Freundlich, U¨ber die adsorption in lo¨sungen, Zeitschrift für Physikalische Chemie. 57 (1906) 385–470.

[23] M.I. Tempkin, V. Pyzhev, Kinetics of ammonia synthesis on promoted iron catalyst, Acta Phys. Chem. USSR. 12 (1940) 327–356.

[24] R. Dhodapkar, N.N. Rao, S.P. Pande, S.N. Kaul, Removal of basic dyes from aqueous medium using a novel polymer, Bioresour. Technol. 97 (2006) 877–885.

[25] K. Porkodi, K.V. Kumar, Equilibrium, kinetics and mechanism modeling and simulation of basic and acid dyes sorption onto jute fiber carbon: eosin yellow, malachite green and crystal violet single component systems, J.Hazard. Mater. 143 (2007) 311–327.

[26] Y. Onal, Kinetics of adsorption of dyes from aqueous solution using activated carbon prepared from waste apricot, J.Hazard. Mater.137 (2006) 1719–1728.

[27] S. Chakraborty, S. Chowdhury, P. D. Saha, Adsorption of Crystal Violet from aqueous solution onto NaOH-modified rice husk, carbohyd. polym. 86 (2011) 1533–1541.

[28] H.Parab, M.Sudersanan, N. Shenoy, T. Pathare, Use of agro-industrial wastes for removal of basic dyes from aqueous solutions, CLEAN-Soil, Air, Water. 37 (2009) 963–969.

[29] L. S, Removal of crystal violet from aqueous solution by sorption into semiinterpenetrated networks hydrogels constituted of poly (acrylicacid– acrylamide–methacrylate) and amylase, Bioresour. Technol. 101 (2010) 2197–2202.

[30] S. D. Khattri, M. K. Singh, Color Removal from Synthetic Dye Wastewater Using a Biosorbent, Water, Air, Soil Pollut. 120 (2000) 283 – 294.

[31] S.Lagergren, Zur theorie der sogenannten adsorption gel¨oster stoffe, Kungliga Svenska Vetenskapsakademiens, Handlingar. 24 (1898) 1–39.

[32] Y.S. Ho, G. McKay, Sorption of dye from aqueous solution by peat, Chem. Eng. J. 70 (1998) 115–124.

[33] W.J. Weber Jr., J.C. Morriss, Kinetics of adsorption on carbon from solution, J. Sanit. Eng. Div. Am. Soc. Civ. Eng. 89 (1963) 31–60.

[34] R.S. Juang, F.C. Wu, R.L. Tseng, The ability of activated clay for the adsorption of dyes from aqueous solutions, J. Environ. Technol. 18 (1997) 525–531.

35] H. Nollet, M. Roels, P. Lutgen, P. Van der Meeren, W. Verstraete, Removal of PCBs from wastewater using fly ash, Chemosphere. 53 (2003) 655–665.

13

Morphology and Mechanical Properties Investigation of Binary Polymer Blends based on PP/SEBS and PP/PC

R.Veys Karami[1], H. Izadi Vasafi[1,*], O. Moini Jazani[2], A. Talaei[3]

[1]*Department of Polymer Engineering, Shahreza Branch, Islamic Azad University, Isfahan, Iran*

[2]*Department of Chemical Engineering, Faculty of Engineering, University of Isfahan, Isfahan, Iran*

[3] *Polymer Engineering Department, Amir Kabir University of Technology, Tehran,Iran*

HIGHLIGHTS

- By increasing PC in PP/PC binary blend, matrix tend to encapsulation of large PC drops.
- Particle size in PP/SEBS samples don't change considerably results from slight coagulation of drops together due to good SEBS interaction with PP matrix.
- PC particles in PP/PC blends can act as nucleation agent in PP crystallinity.
- All binary samples based on PP/SEBS exhibited a remarkable rise in the impact strength compared to pure PP which is attributed to the toughening effect of SEBS phase
- In PP/PC blends, the presence of large rod like composite droplets negatively affects the impact strength.

ARTICLE INFO

Keywords:
Morphology
Polypropylene
Mechanical properties

GRAPHICAL ABSTRACT

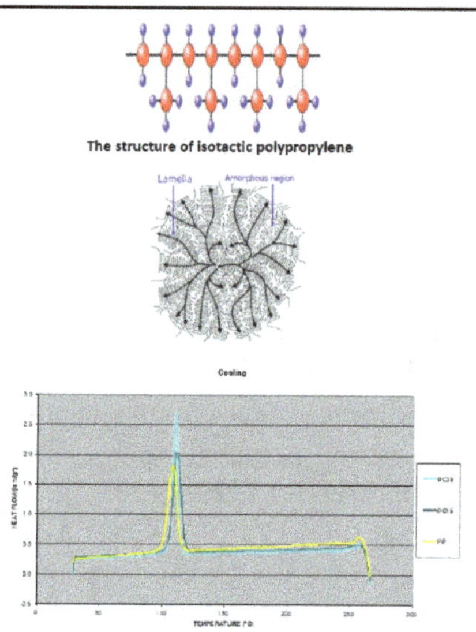

ABSTRACT

In this study, eight binary blends of polypropylene (PP)/ poly [styrene-b-(ethylene-co butylene)-b-styrene] tri-bl°Ck copolymer (SEBS) and polypropylene (PP)/ polycarbonate (PC) were produced at different composition through twin screw extruder. The composition of the minor phases was changed to correlate the morphology and mechanical properties of binary blends (PP/PC and PP/SEBS) through thermal properties. To serve this purpose and to study the composition weight, composition weight ratio of SEBS and PC was change from 10% to 30% in PP/PC and PP/SEBS. With increasing SEBS wt%, toughness grows slightly and in 30wt% of SEBS reached to its maximum value, which uniform morphology created by PP/SEBS plays important role in this matter. On the other hand, with increasing PC wt%, tensile properties have upward trend, but size of holes, due to lack of proper adhesion between two polymeric surface, rose. The results of DCS have shown that SEBS cannot act as nucleation agents (heterogeneous nucleation) but the PC particles in this situation can act as nucleation agent in PP crystallinity. Increasing SEBS percentages leads to immobility of PP segments and hamper the crystallinity. PC particles in PP matrix are heterogeneous nucleation agent, which leads to increasing in crystallinity temperature compared to the pure PP.

* *Corresponding author.*
E-mail address:hizadi@iaush.ac.ir

1. Introduction

1.1. Effective factors in formation and morphology control of the two phase polymeric systems Surface tension (the effect of adding compatibilizer)

The final dimension of disperse phase stems from equilibrium between gravity and surface tension forces. In some researches done by scientists show that if adhesion is appropriate between phases (or low surface tension), Drops can sustain large deformation before fracture, which, in turns, leads to smaller drops morphology. Liang and co-workers [1] have worked on LLDPE/PVC by three compatibilizer. The relation between particle size and surface tension has been studied. They observed the dependence on particle size and surface tension in all three situations which is independence of compabilizer yield. Meanwhile, they mentioned that the relation between surface tension and particle size can be predicted by Taylor theory as well. As a matter of fact, compabilizer can affect deformation and drops fracture by decreasing in surface tension and consequently, deducing hydrodynamic stress which break up droplets with specific size. When a copolymer is used as a compatibilizer, by preventing coagulation leads to modifying and developing of morphology in comparison to drop fracture [2]. Lepers and co-workers [3] carried out a research on PS/EPR, they investigated the effect of surface tension and preventing coagulation by using two compatibilizer (symmetric and asymmetric bl°Ck copolymer). Decreasing in surface tension in asymmetric was more than symmetric bl°Ck copolymer and reduction in the particle coagulation by symmetric copolymer is more than asymmetric. Favis [4] used inomer as a compabilizer in Polyamid/Polyolefin (PE and PP). He observed that in a specific concentration of compabilizer, there is a sharp decline in particle size. This critical concentration depends on the disperse phase area of interface directly.

Zhang and co-workers [5] studied PP/PETG. They produced this blend 30/70 composition by SEBS, SBS and SIS compatibilizer in twin screw extruder. The micrographs represent that in this blend, PETG forms a fibril like structure for all blends and blends compatibilzed by SBS have the smallest particles while SEBS blends have the largest particles. Moreover, compatibilized blends by SIS have average size between two mentioned systems (SBS and SEBS).

Nashar and co-workers [6] have studied LDPE/PET system. They used LDPE-g-AA and LDPE-g-AN as compatibilizer. The blends contain LDPE-g-AA particles with smaller size and homogenous dispersion are compatibilizer. This is due to the interaction of these two compatibilzer with matrix. The blends contain LDPE-g-AA, acrylic acid groups (AA) make hydrogen bonds with PET chains while the connection between PET chains and LDPE-g-AA functional groups contributes to dipole interaction among nitryls and PET segments.

Fillipi and co-workers [7] have investigated the effect of different compatibilzer on LDPE/PA6. They used SEP-g-MA, SEBS-g-MA and HDPE-g-MA as compatibilizer. SEBS-g-MA and SEP-g-MA have dramatic impact on microstructure compared to HDPE-g-MA. Blends contain SEP-g-Ma and SEBS-g-Ma have smaller particles with better dispersion in comparison to blends with HDPE-g-MA as a compatibilizer.

Arefazar and co-workers have studied PET/PC and lantanon acetyl acetate as compatibilizer. Blends with compatibilizer show significant change regarding PC particle size and dispersion in PET matrix [8]. Arefazar and co-worker investigated PET/SBR and SBR-g-MAH as compatibilizer. They observed smaller particle size and better dispersion in PET matrix compared to the PET/SBR, which results in interaction between MA groups in SBR-g-MAH and carboxyl and hydroxyl groups in PET [9, 10].

1.2. Viscosity Ratio

Favis and Chalifoux [11] showed that in PP/PC even in approximately 13 torque ratio (disperse phase torque/ matrix phase torque), there is a significant deformation in disperse particle size in an internal mixer. They concluded that viscosity ratio play major role in morphology and the dimension of the dispersed phase. Furthermore, in 2 to 13 torque ratio, the dimension of dispersed phase increases to three to four times. More deduction in the dispersed particle in lower than 1 torque ratio can be obtained and the minimum particle size can be obtained in 0.25 torque ratio.

Favis and Therrein [12] demonstrated that in PP/PC, fracture of dispersed phase in high viscosity ratio in twin screw extruder with better efficiency happened in comparison to internal mixer and the relation between particle size and viscosity ratio in twin screw extruder is similar to the fracture of the Newtonian fluids in simple shear flow. Kim and co-workers [13] studied the effect of viscosity ratio in binary blend PP/EPR and ternary blend

PP/EPR/PE. The concluded that deformation of drops in matrix strongly accounts for viscosity ratio and to form fibrilar microstructure must be

$$\frac{\eta_d}{\eta_c} < 1$$

Hale and co-workers [14] investigated PBT/ABS and concluded that morphology in this blend strongly influenced by PBT melt viscosity.

1.3.Composition

The effect of composition on particle size in immiscible systems is of prime importance. In a research done by Li and co-workers [15], one of the systems was HDPE/PS which its morphology completely controlled by dispersed phase composition. So that in approximately 18%, 30% and 40% PS, co-continuous ratio is about 25%, 84% and 100% respectively. In 68% PS homogeneously destroyed and phase inversion happened. The dependence of phase structure to phase composition for two types of PP/EPM blends with different rheological behavior has been investigated by Fortenly and co-workers [16]. They showed that samples with similar viscosity of components compared to the samples with high difference viscosity ratio have more co-continuous structure for wide range of compositions.

Wang and co-workers [17] studied the effect of SEBS and SEBS-g-MAH composition on morphology and mechanical properties of PPO/SEBS and PPO/SEBS-g-Ma and observed that by increasing in the composition of SEBS and SEBS-g-Ma in both systems, morphology changes from sea-island in lower compositions to two continues phases in higher compositions. Morphology in both systems is under the influence of disperse phase composition.

1.4. Shear Stress

According to Taylor theory, the particle size of disperse phase has direct relationship to shear stress in mixing. Favis [18] mentioned that any changes in shear stress to two or three times, don't have any desire changes on morphology, which results from the fact that shear stress at interface due to slippage between layers and interfacial adhesion isn't continues. Moreover, in another study by Favis and Therrein [12] on PP/PC in a twin screw extruder shows that shear stress doesn't have dramatic impact on particle size.

1.5.The relation of microstructure and mechanical properties in polymer blends

Wong and co-workers [19, 20] studied PA66/PP modified by SEBS-g-MA in different MA composition. The reported that tensile strength and stiffness rise with increasing SEBS-g-MA in 75/25 of PA66 and PP respectively. Hence, in 50/50 of PA66/PP, with increasing of SEBS-g-MA opposite trend was observed which stems from phase inversion in this system (PP consider as matrix).

Wang and co-workers [21] used DGEBA as compatibilizer in PTT/MB in different weight composition (0 to 5 wt%). They have seen that with adding DGEBA, tensile properties such as tensile strength, Young Modulus and elongation at breaks have modified by increasing weight percentage of this compatibilizer. Moreover, impact resistance increased.

Gonzalez and co-workers [22] studied the influence of phenoxy on mechanical properties of PTT/Phenoxy. With increasing weight percentage, young modulus increases and shows a positive deviation from rule of mixture, so that yield stress rise with increasing in phenoxy composition. Impact resistance shows a negative deviation in similar percentage and rich of phenoxy. Impact resistance have downward trend at first and upward trend in the following. SEMs prove this result.

Yi and co-workers, PP/PET microfilliblar pr°Cessed in a single screw extruder by slit mold. They investigated the effect of increasing of PP-g-GMA as compatibilizer on morphology. They observed that thanks to surface tension by compatibilizer, mechanical properties, tensile strength and impact resistance modified, however young modulus decrease due to lack of stiffness of compatibilizer compared to PET microfibriles.

Jiang and co-workers [24] studied the effect of epoxy resin in PP/SEBS-g-MA. They showed that all mechanical properties increase when cure agent add to epoxy resin during PP/SEBS-g-MA/epoxy pr°-Cessing in comparison to uncured epoxy. This effect is due to creation of specific microstructure surrounding SEBS-g-MA particles. In addition, there is a good balance regarding stiffness and toughness in this system.

Rastin et al. [25] studied a series of binary and ternary polymer blends based on HDPE,polyamide-6 (PA6),poly(ethylene-co-vinyl alcohol) (EVOH) , maleated HDPE and suggested that the core–shell morphology remains unchanged by addition of 0.5, 1.5, 3, and 6 wt.% of HDPE-g-MAH to the ternary blends. But analyzing of impact strength, showed that the addition of 1.5 wt.% HDPEg-MAH to 75/25 MHDPE/EVOH and MHDPE/PA-6 pairs increases about 2.5 and 6 times the impact strength with respect to their corresponding non compatibilized blends, respectively.

Saeb et al. [26] demonstrated that the mechanical properties of high density polyethylene (HDPE),poly (ethylene –co- vinyl alcohol)(EVOH), and polyamide-6 (PA6) binary and ternary blends are strongly influenced by pr°Cessing temperature. In addition, it has been observed that the yield stress increase upon increasing temperature due to enhanced interfacial adhesion between HDPE/EVOH matrix and PA-6 dispersed phase.

Saeb et al. [27] described that for ternary polymer blends based on HDPE/PA-6/EVOH with PA-6 and EVOH as minor components with different weight ratios, yielding behaviour of prepared ternary blends is dependent on the minor component fraction possibly due to the formation of voids at the interface of polymers.

In another research, the morphology of polypropylene (PP)/poly (trimethylene terephthalate) (PTT)/ poly (styrene-b-(ethylene-co-butylene)-b-styrene) (SEBS) 70/15/15 ternary blends prepared by twin screw extrusion was investigated as a function of maleic anhydride-assisted SEBS compatibilizer composition on the morphology and mechanical properties by Moini, et al. [28] It is found that by diminishing SEBS to SEBS-g-MAH weight ratio, the morphology changed from the pure core-shell to a multi-phase system consisting of both core-shell and detached particles. This transition in the phase morphology caused some advantages on the mechanical properties, so that the blend consisting of 50/50 weight ratio of SEBS/SEBS-g-MAH exhibited the maximum value of the impact strength.

Moini et al. [29] have shown that the morphology of the ternary PET/PC/NBR blend is influenced by NBR-g-GMA compatibilizer and by increasing the rubber content, the rod-like structures were disappeared; besides, toughness was increased. Furthermore, by increasing PC content, rod-like structures have seen by morphological study; however, core-shell droplets formed in the blend structure caused enhancing the impact strength and reducing Young's modulus.

In this article, the influence of the composition of PC and SEBS on microstructure and mechanical properties of PP/PC and PP/SEBS through DSC will be investigated.

2. Experimental

2.1. Materials

The following materials were used in this work: (i) An iso-tactic polypropylene homo-polymer (PP), SEETEC H5300 supplied by LG chemical company(Korea) (MFI: 3.5 g/10min, 230 °C, 2.16kg), (ii) Polycarbonate (PC), Makrolon 2858 purchased from Bayer Co(Germany) (MFI: 10 g/10min, 300 °C, 1.2kg), (iii) Poly(styrene-b-(ethylene-co-butylene)-b-styrene) (SEBS) tri-bl°Ck copolymer, Kraton TM G1652 supplied by Shell Chemicals (29% styrene; molecular weight; styrene bl°Ck 7000, EB bl°Ck 37500)

2.2.Blend Preparation

In this study, 8 binary blends were produced at different weight ratio using Brabender co-rotating twin screw extruder (diameter of screw = 2 cm, length/ diameter ratio = 40).The various compositions used for this research are reported in Table1.Prior to pr°-Cessing, the compositions were dried in an oven for at least 17h at 80°C. The barrel of extruder has six temperature-control zones and their temperatures were set at 230-235-240-245-250-255°C (from hopper to die). The screw speed was maintained at 130 rpm.

Table1: Various compositions of binary polymer blends

Code No.	PP(%wt)	PC(%wt)	SEBS (%wt)
PC10	90	10	0
PC15	85	15	0
PC20	80	20	0
PC30	70	30	0
SE10	90	0	10
SE15	85	0	15
SE20	80	0	20
SE30	70	0	30

2.3.Mechanical Properties

After melt blending of designed compounds in twin-screw extruder, the blends were quenched in cooling water bath and pelletized in a granulator. Dried blends were molded to from tensile and impact specimens using an ENGEL injection molding machine. The Barrel temperature profile was 180 °C (hopper) to

240 °C (nozzle) and the mold temperature was maintained at 40 °C. Tensile stress-strain data were obtained using Galdabini testing machine in the rate of 50 mm/min according to the ASTM D-638. Morever Izod impact strength was done for notched specimens according to ASTM D-256 using Zwick pendulum-type tester.

2.4. Morphological Studies

In order to evaluate the effect of particle size and the type of resulted morphology on the mechanical properties of PP/PC and PP/SEBS binary blends, scanning electron microscopy (SEM) micrographs were obtained using AIS-2100 scanning electron microscopy supplied by SERON Company through fracture surface of impact specimens. Before doing scanning electron microscopy, the impact samples were fractured in liquid nitrogen and consequently were etched by cyclohexane for 24h to remove SEBS minor phase. Then, the etched samples were gold sputtered to make the samples conductive.

2.5. Thermal Properties

This test leads to better estimate of compatibility of component in blends. This device (200F3Maia) was made in NETZCH Germany. Approximate weight of samples were 7.5 mg and heating rate was 10 °C/min. Three heating-cooling-heating cycles from 30 to 265 °C were selected. . The percent of crystallinity was calculated using following equation:

$$X_C = \frac{\Delta H_f}{\Delta H_{f^\circ}} \times \frac{1}{\omega_{pp}} \times 100$$

In this equation:

ΔH_f is enthalpy PP

ΔH_{f° is enthalpy isotactic PP (100% crystal) equal to 209 J/gr for PP[31]

ω_{pp} is weight fraction of PP in the blend

3. Results and Discussions

3.1. Morphology investigation of binary blends PP/PC and PP/SEBS

Morphology of PC10, PC15, PC20 and PC 30 belong to PP/PC and SE10, SE15, SE20 and SE 30 belong to PP/SEBS according to the table 2 are in fig. 1. Furthermore, the results of the SEM analysis present in table 3.

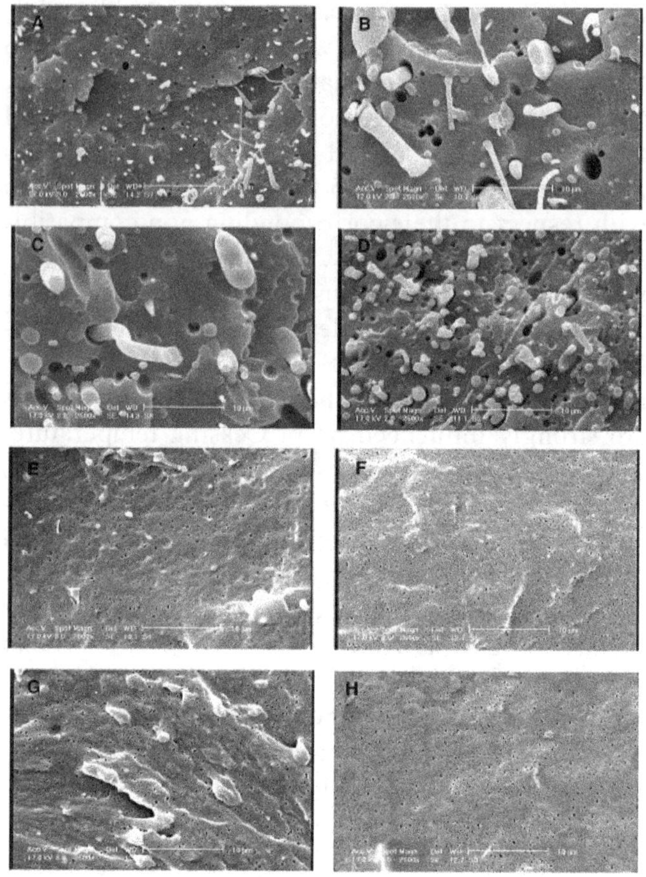

Fig1. SEM micrographs of PP/PC and PP/SEBS at different compositions according to table1.

Relying on SEMs and their analysis, in PP/PC by increasing PC composition to 20% dimension of spherical drops of PC in PP matrix increase due to agglomeration. Moreover, the size of cavities increased due to lack of proper surface adhesion. On the other hand, by increasing PC in this blend, matrix tend to encapsulation of large PC drops, which, in turn, is completely obvious in 15%, 20% and 30% as well as Rod PCs either independence or surrounded exist in that weight fraction. In all samples, hallow cavities implying high interfacial tension can be seen.

PC30 has the highest surrounded drops (either spherical or rod). PP/TPE in separated phases is similar to modified polymeric blends by rubber. Additionally, PP matrix and thermoplastic elastomer has influence on size, particle, shape and its distribution which leads to various morphologies formation especially in high compositions [30]. Referring to SEMs of PP/SEBS it can be inferred that uniform distribution of SEBS drops like black holes which demonstrate drops were extracted by cyclohexane solvent in PP matrix. This is due to the fact that good miscibility of PP with drops owing to Block Ethylene-Butylene

Table2. Results of Image analysis of SEM micrographs according to table1.

Number of rod like PC particles	Number average size of rod like PC particles (μm	Number of indi vidual rod like PC particles	Number average size of individual rod like PC particles (μm	Number of composite droplets	Number average size of composite droplets (μm)	Number of individual PC core particles	Number average size of individual PC core particles	Number of individual SEBS particles	Number average size of individual SEBS particles (μm)	Data type / Sample code
7	0.67514	16	0.40413			50	0.40536	46	0.32439	PC10
9	1.79655	7	1.26314	7	1.05842	20	0.74055	19	0.93258	PC15
6	1.7955			12	1.40008	15	1.17993	12	1.479	PC20
17	1.13282	11	0.87327	19	1.12438	52	0.87062	27	0.69581	PC30
								130	0.2076	SE10
								139	0.15342	SE15
								155	0.21248	SE20
								225	0.1309	SE30

Table3. DSC results of the PP/SEBS and PP/PC binary blend samples

DSC					
XC (%)	wpp	ΔHf (J/gr)	Tm (°C)	Tc (°C)	Sample code
36.68	1	76.67	167.3	108.3	PP
48.79	0.85	86.67	164.766	110.162	PC15
53.55	0.7	78.34	163.738	110.244	PC30
47.003	0.85	83.5	167.128	110.218	SE15
54.45	0.7	79.66	164.768	111.202	SE30

in SEBS tends to PP matrix. With increasing SEBS, the number of SEBS drops increase. Particle size in these samples don't change considerably results from slight coagulation of drops together due to good SEBS interaction with PP matrix, which leads to better stress transfer from matrix to disperse phase and better fracture of drops in matrix and slight coagulation.

3.2. Thermal Properties

According to the table 3, SEBS cannot act as an effective core in PP crystallinity (heterogeneous nucleation), which leads to increasing crystal temperature of PP in both SE15 and SE30. Hence, due to miscibility of PP with SEBS, the high probability of EB block penetration towards PP chains and creation of micelles because of aggregation results in ineffectiveness of SEBS as a nucleation agent in PP crystallinity. On the other hand, due to miscibility and interaction SEBS with PP, the possibility of creation of incomplete crystalline structure increased, which, in turns, leads to low melting temperature of pure PP and this fall increase by increasing in SEBS wt%. Another considerable point is that with increasing SEBS wt% from 15 to 30, heat of fusion declined, which

in SEBS tends to PP matrix. With increasing SEBS, the number of SEBS drops increase. Particle size in these samples don't change considerably results from slight coagulation of drops together due to good SEBS interaction with PP matrix, which leads to better stress transfer from matrix to disperse phase and better fracture of drops in matrix and slight coagulation.

3.2. Thermal Properties

According to the table 3, SEBS cannot act as an effective core in PP crystallinity (heterogeneous nucleation), which leads to increasing crystal temperature of PP in both SE15 and SE30. Hence, due to miscibility of PP with SEBS, the high probability of EB block penetration towards PP chains and creation of micelles because of aggregation results in ineffectiveness of SEBS as a nucleation agent in PP crystallinity. On the other hand, due to miscibility and interaction SEBS with PP, the possibility of creation of incomplete crystalline structure increased, which, in turns, leads to low melting temperature of pure PP and this fall increase by increasing in SEBS wt%. Another considerable point is that with increasing SEBS wt% from 15 to 30, heat of fusion declined, which

which results from better interaction between components with increasing SEBS percentage, which leads to immobility of PP segments and hamper the crystallinity. On contrary, with increasing SEBS, degree of crystallinity decreased. Therefore, degree of crystallinity increased compared to the pure PP. Figure 2 represents the cooling cycle of SE15, SE30 and pure PP. Based on figure, with adding SEBS to PP, peak goes higher and narrower, which results in speeding up the crystallization. The results of table 4 demonstrates that PC particles in PP matrix are heterogeneous nucleation agent, which leads to increasing in crystallinity temperature compared to the pure PP [31]. Another point is that more particle and distribution in matrix, due to nucleation, more PP crystallinity. This trend is completely clear in the table. On contrary, heat of fusion of PC15 and PC 30 rise in comparison to pure PP, which results from increasing in PP crystallinity. Diagram 3 illustrates the cooling cycle of PC15, PC30 and pure PP. Relying on this diagram, with increasing from PC to PP, peak goes higher and narrower, which results in increasing in the rate of PP crystallinity specially in 30wt%. Therefore, PC particles in this situation can act as nucleation agent in PP crystallinity.

3.3. Investigating Mechanical Properties

Table 4 shows the mechanical properties of PP/SEBS and PP/PC. According to this table, it is clear that with increasing SEBS in all PP/SEBS samples, yield stress, young modulus and tensile strength decrease due to elastomeric nature, but elongation at break, regarding elastomeric properties, increases. Due to proper interaction and adhesion between PP and SEBS with increasing in PP, impact resistance increase. This rise in SE30, with the highest SEBS percentage, reaches to its maximum value. In accordance with the PP/PC mechanical properties, it can be inferred that with increasing in PC wt%, tensile

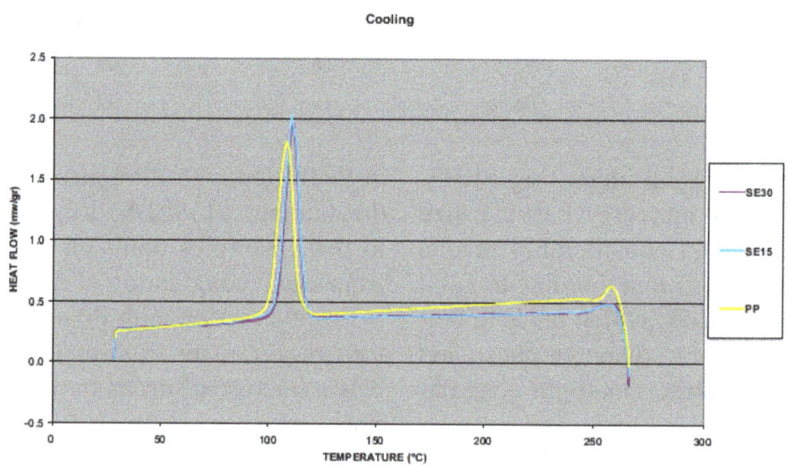

.Fig. 2. Diagram SE15 and SE30 samples cooling cycle

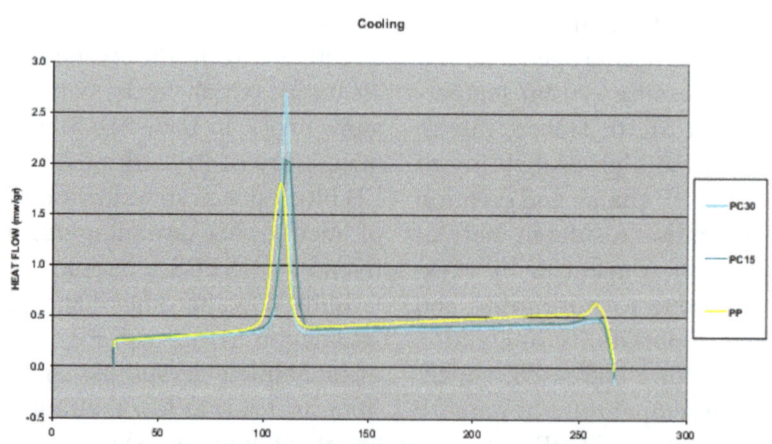

.Fig. 3. Diagram PC15 and PC30 samples cooling cycle

strength increase slightly and yield stress compared to pure PP rises resulting from the nature of high stiffness of PC. Decreasing in yield stress from PC10 to PC30 is a consequence of existence of more structures which matrix surrounding bigger PC particles. In this condition, micro cracks grow in interface causing weakening interface [32]. Since stiffness nature of PC (its high yield stress), with increasing in PC weight percentage leads to rise in yield stress despite lack of interfacial adhesion. Elongation at breaks in PC10 shows a fall which is sharper in higher percentage resulting from weak interaction in these weight percentage is the results of surrounding of bigger PC particles leading to the growth of crakes in interface. SEMs confirms above result easily. Since the stiffness of PC, young modulus increases, reaching to its maximum value in PC30, arise from the presence of more rod like PC individual and surrounded structure in this sample. Tensile strength shows a decline at first arising from undesirable interaction and adhesion between PP and PC, however it slightly increases from PC20. Despite due to an increased in composition and consequently agglomeration in the system the nature of PC stiffness resolves this effect and tensile strength increased. Regarding impact resistance, adding PC to PP matrix due to lack surface interaction to transfer stress, leads to declining in impact resistance of PP. The intensity of this drop from PC10 to PC 15 is more than other samples resulting from surrounded PC particles by PP matrix which is more in higher percentage and causing growth of micro crakes at interface. On the contrary, presence of rod like structures in this system can cause better impact resistance attributing to better orientation of this structures which are perpendicular to crake growth. Therefore, referring to the abovementioned remarks, it can be imply that first factor is much more effective than second factor which is presence of rod like structure and plays a major role.

4. Conclusions

Investigating the effect of disperse phase on morphology and mechanical properties of ternary blends from two binary PP/PC and PP/SEBS blends leads to these results: PP/PC with a heterogeneous morphology stemming from undesirable adhesion between PC and PP at interface. Therefore, disperse phase plays key role in increasing stiffness in this blend. Moreover, PP/SEBS with a desirable morphology wherein SEBS plays important role in increasing toughness in this blend. Finally, DSCs prove that SEBS cannot act as nucleation agents (heterogeneous nucleation) but the PC particles in this situation can act as nucleation agent in PP crystallinity.

References

[1] H.Liang, B.D. Favis, Macromolecules, 32 (1997) 1367.

[2] P.Vanpuyvelde, S.Velankar, P.Moldenaers, Current Opinion in Colloid and Interface Science, 6 (2001) 457.

[3] J.C. Lepers, B. D. Favis, AICHE Journal, 45 (1999) 887.

[4] B.D.Favis, Polymer, 35 (1994) 1552.

[5] X. Zhang, B.Li, K.Wang, Q. Zhang, Q. Fu, Polymer, 50 (2009) 4737.

[6] D. E. EL-Nashar, N. A. Maziad, E. M. Sadek, J Appl Polym Sci, 110 (2008) 1929.

[7] S. Fillipi, L. Minkova, N. Dintcheva, P. Narducci, P. Magagnini, Polymer, 46 (2005) 8054.

[8] P. Zahedi, A. Arefazar, J Appl Polym Sci, 107 (2008) 2917.

Table 4. Mechanical properties of the PP/SEBS and PP/PC binary systems.

Impact strength (J/m)		Tensile strength (MPa)		Young modulus (MPa)		Elongation at break (%)		Yield stress (MPa)		Properties
S.D	average	S.D	average	S.D	average	S.D	average	S.D	average	Samples
0.47	25.66	1.43	19.44	22.60	1167.40	19.96	279.17	0.39	32.30	PP
0.42	24.90	1.79	16.20	9.43	1387.64	3.37	107.23	0.35	34.17	PC 10
0.14	20.00	0.98	12.92	103.49	1402.25	23.38	101.35	0.06	33.99	PC 15
1.68	19.03	0.70	23.86	55.33	1437.72	2.79	29.99	0.10	36.30	PC 20
0.14	17.60	1.29	28.12	0.57	1749.51	4.50	29.03	1.41	37.30	PC 30
5.05	65.13	1.12	18.15	5.01	1081.07	185.04	629.03	0.36	29.04	SE 10
23.33	171.50	0.69	19.03	18.89	978.61	85.80	380.25	0.64	27.25	SE 15
7.94	317.00	0.08	17.06	16.21	965.54	136.72	688.54	0.29	26.18	SE 20
13.58	551.33	0.37	16.70	1.26	898.40	11.72	823.28	0.13	23.15	SE 30

[9] O. Moini Jazani, A. Arefazar, J Appl Polym Sci, 102 (2006) 1615.
[10] O.Moini Jazani, A. Arefazar, Macromol Symp, 263 (2008) 263.
[11] B. D. Favis, J. P. Chalifoux, Polymer, 1761 (29) 1988.
[12] B. D. Favis, D. Therrien, Polymer, 32 (1991) 1474.
[13] B. K. Kim, I.H.DO,J Appl Polym Sci, 60 (1996) 2207.
[14] W. Hale, J. H. Lee, H. Kesskula, D. R. Paul, Polymer, 40 (1999) 3621.
[15] J. Li, P.L. Ma, B.D.Favis, Macromolecules, 35 (2002) 2005.
[16] I. Fortenly, J. Kovar, Eur Polym J, 28 (1992) 85.
[17] X. Wang, W. Feng, H. Li, E. Ruckenstein, Polymer, 43 (2002) 37.
[18] B. D. Favis, J Appl Polym Sci, 39 (1990) 301.
[19] S. C. Wong, Y. W. Mai, Polymer, (1999) 1553.
[20] S. C. Wong, Y. W. Mai, Polymer, 41 (2000) 5471.
[21] K. Wang, Y. Chen, Y. Chang, Polymer, 1 (2009) 1
[22] I. Gonzalez, J. I. Eguiazabal, J. Nazabal, J Appl Polym Sci, 102 (2006) 3246.
[23] X. Yi, L. Xu, Y. L.Wang, G. J. Zhong, X. J, Z. M. Li, Eur Polym J, 46 (2010) 719.
[24] X. L. Jiang, Y. X. Zhang, Chinese Chemical letters, 20 (2009) 877.
[25] H. Rastin, M. R. Saeb, S. H. Jafari, H. A. Khonakdar, B. Kritzschmar, U. Wagenknecht; Macrom. Mat. Eng, 300 (2015) 86.
[26] M.R.Saeb, H.A. Khonakdar, S. H. Jafari, H. Rastin, U. Wagenknecht, . G. Heinrich; Polym-Plast. Tech. & Eng, 54 (2015) 223.
[27] M.R. Saeb, H. A.Khonakdar, S. H. Jafari, N. Jalal-ifar, M. Razban, U.Wagenknecht; . Polym. Bull 71 (2014) 613.
[28] O.Moini Jazani, A. Arefazar, S. H.Jafari, M. R. Pey-manfar, M.R. Saeb, A.Talaei, ., Poly-Plast. Tech. & Eng, 52 (2013) 206.
[29] O.Moini Jazani, A. Arefazar, S. H.Jafari, M. R. Pey-manfar, M.R. Saeb, A.Talaei,B.Bahadori ., Poly- Plast. Tech. & Eng, 52 (2013) 1295
[30] F.O.M.S Abreu , M.M.C.Forte , S.A.Liberman , J Appl Polym Sci , 95 (2005) 254.
[31] Shanshan Dai , L.Ye , Polym Adv Technology , 19 (2008) 1069.
[32] K.R.Srinivasan , A.K.Gupta , J Appl Poly Sci 53 (1994) 1.

14

Effects of catalyst particle size on methanol dehydration at different temperatures and weight hourly space velocities

Leila Khoshrooyan, Ali Eliassi*, Maryam Ranjbar

Chemical Technology Department, Iranian Research Organization for Science and Technology (IROST), Tehran, Iran

HIGHLIGHTS

- Effect of catalyst particle size on methanol conversion was investigated.
- Conversion of methanol in micro-channel reactor was measured.
- Methanol conversion in fixed bed and micro-channel reactor was compared.

GRAPHICAL ABSTRACT

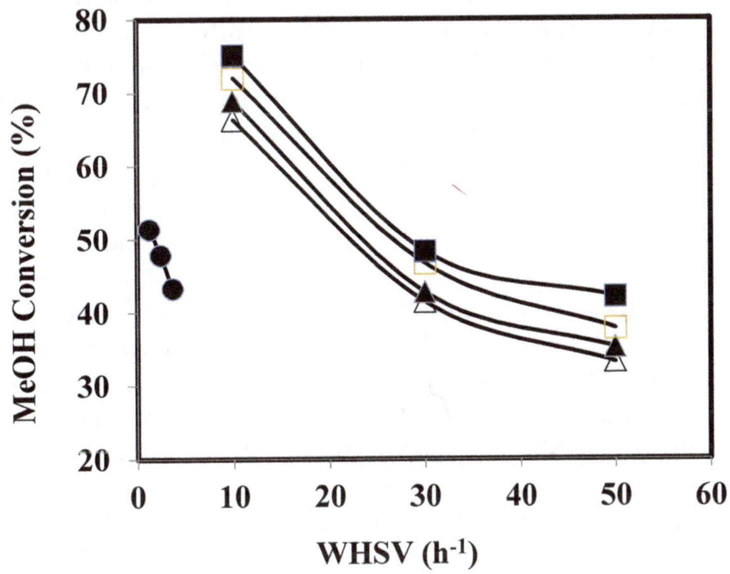

ARTICLE INFO

Keywords:
Methanol Dehydration
Fixed Bed Reactor
Micro-Channel Reactor
Catalyst Particle Size
DME
Gamma Alumina

ABSTRACT

The effect of catalyst particle size on dehydration of methanol to dimethyl ether is investigated using fixed bed and micro-channel reactors at different temperatures and weight hourly space velocities. The experiments were carried out at 290 and 320°C. The space velocities were changed from 10 up to 90h^{-1} and from 1.22 to 3.65h^{-1} for fixed bed and micro-channel reactors, respectively. Considering the catalyst particle size effect on dehydration reaction, the particle size was changed from 0.063 to 1mm. Commercial gamma alumina was used as catalyst in all the experiments. The fabricated micro-channel reactor had 40 channels with 1mm diameter and 6 cm length. The channels were sub-coated with alumina and finally were coated with gamma alumina as dehydration catalyst. The results showed that methanol conversions were increased by increasing the temperature and decreasing the particle size of the catalyst. Furthermore, methanol conversion in micro-channel reactor was less than for fixed bed reactor under the similar WHSVs, due to the special geometrical shape of the micro-channels.

* *Corresponding author.*
E-mail address:alieliassi@yahoo.com

1. Introduction

Dimethyl ether (DME) or CH_3OCH_3, also known as methoxymethane, wood ether, dimethyl oxide or methyl ether, is a non-toxic liquefied gas. At 1 atm and 25°C dimethyl ether is a gas [1]. Moreover, methanol can be replaced by DME as primary substance for production of hydrocarbons such as light olefins, and chemicals such as dimethyl sulphate, methyl acetate [2]. Also, DME can be used as a source of hydrogen for fuel cells [2]. Due to its high cetane number (>55) this compound is considered as an alternative fuel for diesel engines [3].

Generally, two direct and indirect methods are used for dimethyl ether production [4-8]. In direct method, some serial reactions occur such as methanol formation, methanol dehydration and water gas shift (WGS). Methanol is produced in the middle path of the mentioned serial reactions and finally is dehydrated to DME [8, 9]. These reactions can be simplified as follows:

(1)

Indirect method is a simple reaction for DME production, where methanol is dehydrated by using an acidic catalyst such as H-ZSM-5 and gamma alumina (γ-Al_2O_3). Between two mentioned solid acid catalysts, activity of H-ZSM-5 is more than the other one. However, fast deactivation of strong acidic sites of H-ZSM-5 is one of its weaknesses as compared to gamma alumina [10].

Methanol dehydration is a slight exothermic reaction, so that:

(2)

From transport phenomena point of view, catalyst particle size has an important role on global kinetic rate of a heterogeneous reaction. Whereas, there are little information about the effect of catalyst particle size on methanol dehydration. In this work, fixed bed reactor tests were carried out to investigate the effect of gamma alumina grain size (G. S.) on methanol dehydration at two different temperatures and weight hourly space velocities (WHSV). In all of experiments the methanol flow rate was kept constant but temperature and WHSVs were changed. In the second part of this work, a micro-channel reactor was fabricated and its channels were coated by gamma alumina. By using this reactor, the effects of temperature and WHSV were investigated on methanol conversion and the results were compared with the results obtained from the fixed bed reactor.

2. Experimental

2.1. Fixed Bed Reactor

All of the experiments were carried out in an isothermal fixed bed reactor (¾ inch diameter) which was previously used by Eliassi et al. [11]. In this system pure methanol was pumped from the methanol feed tank at a rate of 63cc/h to an evaporator. The evaporated methanol was sent to a super-heater before flowing to the isothermal fixed bed reactor. A thermo well and a thermocouple system was used in order to control the reactor temperature. Reactor products were passed through a cooler to cool down to the room temperature. Then the liquid and gas products were separated into the trap. Experiments were carried out at 290 and 320°C under the constant atmospheric pressure.

2.2. Micro-Channel Reactor

A micro-channel reactor is an apparatus which has a sandwich construction and contains several channels. Usually, diameter of these channels are less than 1mm and after that a catalyst layer was coated on these channels, the reactor can be used for considering the various chemical reactions [12-13]. In this work, a micro-channel reactor was made by using titanium plate, which 40 parallel micro-channels in this plate were made by milling and lathing process. The diameter and length of each channel were 1 mm and 40 mm, respectively. Fig. 1. shows perspective picture of the made micro-channel plates.

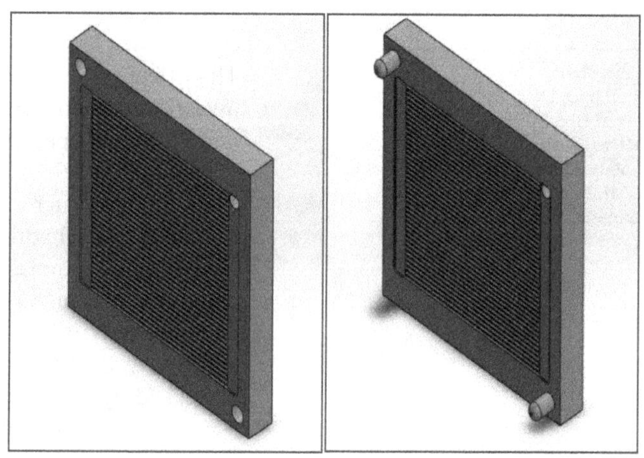

Fig. 1. Perspective of the fabricated micro-channel plates

Fig. 2. Micro-channel reactor with jaw plates, connections, heater and thermocouples

The feed was injected to the reactor through a hole near the vertex of the titanium plate and the products exit from another hole which is placed on the opposite side of the entrance hole.Fig. 2 shows the fabricated micro-channel reactor with jaw plates, connections, heater and thermocouples. The micro-channel reactor set up is made of the following components: syringe pump (Irman tech, Iran) for methanol injection, heater, thermocouples, temperature controller and gas-liquid separation vessel.

Micro-channel coating is an important step for fabrication of these types of reactor. It is required that the coated catalyst forms a uniform and stable layer on the reactor channels. In this work, a thin layer of gamma alumina was coated on the channels by using catalyst slurry mixture. Usually, for increasing specific surface area and mechanical resistance, a nano-porous oxide used as sub-layer on the channels [14]. For this purpose, an alumina sol was prepared via sol-gel method by aluminum iso-propoxide (AIP) and iso-propanol as a solvent. To prepare the sol, initially AIP and isopropanol was mixed for 30 minutes at 65-70°C by a sonicator. Acetic acid, propanoic acid and double distilled water were used to control the pH. Also, ethyl acetate was used as a surfactant for increasing sol adhesion to the micro-channels. Then the sol was applied on the micro-channels by injection method and finally the plates were dried in an oven. In the second step, the powder of gamma alumina was used to preparation of slurry of the catalyst. The particle size of the gamma alumina powder was less than 40 microns. Similar to the previous mentioned coating procedure, the prepared slurry was injected to the channels and the plates were placed in the furnace for 3 hours for calcination at 350°C.

2.3 Materials

Commercial extruded gamma alumina was prepared from BASF, Germany (Kat D10-10 S4), aluminum iso-propoxide (AIP) and iso-propanol from Merck and methanol from Iran Petrochemical Company.

2.4 Experimental Conditions

In the fixed bed reactor, the experiments were performed by using four different catalyst particle sizes (P.S.) less than 0.063 up to 1mm, namely: 0.5<P.S.<1mm, 0.125<P.S.<0.5mm,0.063<P.S.<0.125 mm and P.S.<0.063. Also, the reactor temperature was fixed on 290 and 320°C. All of experiments were carried out by 63cc/h methanol as the feed of the reactor, but WHSV (ratio of methanol mass flow rate to the mass of the used catalyst) were changed from 10 up to 90 h^{-1}.

The experiments were repeated three times and the average of the obtained results were reported as the measured quantity.

The experiments in micro-channel reactor were carried out at 290 and 320°C and methanol flow rates in micro-reactor were fixed on 1, 2 and 3ml/h. Also, WHSV values were 1.22, 2.43 and 3.65 h^{-1}.

3. Results and discussion

Figures 3- 6 show the changes of methanol conversion versus WHSV at different temperature and catalyst particle size in fixed bed reactor.

It is worth noting that all of experiments in fixed bed reactor were carried out at constant methanol flow rates and WHSVs were changed just by changing amount of the loaded catalyst. Methanol conversion (X) and WHSV are defined as:

(3)

(4)

where, F_{in}, F_{out} and m_c are the methanol mass flow rates at inlet and outlet of the reactor and the mass of the loaded catalyst, respectively.

Figures 3- 6 clearly show that by increasing temperature, the methanol conversion increased rapidly.

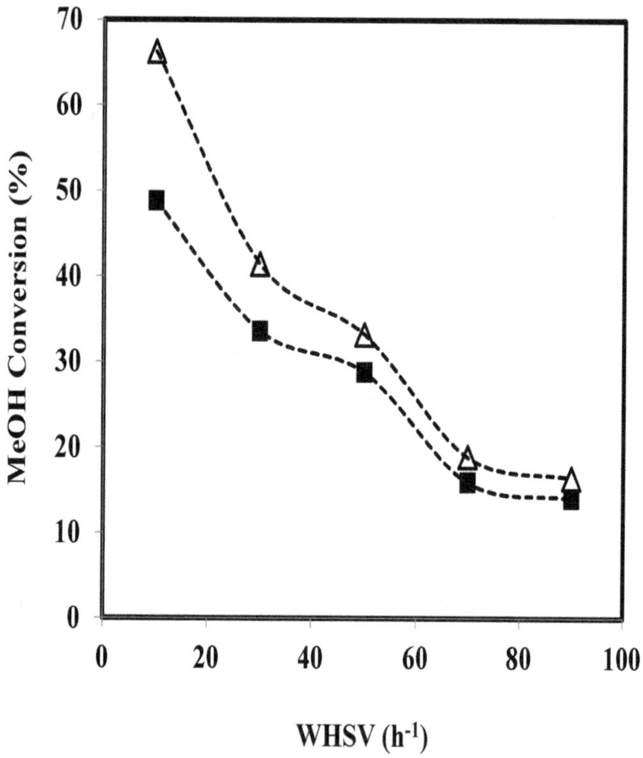

Fig. 3. Methanol conversion versus WHSV at 290°C (■) and 320°C (Δ) in the fixed bed reactor for catalyst particle size of 0.5<P.S. .<1mm. Dotted lines show the trend of changes

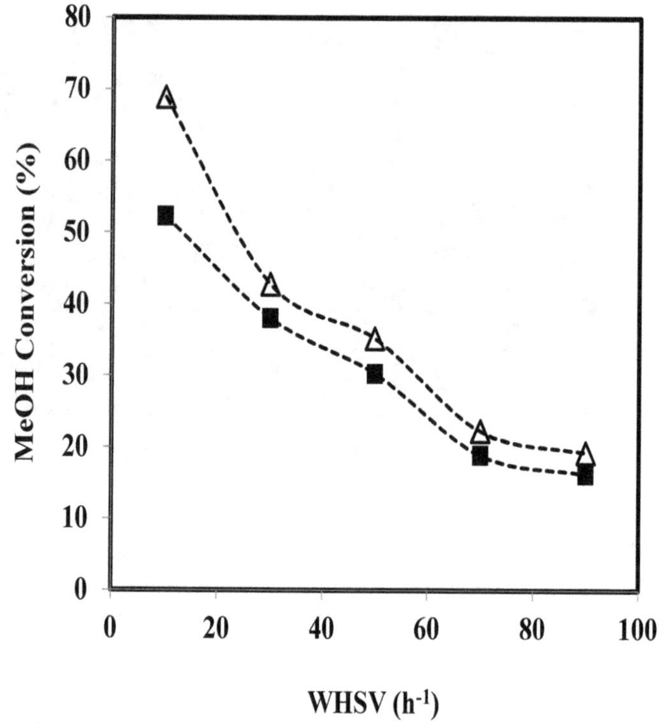

Fig. 4. Methanol conversion versus WHSV at 290°C (■)and 320°C (Δ) in the fixed bed reactor for particle size of 0.125<P.S. <0.5mm. Dotted lines show the trend of changes..

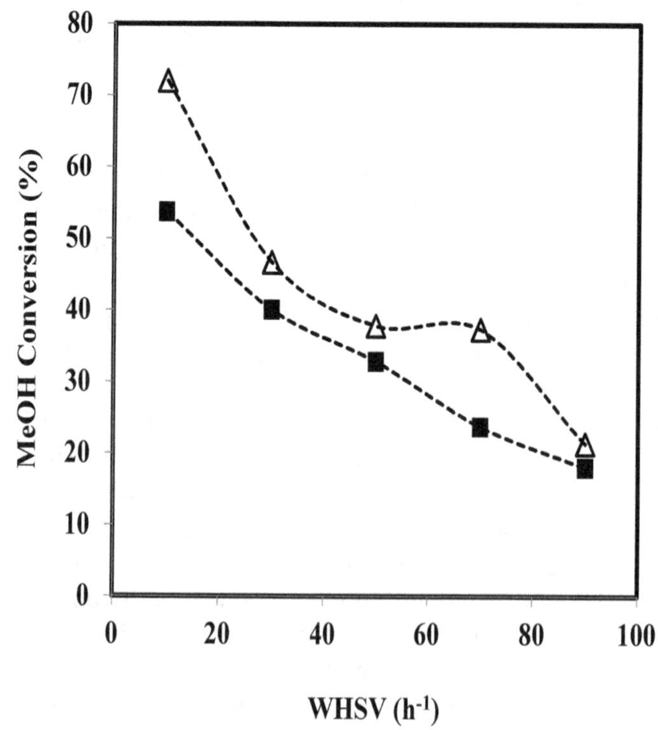

Fig. 5. Methanol conversion versus WHSV at 290°C (■)and 320°C (Δ) in the fixed bed reactor for catalyst particle size of 0.063<P.S. <0.125mm. Dotted lines show the trend of changes.

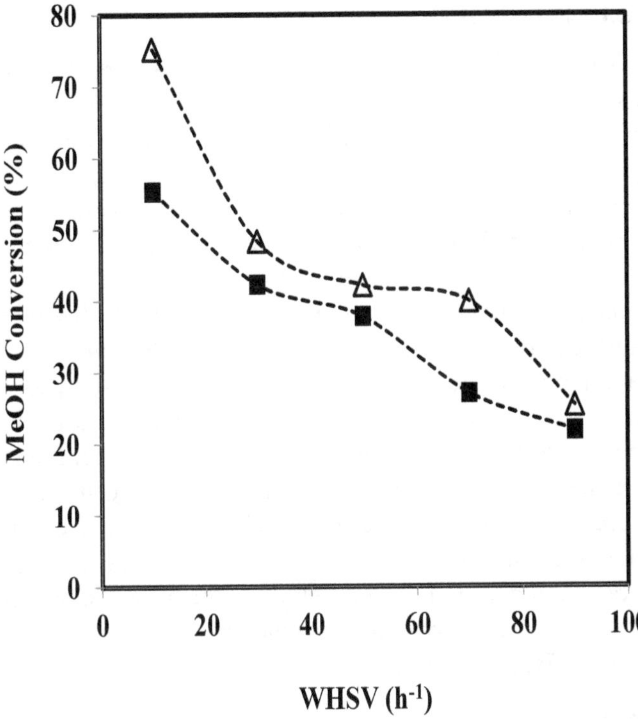

Fig. 6. Methanol conversion versus WHSV at 290°C (■) and 320°C (Δ) in the fixed bed reactor for catalyst particle size of P.S. <0.063mm. Dotted lines show the trend of changes

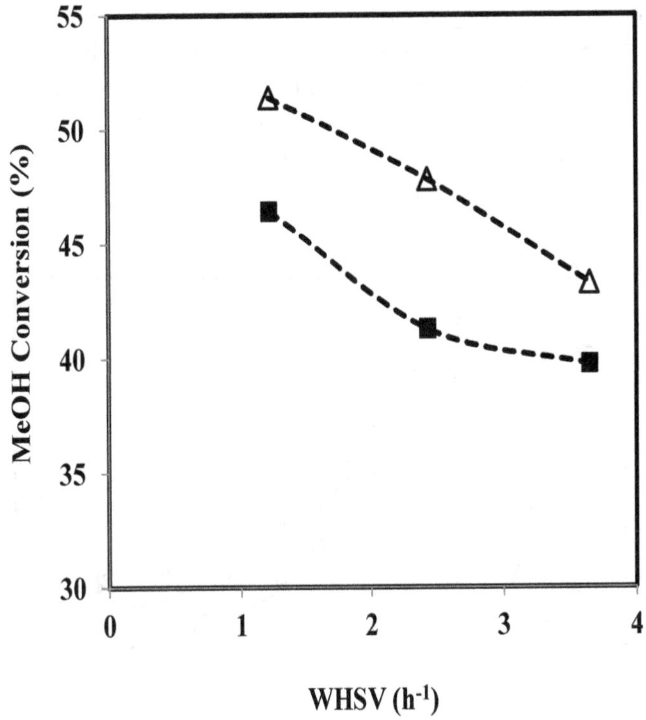

Fig. 7. Methanol conversion changes versus WHSV in micro-channel reactor at 290°C(■) and 320°C (Δ). Dotted lines show the trend of changes

Also, these figures show that the methanol conversion increased by decreasing of the catalyst particle size. For example, according to Figure 6, at 290°C and WHSV=10h^{-1}, the conversion is reached to 55.25% , when the catalyst particle size is less than 0.063mm. While, according to Fig. 3, for 0.5 < P.S. < 1mm and at the same conditions (temperature and WHSV) the methanol conversion is reached to 48.77%. In the other word, in the second case, the methanol conversion was decreased about 12%. With notice to this matter that the methanol flow rates were the same in the both experiments, the observed changes can be related to two different phenomena. One of them is that the available surface of the catalyst particles for the reactant molecules were increased for the smaller catalyst particles. The second phenomenon can be related to increasing mass transfer coefficient by decreasing the catalyst particle size, because, the flow pattern of the gases (reactants and/or products) can be affected by catalyst particle sizes in the fixed bed reactor. In the second series of experiments, the micro-channel reactor was used. In these experiments, the mass of the coated catalyst was nearly constant and WHSVs were changed by changing the methanol flow rates. The obtained results are given in Figure 7. By using this reactor, the average increasing of methanol conversion was about 5% by changing temperature from 290 to 320°C.

Figure 8 shows methanol conversion versus WHSV for fixed bed reactor in compared with micro-channel reactor at 290°C. According to this figure, at the same WHSVs, methanol conversion is higher for fixed bed reactor than micro-channel reactor. This fact can be related to the geometrical shape of the channels, so that the pressure drop through different channels was not to be the same and all of the channels were not available for the reactant molecules and finally the conversion was decreased. This claim can be approved by considering the status of the channels after the reaction. Figure 9 shows a picture of the micro-channel reactor after using the dehydration reaction. With notice to the color changes of the catalyst at different channels, it can be seen that more than half of the channels were not available for the reaction. In the other word, the real WHSV for this reactor is more than that was reported initially and this fact was caused to methanol conversion decreasing. It seems that by changing the geometrical shape of the channels and decreasing the pressure drop, the methanol conversion will be increased. Non-uniform coating of the catalyst on the surface of the channels is the other fact for different pressure drop in different channels. Therefore, by improving the coating techniques the conversion can be increased. Also, mass transfer coefficient is another important factor in reaction rate of a heterogeneous reaction and it can be influenced by flow rate of the reactants in the reactor. Since the flow rate of methanol in fixed bed and micro-channel reactors were different, therefore, mass transfer coefficients in two reactors are not the same, as a results, methanol conversions in two reactors were different, whereas the WHSVs were the same.

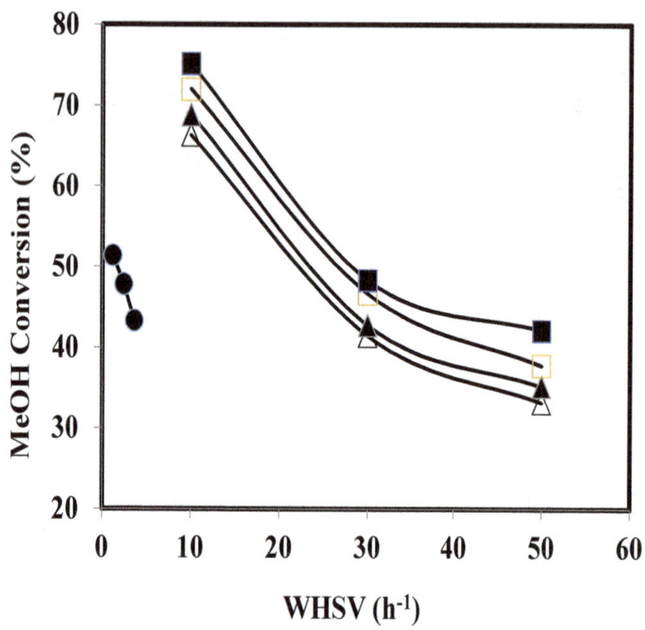

Fig. 8. Methanol conversion versus WHSV for fixed bed reactor in comparison with micro-channel reactor at 290°C. Micro-channel reactor (●). Fixed bed reactor: P.S.<0.063mm ,0.063< P.S. <0.125mm (□),0.125< P.S. <0.5mm (▲),0.5< P.S. <1mm (Δ).

Fig. 9. White channels were not used during methanol dehydration

4. Conclusions

Effects of commercial gamma alumina particle size as a catalyst for dehydration of methanol to DME were investigated. For this purpose the reaction was performed in a fixed bed reactor at two different temperatures and different weight hourly space velocities. The results showed that methanol conversion was increased by decreasing catalyst particle size. This phenomenon can be interpreted by increasing the effective surface area of the catalyst by its size reduction. Also, methanol conversion in fixed bed reactor was more than for micro-channel reactor. This phenomenon can be due to the: (a) non-uniform coating of the catalyst on the surface of the channels and non-uniform pressure drop in different channels (b) some of the channels were not available for reactants during the dehydration reaction for geometry of the micro-channel reactor (c) flow rate of the reactants in the channels was different in compare with the fixed bed reactor.

Acknowledgment

Authors thanks Iranian Research Organization for Science and Technology (IROST) for their helps and special thanks to Ms. Niloofar Naseri for doing some analysis.

References

[1] L. C. L. Agostinho, C. Barbosa, L. Nascimento, J. Rodbari, Catalytic Dehydration of Methanol to Dimethyl Ether (DME) Using the Al62,2Cu25,3Fe12,5 Quasicrystalline Alloy, Journal of Chemical Engineering & Process Technology 4, 164 (2013) 7-8.

[2] J. Ereña, I. Sierra, M. Olazar, A. G. Gayubo, A. T. Aguayo, Deactivation of a CuO-ZnO-Al_2O_3/γ-Al_2O_3 catalyst in the synthesis of dimethyl ether, Industrial & Engineering Chemistry Research 47 (2008) 2238-2247.

[3] F. Yaripour, F. Baghaei, I. Schmidt, J. Perregaard, Synthesis of dimethyl ether from methanol over aluminium phosphate and silica-titania catalysts, Catalysis Communications 6 (2005) 542-549.

[4] F. Dadgar, R. Myrstad, P. Pfeifer, A. Holmen, J. V. Hilde, Direct dimethyl ether synthesis from synthesis gas: The influence of methanol dehydration on methanol synthesis reaction, Catalysis Today 270 (2016) 76-84.

[5] D. Macina, Z. Piwowarska, K. Tarach, K. Góra-Marek, J. Ryczkowski, L. Chmielarz, Mesoporous silica materials modified with alumina polycations as catalysts for the synthesis of dimethyl ether from

methanol, Materials Research Bulletin, 74 (2016) 425-435.

[6] Z. Hosseini, M. Taghizadeh, F. Yaripour, Synthesis of nanocrystalline γ-Al_2O_3 by sol-gel and precipitation methods for methanol dehydration to dimethyl ether, Journal of Natural Gas Chemistry 20 (2011) 128-134.

[7] Q. Tang, H. Xu, Y. Zheng, J. Wang, H. Li, J. Zhang, Catalytic dehydration of methanol to dimethyl ether over micro–mesoporous ZSM-5/MCM-41 composite molecular sieves, Applied Catalysis A: General 413 (2012) 36-42.

[8] Y. Tavan, R. Hasanvandian, Two practical equations for methanol dehydration reaction over HZSM-5 catalyst–Part I: Second order rate equation, Fuel 142 (2015) 20-214.

[9] F. Yaripour, Z. Shariatinia, S. Sahebdelfar, A. Irandoukht, The effects of synthesis operation conditions on the properties of modified γ-alumina nanocatalysts in methanol dehydration to dimethyl ether using factorial experimental design, Fuel 139 (2015) 40-50.

[10] B. Sabour, M. H. Peyrovi, T. Hamoule, M. Rashidzadeh, Catalytic dehydration of methanol to dimethyl ether (DME) over Al-HMS catalysts, Journal of Industrial and Engineering Chemistry 20 (2014) 222-227.

[11] F. Raoof, M. Taghizadeh, A. Eliassi, F. Yaripour, Effects of Temperature and Feed Composition on Catalytic Dehydration of Methanol to Dimethyl Ether over Gamma-Alumina, Fuel 87 (2008) 2967-2971.

[12] P. L. Mills, D. J. Quiram, J. F. Ryley, Microreactor technology and process miniaturization for catalytic reactions—a perspective on recent developments and emerging technologies, Chemical Engineering Science 62 (2007) 6992-7010.

[13] A. I. Stankiewicz, J. A. Moulijn, Process intensification: transforming chemical engineering, Chemical Engineering Progress 96 (2000) 22-34.

[14] K. Haas-Santo, M. Fichtner, K. Schubert, Preparation of microstructure compatible porous supports by sol–gel synthesis for catalyst coatings, Applied Catalysis A: General 220 (2001) 79-92.

15

The measurement of droplet size distribution of water-oil emulsion through NMR method

Arash Amani[1], Ali Reza Solaimany Nazar[1,*] , Hasan Sabzyan[2], Gholamhassan Azimi[2]

[1]*Department of Chemical Engineering, University of Isfahan, Isfahan, Iran*

[2]*Department of Chemistry, University of Isfahan, Isfahan, Iran*

HIGHLIGHTS

- Average water droplet size in water-oil emulsions were measured by applying NMR technique.
- Effect of type and concentration of demulsifier on average water droplet size was investigated.
- Influence of water volume ratio, water salinity and mixing speed was also evaluated.
- Water/oil volume ratio and water salinity increased the average droplet size.

GRAPHICAL ABSTRACT

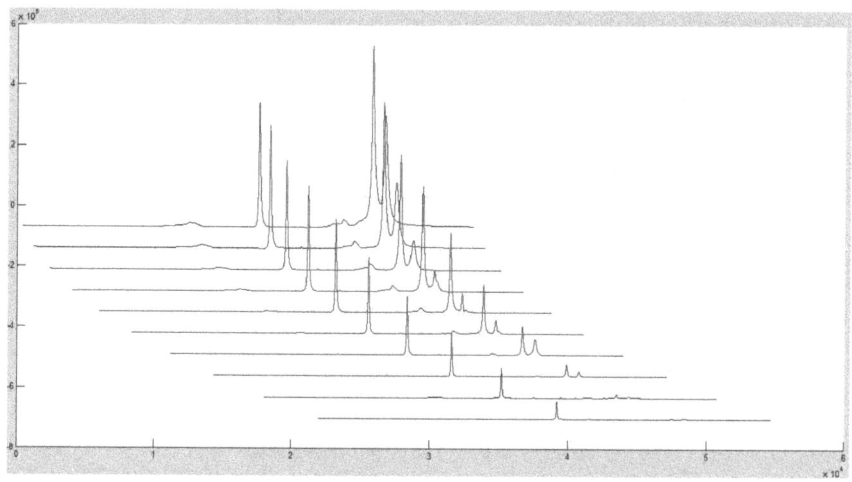

ARTICLE INFO

Keywords:
NMR measurement
Water-oil emulsion
Droplet size
Demulsifier

ABSTRACT

The effects of water/oil volume ratio, type and concentration of demulsifier, water salinity and mixing speed on the average water droplets size in water-oil emulsion are evaluated at different times through NMR measurements. The type and concentration of demulsifier have the greatest effects on the average droplets size with 38% and 31.5%, respectively. The water/oil volume ratio, water salinity and mixing speed are significant factors with 13.1%, 7.5% and 5.71%, respectively. The commercial demulsifier Break 6754 has the greater influence on the average droplets size compared to the acrylic acid. The water droplets size increases upon increasing the concentration of demulsifier, the water volume ratio and the salinity of water and decreases upon increasing the mixing speed.

* *Corresponding author.*
E-mail address: asolaimany@eng.ui.ac.ir

1. Introduction

The water in oil emulsions are formed during the production process, transportation and crude oil refining [1]. Generally, formation of emulsionsis undesirable in oil and petrochemical industry. The dispersed water occupies some of the volume of the oil processing equipment and pipelines and increases the operational costs of processing. Moreover, the physical properties of oil changes substantially due to the emulsion formation [2]. The mechanism of emulsion formation and its stability is almost the same in different industries; therefore, the study of emulsion behavior and factors affecting its stability is beneficial. The droplet size distribution of the emulsion is an important and effective factor on the emulsion stability. There exist several techniques to measure the size of the emulsion droplets having their advantaged and disadvantaged [3,4]. Economical consideration, reliability and the ease of the measurement method are the criteria for the technique selection.

Microscopic imaging method can be used to determine the shape, size, lateral surface and the volume of particles and droplets within a system. The advantage of this method is direct imaging of particles, while in other method, particles sizesare derived indirectly using different size distribution functions with parameters which values are optimized based on measurements on known samples. Direct microscopic imaging method has its own limitations, especially for turbid and dark color samples, as well as opaque concentrated emulsions appearing as cloudy, where do not allow the light to pass through the sample.

Nuclear magnetic resonance (NMR) has been used to measure mainly because of its simple preparation of sample step. The entire sample can be analyzed within a relatively short experiment time. Also there is no limitation for the turbid, dark color or concentrated emulsion samples. NMR has so far been used to measure the weak field coefficient [5], determine the droplet size distribution in oil-in-water and water-in-oil emulsions [6,7]. Houlingsworth et al. [8] proposed a method for measuring droplet size distribution by NMR which reduces the experiment time effectively (from 5-20 minutes to 3-10 seconds per sample). By using this method they succeed to measure silicon oil emulsion droplet size distribution dynamically for water-in-silicon oil samples. This method also allows measurements in non-equilibrium systems.

Van der Tuuk et al. [9] combined the previous methods including Carr-Purcell-Meimboom-Gill (CPMG) pulse sequence, pulsed field gradient (PFG) and simulated echo (STE) and proposed a new method for determining the droplet size distribution by NMR. In this method also the short observation time model [10] and slightly modified method of Parker and Rees are combined to develop a new method for calculating the droplet size distribution of crude oil emulsions.

In this article, NMR relaxation time and measurements of diffusion coefficient of the dispersed phase are used to determine the water-in-oil samples droplet size distributions. The effect of type and concentration of demulsifier, volume water/oil ratio, water salinity and mixing speed on the average droplets size are investigated.

2. Theory

The droplet size distribution is determined through two different methods depending on the sizes range of droplets.

2.1. Large droplet size

If the square root of the average of molecular penetration length inside the droplet is much smaller than the radius of the drop i.e. $(tD_0)^{\frac{1}{2}} \ll \bar{R}$ the short observation time model is used [10]. In short observation time, only a small fraction of molecules within the droplet feels droplet wall. Ignoring this fraction of molecules, Mitra et al. [10] showed that the deviation of the observed diffusivity at time t from the liquid bulk diffusivity within the droplet is expressed as:

$$\frac{D(t)}{D_0} = 1 - \frac{4}{9\sqrt{\pi}}\sqrt{D_0 t}\left(\frac{s}{v}\right) + Q(P,R,T) \tag{1}$$

where, D(t) is the diffusion coefficient of the molecules trapped in the droplets at time t, D_0 is bulk diffusion coefficient of the bulk of dispersed phase, $\left(\frac{s}{v}\right)$ is the surface to volume ratio of the droplets, and Q(P,R,T) is a parameter related to the curvature of the droplet surface. For short times, the Q(P,R,T) parameter is small and thus the $\frac{D(t)}{D_0}$ ratio is related to $\left(\frac{s}{v}\right)$ directly. If values of all other parameters are known the $\left(\frac{s}{v}\right)$ can be obtain from Eq. 1. For a distribution of droplets considered as the average surface-to-volume ratio of all droplets $\left(\frac{S}{V}\right)_{ave}$ and thus, D(t) can be regarded as the average diffusion coefficients of water molecules inside all droplets ($D(t)_{ave}$).

A simple relation connects relaxation time (T_2) to the surface to volume ratio of the droplets [11], i.e.

$$\frac{1}{T_2} = \rho(\frac{S}{V}) \quad (2)$$

where, ρ is the surface relaxation parameter. Averaging above equation for all droplets results in a relation between the average surface to volume ratio of droplets to the average relaxation time, $\left(\frac{1}{T_2}\right)_{ave}$ obtained from relaxation time distribution as follow:

$$\left(\frac{1}{T_2}\right)_{ave} = \rho\left(\frac{S}{V}\right)_{ave} \quad (3)$$

Distribution of the T_2 values is obtained from CPMG inversion recovery (IR) measurements in which the NMR signal attenuation (M^{abs}) is followed with time, and the T_2 values at each NMR frequency is derived from fitting CPMG results to the IR equation [12].

$$\frac{I}{I_0} = exp(-\frac{\tau}{T_2}) \quad (4)$$

The $\left(\frac{1}{T_2}\right)$ values are averaged over the T_2–frequency distribution furthermore to obtain the $\left(\frac{1}{T_2}\right)_{ave}$, the $\left(\frac{S}{V}\right)_{ave}$ value is obtained. By inserting these two values into Eq. 3, the surface relaxation parameter, ρ, can be calculated. Since the ρ value is constant and does not depend on the droplet size, this value can be used in Eq. 2 to relate the $\left(\frac{1}{T_2}\right)$ to $\left(\frac{S}{V}\right)$ values for any range of droplet size; therefore, the droplet size distribution can be obtained from the T_2–frequency distribution.

2.2. Small droplet size

For small droplet size, the average length diffusion of water molecules is larger than the droplet size. At this condition the measured NMR signal is simplified to:

$$\frac{I}{I_0} = \exp\left(-\frac{1}{5}\gamma^2\delta^2G^2R^2\right) \quad (5)$$

where, G is the applied magnetics field gradient, γ is the gyromagnetic ratio of the nucleus, δ is the width of the gradient pulse and R is the droplet radius. This equation can be used to derive the average value of droplet size distribution, $\overline{R}$, by following the NMR signal. The average radius can be used to calculate $\left(\frac{S}{V}\right)_{ave}$ assuming perfect sphere shape for droplets using

$$\left(\frac{S}{V}\right)_{ave} = \frac{3}{\overline{R}} \quad (6)$$

The $\left(\frac{1}{T_2}\right)_{ave}$ is obtained through the same procedure described.

3. Materials and methods

3.1. Materials

In this study, a mixture of 50% crude oil and 50% diesel is used for sample preparation. The crude oil is mixed with dieselreduce its high viscosity which prevented its delivery into the NMR tube. Note that the dilution diesel sample is the product of the same crude oil fed in the Isfahan refinery. The viscosity and density of the crude oil and diesel used are reported in Table 1. Viscosity measurement is carried out in accordance with ASTM D7042 using a Stabinger viscometer (SV300 model). Aqueous phase re-distillated deionized water is used to prepare the sample with purity 99.99%. NaCl is provided by Merck, Germany. In this study two different demulsifier types are used that the first is Break 6754 commercial demulsifier produced by Ahwaz oil Company, Iran and the second demulsifier is acrylic acid with a laboratory purity of Merck, Germany.

3.2. Methods

3.2.1. Water in oil preparation

Small amounts of clay and the other insoluble solid particles can be suspended during the production and transportation of crude oil. These solids can affect the water droplets distribution in water-oil emulsions; therefore, the solids must be removed from the oil samples. To ensure a complete separation of the particles at first the oil samples are centrifuged by a centrifuge machine at 8000 rpm for 30 minutes and next is filtrated by syringe filters made of regenerated cellulose.

The procedure of water in oil samples preparation consists of three steps: 1- An aqueous solution with

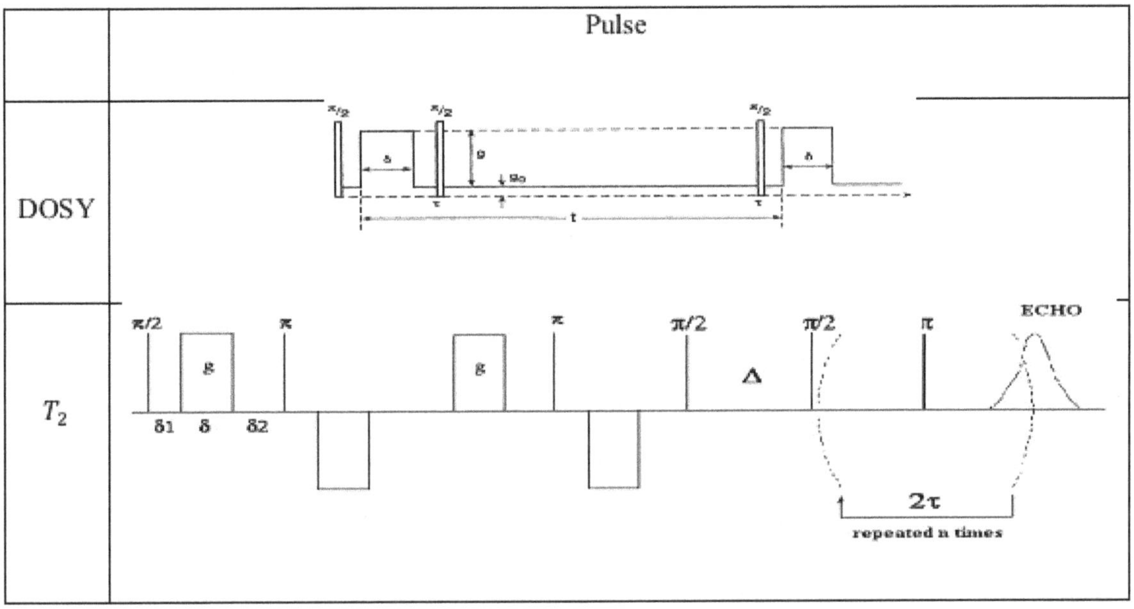

Fig.1. The pulse sequence used for the DOSY and T_2 measurements.

Table1.
API degree, density and kinematic viscosity (all at 25 °C) of the crude oil and the diesel samples used in this study.

Sample	Degree API	Density (g/ml)	kinematic viscosity (mm^2/s)
Crude oil	28.22	0.886	53.4
Diesel	61.28	0.734	3.48

consists of three steps: 1- An aqueous solution with a desired dissolved NaCl in distillated water is prepared. 2- A mixture of 50% crude oil and 50% diesel sample is mixed with the aqueous solution 3- A known volume of a sample demulsifier is added to the sample. In all experiments a mechanical stirrer is applied during mixing processes, for five and one minutes at steps 2 and 3, respectively.

3.2.2. NMR measurements

A 400MHz FT NMR product of Bruker Company is applied in this study. The water in oil emulsion samples are used in the NMR experiments immediately after preparation.The pulse sequence related to DOSY and T2 measurements are presented in Fig. 1. A 5 ml of the prepared sample is charged inside the NMR tube and the sample placed inside the machine for performing the spectroscopy measurements. A period about 5 minutes is needed to reach a thermal equilibrium inside the sample and the spectroscopy on the samples is started. The spectrums of the sample emulsion are shown in Fig. 2 at different times (τ). A typical T_2 distribution obtained in the whole frequency range of the NMR spectra is shown in Fig. 3. In this figure the vertical axis represents the ratio of the numbers of T_2 with a same size to the total counted T_2 and the horizontal axis is chosen somehow that the T_2 distribution has normal shape.

3.3. Experimental Design

The experimental design and the analysis of experiments are adopted through Taguchi method [13] and the average size of water droplets is considered as the experiment response after a long time to achieve the final average size. The experiments are designed using L18 standard orthogonal array. The studied factors and their levels are presented in Table 2. The experiments are conducted as per the experimental layout given in Table 2.

4. Results and discussion

The effects of type and concentration of demulsifier, the water volume ratio, the water salinity, and the mixing speed on the average water droplets size are investigated.

4.1. Analysis of Variance (ANOVA)

The analysis of variance (ANOVA) is used to recognize the significant factors among all the process parameters which are affecting the output quality characteristics using the quantities such as degrees

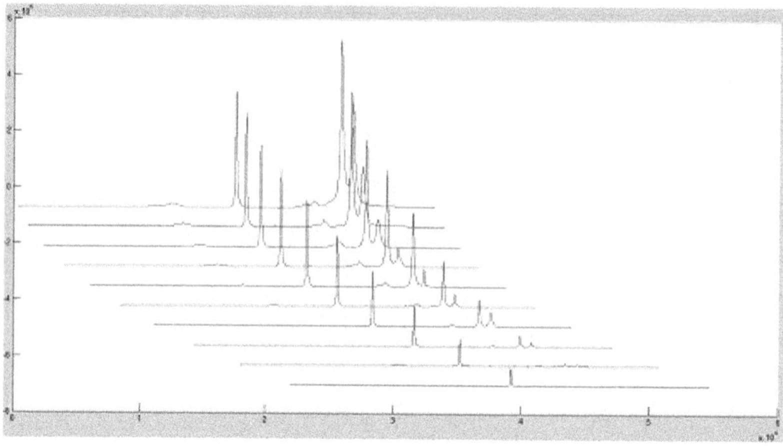

Fig. 2. Typical NMR spectra at different times.

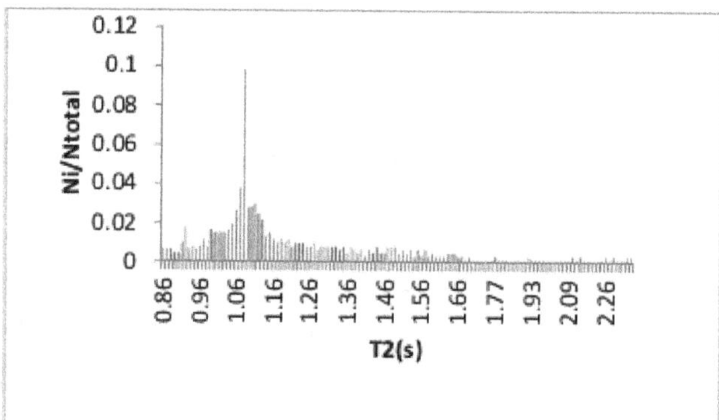

Fig. 3. T_2 distribution related to water/oil emulsion at 30% water/oil volume ratio.

Table 2.
NMR Experimental results with design tests in Taguchi method at different levels.

Number of experiment	Type of demulsifier (TD)	Demulsifier concentration (CD) (ppm)	Water/Oil volume ratio (W) (%)	Mixing speed (N) (rpm)	Salinity (S) (g/lit)	Average water droplet size (micrometer)
1	1	10	10	800	25	155
2	1	10	20	1000	35	180
3	1	10	30	1200	45	214
4	1	50	20	800	25	229
5	1	50	30	1000	35	259
6	1	50	10	1200	45	193
7	1	100	10	800	35	260
8	1	100	20	1000	45	281
9	1	100	30	1200	25	239
10	2	10	30	800	45	170
11	2	10	10	1000	25	98
12	2	10	20	1200	35	109
13	2	50	30	800	35	203
14	2	50	10	1000	45	155
15	2	50	20	1200	25	137
16	2	100	20	800	45	229
17	2	100	30	1000	25	196
18	2	100	10	1200	35	144

of freedom, sum of squares, variance, F-ratio and percent contribution [13]. Table 3 shows the computed results of the ANOVA with 95% confidence level which meanstheresponse value witha p-value lesser than 0.05 is acceptable and indicates that the factor has a significant effect on the average droplets size. The F-ratio and the percent contributions of the various parameters as quantified under the respective columns of Table 3 reveal that type and concentration of demulsifier, water/oil volume ratio, water salinity and mixing speed have significant effects on the average droplets size. The type and concentration of demulsifier factors contribute 38% and 31.5% on the total value, respectively while mixing speed and salinity contribute 5.71% and 7.5%, respectively and the water/oil volume ratio has contributed 13.1% on the total value.The interactions between type of demulsifier and concentration of demulsifier have no significant effect on the response in the studied range of levels.

4.2. Effect of type of demulsifier on the average droplets size

Among all the five factors, the most significant factor is the type of demulsifier, and this factor contributes to 38% on the average droplets size. The counter line diagram of the type of demulsifier on the average droplets size is shown in Fig. 4. The demulsifier has greater influence on the resistant layers around droplets; these layers are destroyed during droplets collisions favoring faster droplets coagulation. Since the demulsifier molecules migrate to the layers between droplets and oil media and adsorb on the interfacial surfaces, the layers are destructed by reaction or dissolution mechanisms within the aqueous or oil phases; hence, a rapid coagulation of droplets is resulted. Due to the lack of information about the formula and the other properties of the commercial demulsifier it is not possible to compare the demulsifier performances through their molecular structures.

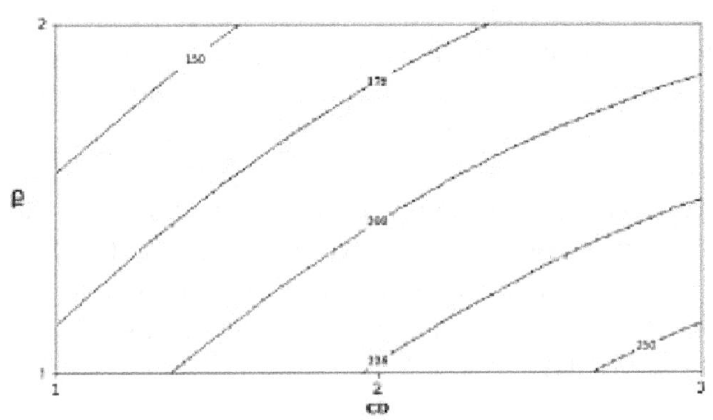

Fig. 4. Contour diagram of average water droplet size in terms of demulsifier types and concentration of demulsifier.

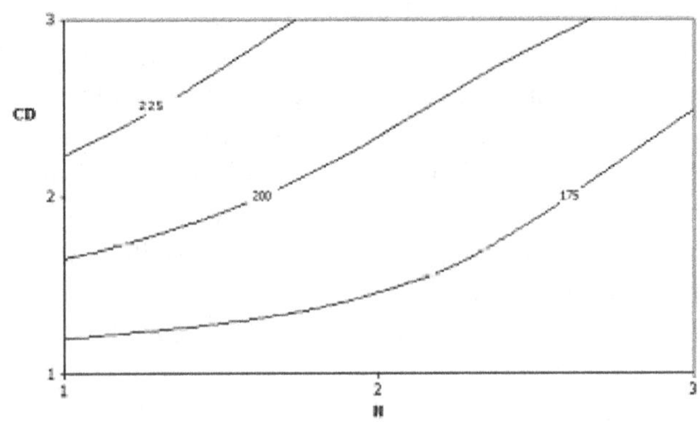

Fig. 5. Contour diagram of average water droplet size in terms of the concentration of demulsifier and mixing speed.

Table 3.
Variance estimation of average water droplet size as response

Factor	Degrees of freedom	Mean squared error	Sum of squared errors corrected	F ratio	P value	Contribution percent
TD	1	17986.7	17986.7	151.01	0	38
CD	2	15075.4	7537.7	63.28	0	31.5
W	2	3762.1	1881.1	15.79	0.004	13.1
N	2	2950.8	1457.4	12.39	0.007	5.71
S	2	6401.8	3200.9	26.87	0.001	7.5
TD*CD	2	130.1	65.1	0.55	0.605	-
Residual error	6	714.7	119.1	-	-	4.19
Total	17	47021.6	-	-	-	100

4.3. Effect of demulsifier concentration

A key factor affecting the average droplets size is the demulsifier concentration. The significant effect of this factor on the output response is shown in Fig. 5. With increasing the demulsifier concentration, the demulsifier molecules numbers between the interface of water droplets increase and the protective layers around the droplets weaken rapidly and break. Therefore, the coagulation rate of droplets increases and larger droplets are formed; hence, the average diameter of droplets increases.

4.4. Effect of salinity on the average droplets size

The water salinity effect on the average water droplet size is evaluated and as shown in Fig. 6, the average droplets size increases with an increase of the water salinity. As water density increases with increasing salinity, the differential density between water droplets and the continuous phase oil is increased; therefore, assists the separation of the oil and water phases and enhances the water droplets coagulation process. Moreover, the existence of small amount of salt or other dissolved solids in the water reduces drastically surface tension which is affective on the droplets coagulation. From another point of view, an increase in the salinity yields to increase the concentration of ions with opposite charges on the surface of water droplets; hence, the faster coagulation of droplets is achieved.

4.5. Effect of water/oil volume ratio on the average droplets size

The result of water/oil volume ratio effect on the average droplets size is shown in Fig. 7. It can be observed from the figure that changes in this factor value will cause a significant effect on the output response.This effect is attributed to an increase in the droplets numbers per unit volume at higher volume ratio of water; consequently, increases the collision probability among the droplets and the larger droplets are formed.

4.6. The effect of mixing speed on the average droplets size

The effect of mixing speed on the average droplets size is shown in Fig. 8. With increasing mixing speed the average droplets size decreases. An increase in the mixing speed causes the droplets become smaller due to the applied shear and both the coagulation induced by Brownian motion and sedimentation are reduced.

4.7. Optimum conditions

One of the purposes of statistical analysis and experimental design is to determine the optimum operation conditions for the process. Here, the purpose is to achieve a larger average size of the droplets. The optimum conditions for the average droplets size according to the mean effect diagrams are illustrated in Fig. 9. The appropriate demulsifier is Break 6754. The optimum conditions of demulsifier concentration, water volume ratio, mixing speed, and salinity are 100 ppm, 30%, 800 rpm and 45gr/liter, respectively.

5. Conclusions

In this study the average water droplet size in

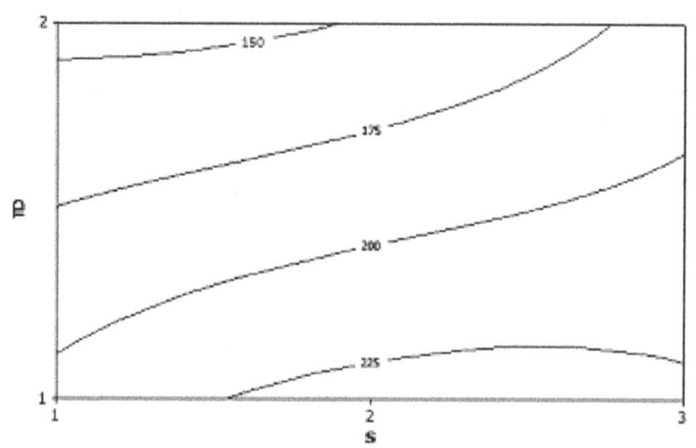

Fig. 6. Counter diagram of average water droplet size change in terms of the demulsifier types and water salinity.

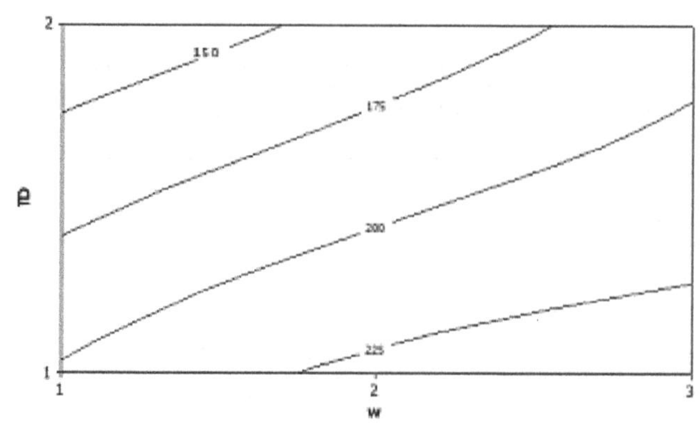

Fig. 7. Counter diagram of average water droplet size in terms of demulsifier types and water/oil volume ratio.

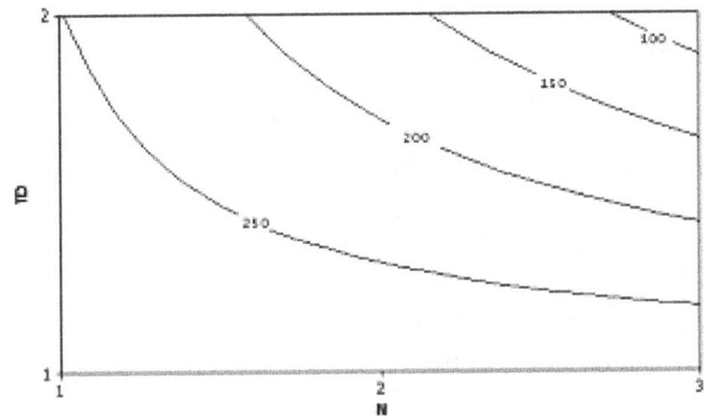

Fig. 8. Counter diagram of average water droplet size in terms of demulsifier types and mixing speed.

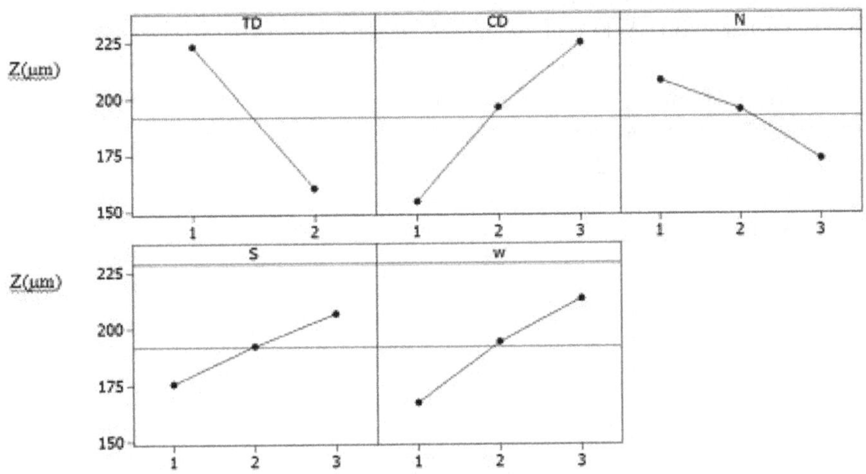

Fig. 9. The main effects diagrams

water-oil emulsions are measured by applying NMR technique in order to evaluate the dependency of the emulsions stability in terms of the operating factors. The effect of type and concentration of demulsifier, water volume ratio, water salinity and mixing speedon the average water droplet size are studied. The demulsifier Break 6754 in compare to acrylic acid has a more impact on the average droplets size and leads to larger average droplets size. The average water droplet size increases with increasing the concentration of demulsifier. Increasing the water/oil volume ratio and water salinity increase the average droplets size, while the mixing speed has an opposite effect.

References

[1] I. B. Ivanov, P. A. Kralchevsky, Stability of emulsions under equilibrium and dynamic conditions, Colloids Surf. A: Physicochem. Eng. Asp., 128 (1997) 155-175.

[2] M. Fingas, B. Fieldhouse, M. Bobra, E. Tennyson, The physics and chemistry of emulsions. Environment Canada and Consult chem, Ottawa, Canada and US Minerals Management Service, Herndon, Virginia, 1993.

[3] D. B. Curtis, M. Aycibin, M. A. Young, V. H. Grassian, P. D. Kleiber, Simultaneous measurement of light-scattering properties and particle size distribution for aerosols: Application to ammonium sulfate and quartz aerosol particles, Atmos. Environ. 41 (2007) 4748-4758.

[4] S. Chodankar, V. K. Aswal, P. A. Hassan, A. G. Wagh, Structure of protein–surfactant complexes as studied by small-angle neutron scattering and dynamic light scattering, Physica B: Condens. Matter. 398 (2007) 112-117.

[5] D. S. Parker, W.J. Kaufman, D. Jenkins, Floc breakup in turbulent flocculation processes, J. Sanitary Eng. Div. 98 (1972) 79-99.

[6] P. S. Denkova, S. Tcholakova, N. D. Denkov, K. D. Danov, B. Campbell, C. Shawl, D. Kim, Evaluation of the precision of drop-size determination in oil/water emulsions by low-resolution NMR spectroscopy, Langmuir, 20 (2004) 11402-11413.

[7] C. P. Aichele, M. Flaum, T. Jiang, G. J. Hirasaki, W. G. Chapman, Water in oil emulsion droplet size characterization using a pulsed field gradient with diffusion editing (PFG-DE) NMR technique, J. Colloid interface Sci. 315 (2007) 607-619.

[8] K.G. Hollingsworth, A.J. Sederman, C. Buckley, L.F. Gladden, M.L. Johns, Fast emulsion droplet sizing using NMR self-diffusion measurements, J. Colloid Interface Sci. 274 (2004), 244-250.
[9] N. van der Tuuk Opedal, G. Sørland, J. Sjöblom, Methods for droplet size distribution determination of water-in-oil emulsions using low-field NMR, Diffusion Fundamentals 7 (2009) 1-29.
[10] P.P. Mitra, P.N. Sen, L.M. Schwartz, Short-time behavior of the diffusion coefficient as a geometrical probe of porous media, Phys. Rev. B, 47 (1993) 8565.
[11] M.D. Hurlimann, K.G. Helmer, L.L. Latour, C.H. Sotak, Restricted diffusion in sedimentary rocks. Determination of surface-area-to-volume ratio and surface relativity, J. Magn. Reson. A 111 (1994) 169-178.
[12] K.S. Mendelson, M.H. Cohen, The effect of grain anisotropy on the electrical properties of sedimentary rocks, Geophys. 47 (1982) 257-263.
[13] R. Roy, Design of experiments using the Taguchi approach, New York: Wiley, 2001.

16

Grinding-aid effect on the colour properties (Ry, whiteness and yellowness) of calcite in stirred media milling

Oner Yusuf Toraman

Mining Engineering Department, Faculty of Engineering, Omer Halisdemir University, 51240 Nigde, Turkey

HIGHLIGHTS

- Dry fine grinding of calcite was performed in a batch type stirred media mill.
- The effects of grinding aids on fineness and color properties were investigated.
- Ethanol and methanol are noticeably effective with regard to colour properties.

GRAPHICAL ABSTRACT

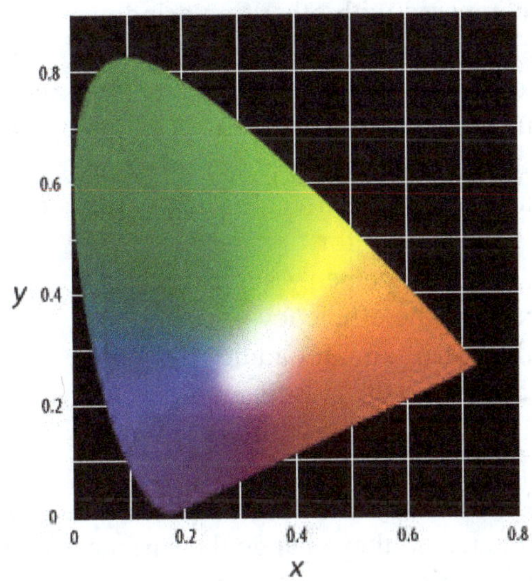

ARTICLE INFO

Keywords:
Stirred media mill
Fine grinding
Calcite
Grinding aid
Colour properties

ABSTRACT

This study investigates the influence of some chemical additives such as methanol, ethanol, sodium oleat, chloroform and sodium hexametaphosphate (SHMP) on the dry fine grinding of calcite (X_{50}= 33 μm) using a stirred media mill. The experiments were carried out by a batch operation, and the change in colour properties (Ry, whiteness and yellowness) of calcite powder. The results showed that the chemical additives promote the fine grinding of calcite obtained with ethanol and methanol at a range of 0.5%. Ry and whiteness values of the ground calcite products very slightly increased from 94.10 and 87.04 to 94.76 and 87.75 respectively with grinding aid (ethanol) increased from 0% to 1%. Ry value was affected slightly adversely with sodium oleat, chloroform and hexametaphosphate indicating that the quality of colour of calcite deteriorates. It was also found that ΔRy increases with increasing amount of grinding aids from 0% to 0.5% for methanol and ethanol, indicating that the quality of colour of calcite heals.

* *Corresponding author.*
E-mail address: otoraman@ohu.edu.tr

1. Introduction

It is known that grinding is an important industrial operation that is used for the size reduction of materials, production of large surface area and/or liberation of valuable minerals from their matrices but it is also one of unit operations with the lowest energy efficiency [1]. The need for fine/ultrafine particles has increased in the field of preparing raw powders and high value added products in many industries such as mineral, ceramic, pigments, paint and pharmaceutical. On the other hand, stirred bead mills have been used in recent years for grinding particles to micron and submicron sizes due to their easy operation, simpler construction, higher grinding rate and lower energy consumption compared with other fine grinding machines [2]. It is also known that particle aggregation/agglomeration causes poor flowability of dry material to be ground in a mill. Moreover, grinding media and liner coating result in a poor dry grinding efficiency due to the cushioning effect [3]. Grinding aid has been used successfully for decades in many industries such as mineral, cement, ceramics, pigments etc. It can improve the efficiency of the grinding remarkably with a small amount addition should more positively be applied to the grinding operations, especially to dry ultrafine grinding with higher energy consumption [1]. Moreover, it can reduce the surface free energy of powder and prevent fine particles from closing each other. And they change the surface charge distribution by shielding or neutralizing the particles surface partial charge and prevent fracture surface from healing and promoting the cracks to extend easily [4]. When it is used a grinding aid, it must be selected an appropriate one that has no detrimental effect on downstream processing or the final product [5]. If it is actually used a grinding aid at the present technical level, it must be empirically determined the variety and quantity of the grinding aid based on experimental data [6]. In most of the studies on grinding aids, the effect of moisture [7] and grinding aids [8-12] have been discussed to get the fine powders in wet grinding method for calcite/limestone, but there are only a few reports in dry conditions for calcite/limestone [5,13]. In these studies, product analyses consist of surface area, fineness, particle size distribution, crystalline structure and specific energy consumption. Neverthless, effects of grinding aids on colour properties of ground products were not found in the literature.

The main purpose of this study was to systematically investigate the effects on colour parameters (Ry, whiteness and yellowness) of some chemical additives such as methanol, ethanol, sodium oleat, chloroform and sodium hexametaphosphate on dry fine grinding of calcite powder using a laboratory scale stirred media mill.

2.Materials and Methods

2.1.Materials

Powder sample used in this study was calcite ($CaCO_3$) (X_{50}=33 μm) from Micron'S Co. (Nigde, Turkey) and its density was 2700 kg/m^3. Chemical properties of sample are shown in Table 1. The grinding media selected for the tests was 3.5-4.0 mm alumina (Al_2O_3) beads. Four kinds of additives were used as grinding aids, as shown in Table 2. These additives were special grade reagents (Merck and Sigma-Aldrich Corporation, St. Louis, MO, USA) and used without further purification. Summary of experimental conditions is also shown in Table 3.

2.2.Methods

Grinding tests were carried out in a vertical type stirred media mill Standard-01 Model manufactured by Union Process (U.S.A.) which was reported in our previous study [14]. Sympatec HELOS (H0983) laser diffraction analyser (Sympatec GmbH, Clausthal-Zellerfeld, Germany) was used for the analysis of the feed and the ground products. Each test was repeated three times and the values reported are a mean average. The colour parameters (Ry, whiteness and yellowness) of ground products were measured using a Datacolour Elrepho SF450X spectrophotometer in the study.

CIE considered the tristimulus values for red, green, and blue to be undesirable for creating a standardized colour model. Instead, they used a mathematical formula to convert the RGB data to a system that uses only positive integers as values. The reformulated tristimulus values were indicated as XYZ (Fig.1). These values do not directly correspond to red, green, and blue, but are approximately so. The curve for the Y tristimulus value is equal to the curve that indicates the human eye's response to the total power of a light source. For this reason the value Y is called the luminance factor and the XYZ values have been normalized so that Y always has a value of 100.

Tristimulus value Y (*ISO 5631- Y-value C/2°*) in the CIEXYZ-system of a layer of material of such a

Table 1. Chemical composition of the calcite sample (wt%)

$CaCO_3$	$MgCO_3$	Fe_2O_3	SiO_2	Al_2O_3	Total
99.5	0.2	0.01	0.01	0.02	99.74

Table 2. Physical properties and addition amount of grinding aids used

Additives	Density (g/cm³)	Chemical formula	Molecular weight (g/mol)	Additive dosage (wt%)
Methanol	0,792	CH_3OH	32,0	0.25, 0.5, 1, 2
Ethanol	0,789	C_2H_5OH	46,1	0.25, 0.5, 1, 2
Sodium oleate	0,900	$C_{18}H_{33}NaO_2$	304,4	0.25, 0.5, 1, 2
Cloroform	1,49	$CHCl_3$	119,4	0.25, 0.5, 1, 2
Sodium hexametaphosphate (SHMP)	2,48	$Na_6P_6O_{18}$	611,7	0.25, 0.5, 1, 2

Table 3. Summary of experimental conditions

Item	Experimental conditions
Bead filling ratio	0.70
Sample quantity	40 g
Sample filling ratio	0.05
Rotation speed of stirrer	600 rpm
Grinding time	10 min
Internal volume of grinding pot	750 ml
Temperature	Room temperature
Material of grinding media	Alumina
Grinding media size	3.5-4 mm

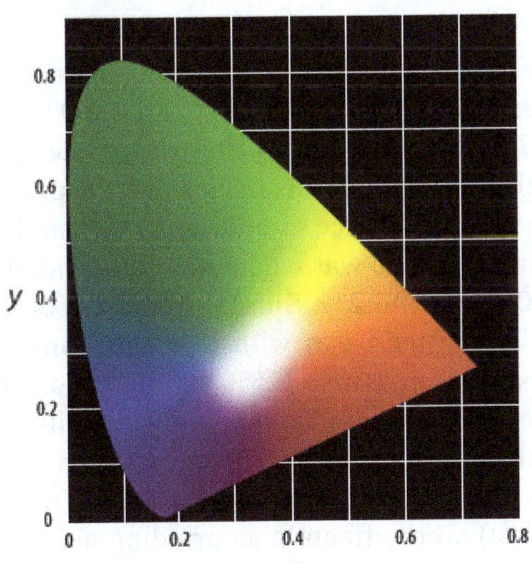

Fig. 1. The tristumulus values XYZ [15]

thickness that there is no change in Y when the thickness is increased. The illumination is here CIE illuminant C [16].

3. Results and Discussion

3.1. Effect of additive on the fineness (X_{50}) of calcite

The effect of grinding aid on fine grinding performance has been explained mainly by two kinds of mechanism. One is based upon the alteration of surface and mechanical properties of individual particles, such as reduction of surface energy and modification of surface hardness, and the other is the change in arrangement of particles and their flow in suspensions [6]. Table 4 shows the summary of experimental results on product size (X_{50}, X_{97}). The result with (0.25-2%) and without (0%) an additive are shown in this table. The use of grinding aid in the mill indicated enough beneficial effect on product particle size. The observed beneficial effect is due to the adsorption of grinding aid on fine calcite particles by influencing the mass transport. It can be seen from Table 5, the median diameters (X_{50}) of ground products are decreased with the increase of additive dosage from 0.25% to 1% for methanol, ethanol and sodium oleat. But, all X_{50} values are increased with chloroform and sodium hexametaphosphate. Especially, the obvious function of methanol and ethanol was both of increasing the content of fine powder and decreasing of average particle size (X_{50}). The results indicate that the best addition of grinding aids were 0.5% of ethanol and methanol, and both of them decrease the average particle size from 33.02 to 2.71 and 2.79 μm (Fig 2).

3.2. Effect of additive on the colour properties of calcite

Table 6 shows the summary of results on colour properties of product. Notably, the Ry and whiteness values of the ground calcite products very slightly increased from 94.10 and 87.04 (Run2) to 94.76 and 87.75 (Run9) respectively as grinding aids (ethanol) increased from 0% to 1%. This means that these two additives have even partially positive effect to the colour properties of calcite. Ry is affected slightly adversely with sodium oleat, chloroform and sodium hexametaphosphate indicating that the quality of

Table 4. Summary of experimental conditions

Grinding aid	Run no	Additive dosage, wt% of sample
Feed material	1	
No grinding-aid	2	0
Methanol	3	0.25
	4	0.5
	5	1
	6	2
Ethanol	7	0.25
	8	0.5
	9	1
	10	2
Sodium oleat	11	0.25
	12	0.5
	13	1
	14	2
Chloroform	15	0.25
	16	0.5
	17	1
	18	2
Sodium hexametaphosphate	19	0.25
	20	0.5
	21	1
	22	2

Table 5. Particle size of the feed sample and products

Run no	X_{50}/X_{97} (μm)
1	33.02/139.11
2	3.50/42.67
3	3.10/30.49
4	2.71/28.62
5	3.22/25.93
6	6.82/104.36
7	3.09/29.76
8	2.79/24.12
9	3.30/28.26
10	5.68/122.54
11	4.27/25.20
12	4.78/30.38
13	3.91/25.97
14	4.18/43.41
15	3.57/70.87
16	3.92/70.68
17	4.04/70.52
18	3.63/70.97
19	3.30/57.31
20	3.62/69.82
21	4.04/70.41
22	3.27/68.64

colour of calcite deteriorates. The maximum value of Ry was measured in Run8 (94.62) and Run9 (94.76).

Table 7 shows the colour difference values (ΔRy). If this value is positive, this means that the product is lighter than the reference sample or vice versa. The total colour difference was calculated for grinding aids of 0.25, 0.50, 1, and 2%. Similar trends are observed for ΔRy, which increases with increasing amount of grinding aids from 0% to 0.5% for methanol (Run4) and ethanol (Run8), indicating that the quality of colour of calcite heals (Fig 3). The results with alcohols additives indicated that both two alcohols (methanol and ethanol) were effective as grinding aid and that the ΔRy increased positively with an increase in the addition amount of alcohol from 0% to 0.5% at the same grinding time. This is because it has also been obtained best fineness for this two alcohol as mentioned in section 3.1. It is obvious that the fineness of calcite is increased with grinding and the whiteness is also increased from 83.09 to 87.04 without grinding aid. This whiteness value is reaching up to 87.87 for 0.5% ethanol aid (Table 6). Consequently, ΔRy increases with increasing amount of grinding aids from 0% to 0.5% for methanol and ethanol, indicating that the quality of colour of calcite heals.

Table 6. Colour properties (Ry, whiteness and yellowness) of the feed sample and products

Run no	Ry	Whiteness	Yellowness
1	86.88	83.09	6.08
2	94.10	87.04	3.02
3	93.75	86.70	3.00
4	94.53	87.71	2.91
5	94.03	87.25	2.90
6	88.70	86.73	3.42
7	92.67	85.52	3.06
8	94.62	87.87	2.89
9	94.76	87.75	3.01
10	90.79	81.02	4.24
11	93.78	86.78	3.02
12	93.36	86.39	2.99
13	93.48	85.37	3.47
14	92.49	83.59	3.79
15	92.52	83.14	4.05
16	91.82	82.14	4.18
17	92.36	82.98	4.05
18	92.03	82.48	4.14
19	93.60	86.19	3.28
20	92.35	82.57	4.24
21	92.37	82.96	4.08
22	92.41	83.04	4.05

Table 7. ISO-5631 Ry-value of the products after milling with grinding aid

Run no	Ry	ΔRy Before and after milling with grinding aid (%)
3	93.75	-0.35
4	94.53	0.43
5	94.03	-0.07
6	88.70	-5.40
7	92.67	-1.43
8	94.62	0.52
9	94.76	0.66
10	90.79	-3.31
11	93.78	-0.32
12	93.36	-0.74
13	93.48	-0.62
14	92.49	-1.61
15	92.52	-1.58
16	91.82	-2.28
17	92.36	-1.74
18	92.03	-2.07
19	93.60	-0.5
20	92.35	-1.75
21	92.37	-1.73
22	92.41	-1.69

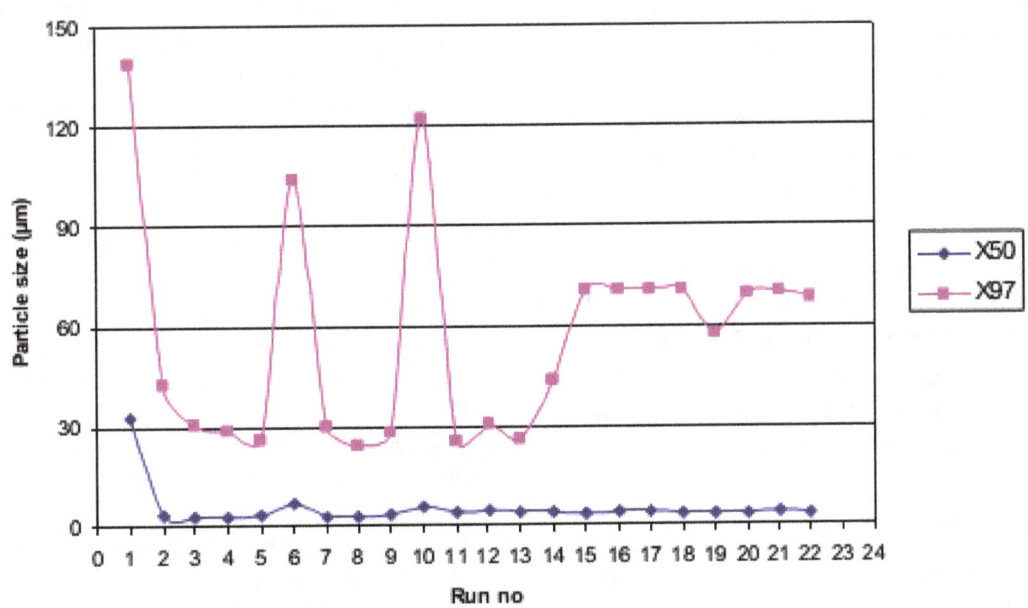

Fig. 2. Particle size (X_{50} and X_{97}) of the feed sample and products

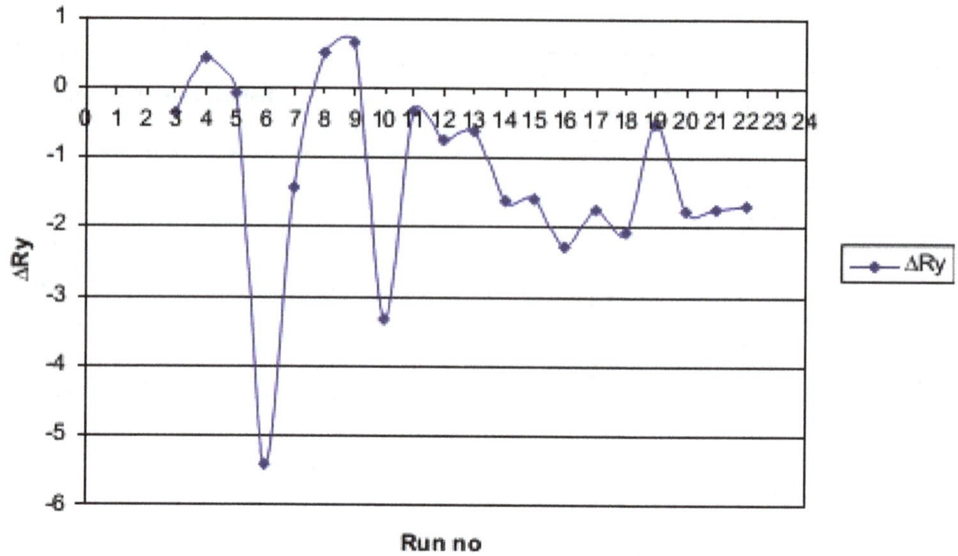

Fig. 3. ΔRy-value of the products before and after milling with grinding aid

4.Conclusions

The effects of grinding aid on the fineness (X_{50}) and some colour properties such as Ry, whiteness and yellowness were examined. The followings were found out:

1. The median diameter (X_{50}) at the methanol and ethanol addition amount of 0.5% was about 0.3 and 0.25 times as small as the particle size without an additive, and about 12.5 and 13 times smaller than that before grinding, respectively.

2. Ry and whiteness values of the ground calcite products very slightly increased from 94.10 and 87.04 to 94.76 and 87.75 respectively as grinding aids (ethanol) increased from 0% to 1%.

3. Ry value is affected slightly adversely with sodium oleat, chloroform and sodium hexametaphosphate indicating that the quality of colour of calcite deteriorates.

4. ΔRy increases with increasing amount of grinding aids from 0% to 0.5% for methanol and ethanol, indicating that the quality of colour of calcite heals.

5. As a result, ethanol and methanol used as grinding aid are noticeably effective as grinding aids with regard to colour properties of calcite.

Acknowledgement

The author wish to thank Micron'S Company (Nigde, Turkey) for supporting this work.

References

[1] M. Hasegawa, M. Kimata, M. Yaguchi, Effect of behavior of liquid additive molecules in dry ultrafine grinding of limestone, KONA Powder Part. J. 24 (2006) 213-221.

[2] H. Choi, W. Lee, S. Kim, Effect of grinding aids on the kinetics of fine grinding energy consumed of calcite powders by a stirred ball mill, Adv. Powder Technol. 20 (2009) 350-354.

[3] Y. Wang and E. Forssberg, Dispersant in stirred ball mill grinding, KONA Powder Parti. J. 13 (1995) 67-77.

[4] J. Zhao, D. Wang, X. Wang, S. Liao, H. Lin, Effect of grinding aids on the particles characteristics of cement and analysis of action mechanism, Adv. Mater. Res. 936 (2014) 1404-1408.

[5] M. Hasegawa, M. Kimata, M. Shimane, T. Shoji, M. Tsuruta, The effect of liquid additives on dry ultrafine grinding of quartz, Powder Technol. 114 (2001) 145-151.

[6] D. W. Fuerstenau, Grinding aids, KONA Powder Part. J. 13 (1995) 5-18.

[7] A. H. Shinohara, K. Sugiyama, E. F. Kasai Saito, Y. Waseda, Effects of moisture on grinding of natural calcite by a tumbling ball mill, Adv. Powder Technol. 4(4) (1993) 311-319.

[8] J. Zheng, P. Harris, P. Somasundaran, The effect of additive on stirred media milling of limestone, Powder Technol. 91 (1997) 173-179.

[9] R. R. Klimpel, The selection of wet grinding chemical additives based on slurry rheology control, Powder Technol. 105 (1999) 430-435.

[10] R. Greenwood, N. Rowson, S. Kingman, G. Brown, A new method for determining the optimum dispersant concentration in aqueous grinding, Powder Technol. 123 (2012) 199-207.
[11] H. K. Choi, W. S. Choi, Ultra fine grinding mechanism of inorganic powders in a stirred ball mill-The effect of grinding aids-, Korean J. Chem. Eng. 20(3) (2003) 554-559.
[12] H. Choi, W. Lee, D. U. Kim, S. Kumar, S. S. Kim, H. S. Chung, J. H. Kim, Y. C. Ahn, Effect of grinding aids on the grinding energy consumed during grinding of calcite in a stirred ball mill, Miner. Eng. 23 (2010) 54-57.
[13] O. Y. Toraman, Effect of chemical additive on stirred bead milling of calcite powder, Powder Technol. 221 (2012) 189-191.
[14] O. Y. Toraman, D. Katırcıoglu, A Study on the Effect of Process Parameters in Stirred Ball Mill, Advanced Powder Technol. 22(1) (2011) 26-30.
[15] Technical guides colour models CIEXYZ (http://dba.med.sc.edu/price/irf/Adobe_tg/models/ciexyz.html
[16] Scandinavian Pulp, Paper and Board Testing Committee (SCAN-P 89:03), Preparation of tablets for the measurement of ISO brightness, Y-value and colour (C/2°), 4 p. SE-114 86 Stockholm, Sweden, 2003.

Hollow alumina nanospheres as novel catalyst for the conversion of methanol to dimethyl ether

N. Rostamizadeh[1], M. S. Sadjadi[1,*], S. A. A. Sadjadi[2]

[1] *Department of Chemistry, Science and Research Branch, Islamic Azad University, Tehran, I. R. Iran*

[2] *Institute of Water and Energy, Sharif University of Technology, P.O. Box 11365-8639, Tehran, I. R. Iran*

HIGHLIGHTS

- Hollow and porous alumina nanospheres were prepared by integrating techniques of microemulsion and templating.
- The prepared samples were used as catalyst for conversion of methanol to DME.
- Effect of acidity, temperature and WHSV on performance of the catalysts wasinvestigated.

GRAPHICAL ABSTRACT

Dehydration of methanol to dimethyl ether over **hollow alumina nanospheres**

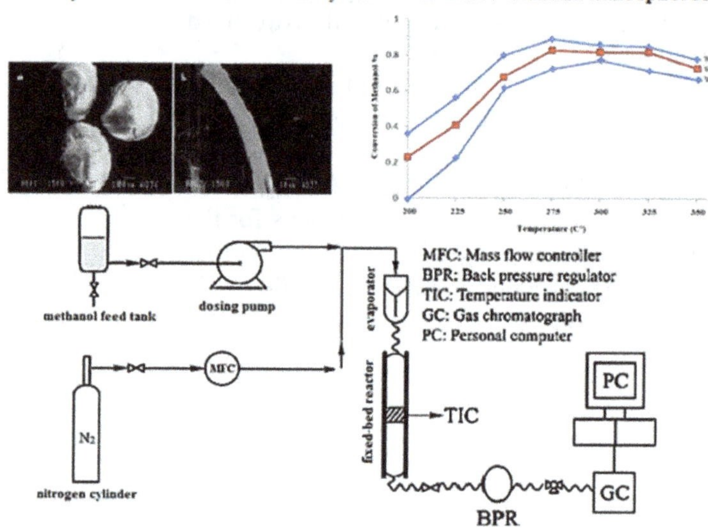

ARTICLE INFO

Keywords:
Methanol
Dehydration
Dimethyl ether
Hollow
Alumina nanospheres

ABSTRACT

This paper investigates hollow and porous alumina nanospheres that were previously synthesized to be used for the dehydration of methanol to dimethyl ether (DME). As hollow nanostructures possess characteristics such as low density and high surface to volume ratio, their catalytic activity between hollow and porous structure is compared. For this purpose, three most important parameters (acidity, temperature and weight hourly space velocity (WHSV)) affecting the performance of these catalysts were investigated. The catalysts were characterized by scanning electron microscopy (SEM), BET, X-ray diffraction (XRD), and the temperature programmed desorption of ammonia (NH_3-TPD) techniques. Results show that the optimum operating condition for hollow alumina nanosphere can be achieved at temperature of 275 °C and WHSV of 20 h^{-1} compared with operating condition for porous alumina at temperature of 325 °C and WHSV of 20 h^{-1}.

* *Corresponding author.*
E-mail address:m.s.sadjad@gmail.com

1. Introduction

Application of dimethyl ether (DME) as a fuel and fuel additive for use in diesel engines has been investigated due to its high volatility, high cetane number, lower NOx and smoke emissions and lower engine noise. In addition to alternative fuel, DME can also be used as feed stock to many products such as short olefins (ethylene and propylene), gasoline, hydrogen, acetic acid and dimethyl sulphate [1-4]. It can also be used a substitute for Liquefied Petroleum Gas (LPG) because of having similar properties. Its major technical advantages coupled with its environmentally-friendly nature of the DME have attracted the attention of various planners to design and construct the DME plants [5-9]. DME can either be produced by direct synthesis from syngas ($CO/H_2/CO_2$) in a single step [10-14], or indirect synthesis by conventional bimolecular catalytic dehydration of methanol over various solid acids [15–18]. As conditions of the two consequent reactions are similar, these can be conducted simultaneously in the same reactor over a bifunctional catalyst. The direct method, known as syngas to dimethyl ether (STD) process is a single-step vapour phase having three reactions. The combination of reactions affects the methanol production equilibrium. One of the disadvantage of this process is that high operating conditions cause more by-product synthesis, which in return, requires more complex distillation in order to separate the reactor effluent to achieve high purity DME [19-20]. For these fundamental reasons, more focus has recently been made on methanol dehydration process and dimethyl ether purification.

Dehydration of methanol to DME can be achieved by employing solid-acid catalysts such as γ-alumina and zeolite. The catalytic activity of solid-acid for methanol dehydration is linked to Lewis acid-Lewis base pair formed during calcination [7-11]. A mechanism involving an acid-base pair in methanol dehydration is proposed by Padmanabhan and Eastburn (1972):

Among solid acid catalysts, γ-alumina is commercially preferred due principally to its fine particle size, high surface area, excellent thermal stability, high mechanical resistance and low cost catalyst to mention a few examples [4-6].

Much of the research work in the last decade has been devoted for preparation dense nanoparticles. Hollow nanomaterials have become the focus of investigations because of their great application potentials such as the confined nanoreactors in drug delivery and catalytic purpose [21-23]. Among these category materials, hollow γ-Al_2O_3 nanosphere has been favoured because of its very large specific surface area and low density makes it appropriate for catalytic usage [24-25].

Two samples of alumina nanosphere are used in this study as the catalysts for the dehydration of methanol to DME. The principal effects of preparation method on the morphology, acidic properties and catalytic activity of γ-alumina samples have been investigated.

2. Experimental

2.1. Catalyst preparation and characterization

Two catalyst samples, hollow and porous alumina nanospheres, were prepared by integrating two techniques of microemulsion and templating process. These were based on the procedure described in our previous work [26]. In line with this, the alginate hydrogels was prepared by microemulsion method, and was subsequently used as templates for the formation of core-shell structures of aluminum-alginate. Hollow and porous alumina nanospheres were then obtained in the presence and absence of ammonia referred to here as Al-B and Al-A catalysts respectively.

The powder XRD pattern was recorded on PW1800 Philips Diffractometer using Cu Kα radiation (λ=1.5418Å) to characterize crystalline structure of the prepared samples.

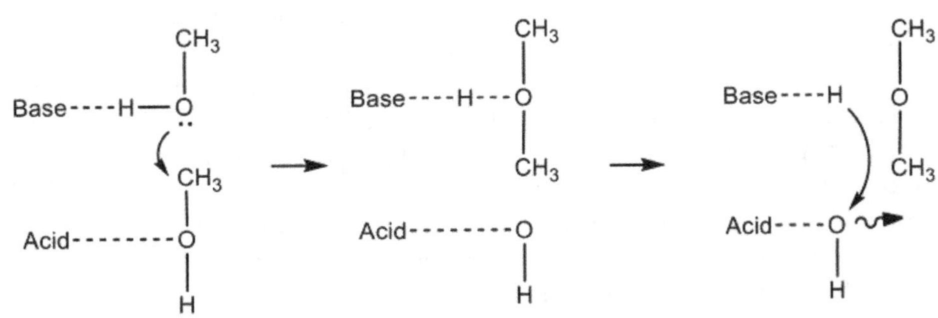

A scanning electron microscope (S-360) was used to show hollow structure, shell thickness and finally surface morphology. The BET technique measures surface area, pore size and pore volume of the products, using Quanta chrome instrument (Model Nova 2000). The analysis of temperature programmed desorption of ammonia (NH_3-TPD) was performed in order to determine the acidic sites on the catalysts using BEL- CAT (type A, Japan) instrument. The methodology involved initially flushing about 0.1 g of the sample with a He under the flow rate of 50 ml/min for 2h at 500 ° C. It was then cooled to 100 ° C and subsequently saturated with NH_3. After exposing NH_3, the sample was purged with He until the initial excess of NH_3 is removed. The sample was then heated under the flow of the carrier gas from 100 °C to 500 °C and the amount of ammonia in effluent was measured via thermal conductivity detector (TCD).

2.2. Catalytic performance tests

The schematic diagram of the used set up is shown in Fig. 1. The gas phase dehydration of methanol was carried out in a stainless steel fixed-bed microreactor (inner diameter = 0.8 cm and length = 20 cm). Prior to experiments, the catalyst was pretreated for 2h at 350 °C under a N_2 flow. The liquid methanol was fed into the reactor using a dosing pump. The reaction was carried out with N_2 as a carrier gas at atmospheric pressure, in the temperature range of 200–350 °C and in the weight hourly space velocity range of 20-50 h^{-1}. Reactor consists of two heating zones. The first is to raise the feed temperature to the desirable level and the second to maintain the reactor surrounding at the proper temperature in order to minimize heat losses and simulate an adiabatic reactor. The effluent of the reactor was analyzed on a gas chromatograph (GC) equipped with a flame ionization detector connected with a capillary column.

3. Results and discussion

3.1. Morphological study

Fig. 2a,b shows SEM images of hollow γ-alumina nanosphere prepared in the alkaline condition (Al-B) and then calcined at 650 °C. The shell thickness image of the sample after nanosphere breaking technics (Fig. 2b) confirms successful formation of hollow γ-alumina nanosphere with ca. 400-500 nm in

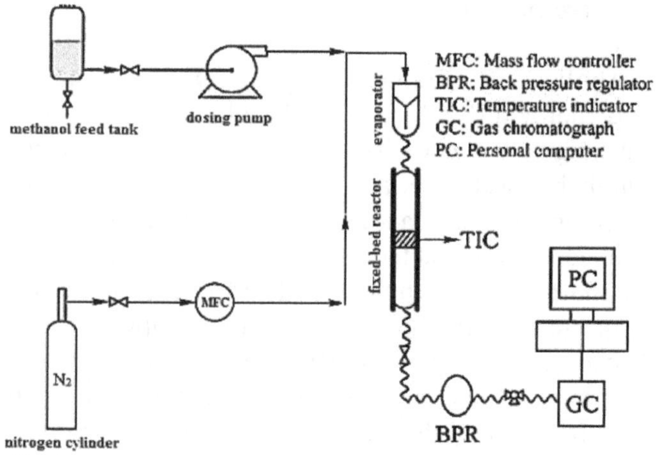

Fig. 1. Schematic view of experimental setup.

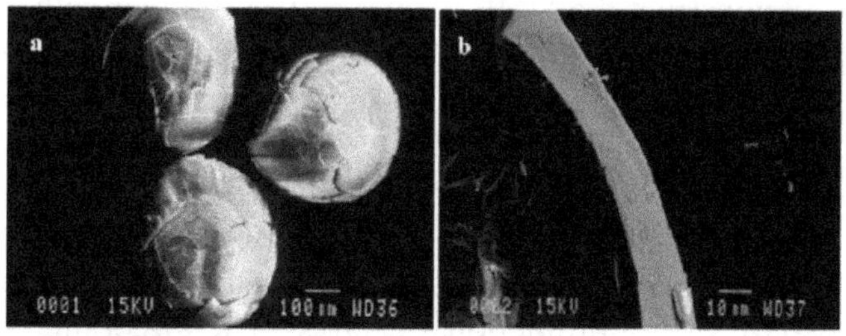

Fig.2. SEM images of hollow γ-alumina nanosphere prepared in alkaline medium, a) after calcination at 650 °C for 3 hours; b) shell thickness image after breaking of nanosphere

diameter, and 20 nm in thickness.

Fig. 3c,d represents SEM images of the samples prepared in the aforementioned techniques but in acidic condition (Al-A). The powder obtained after breaking of nanosphere (Fig.3d) indicates formation of γ-alumina nanosphere with ca. 450 nm in diameter. Based on this observation, yielding a large volume nanosphere and powder after breakage suggest that the synthesized sample in acidic condition may not favored formation of hollow γ-alumina nanosphere.

3.2. BET study

From data obtained using BET technique as seen in Table 1, get about surface area, average pore size and total pore volume of the samples. The specific surface area for the samples (Al-B) and (Al-A) was found to be 226.02m^2/g and 196.15m^2/g respectively.

The N_2 adsorption and desorption isotherm is of type IV, as is seen from Fig. 4a,b. According to IUPAC classification [27], the Al-B sample has a hysterics loop of type 1 (H1) that indicates a narrow pore size distribution with uniform size and shape (Fig.

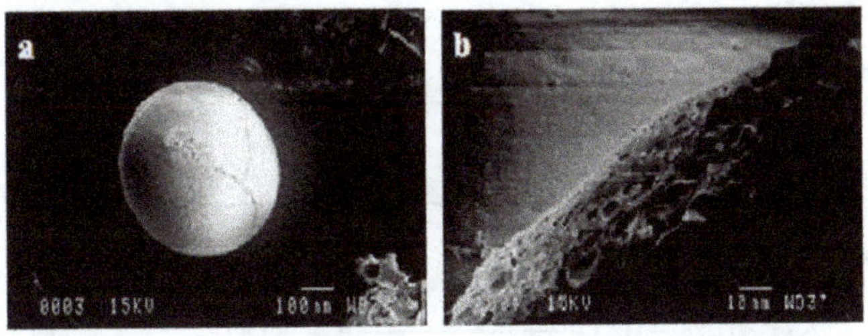

Fig. 3. SEM images of γ-alumina nanosphere prepared in acidic medium , a) after calcination at 650 °C for 3 hours; b) porous powder image after breaking of nanosphere

5a). The Al-A sample has a hysterics loop of type 2 (H2) that suggests a broad pore size distribution (Fig. 5b).

3.3. X-ray structural study

Fig.6. represents the diffraction pattern of as prepared alumina nanosphers obtained after calcination at 650 °C compared with those of standard γ-alumina. The three peaks at 2Θ=37.8, 2Θ=45.7 and 2Θ=66.9 are assigned to (311), (400) and (440) reflection plans of γ-Al_2O_3. JCPDS file number 29-0063 and JCPDS file no.10-0425 can be assigned to cubic γ-Al_2O_3. Cubic γ–alumina has a defect cubic spinel structure in which Al (III) ions occupy both octahedral and tetrahedral positions where the relative partial occupancy in each position is still a matter of dispute [28, 29]. According to the Scherrer equation, the crystalline size of the prepared nanosphere is ca. 14.18 nm.

3.4. Temperature-programmed experiments

NH_3-TPD measurements are performed to determine the acid strength and the amounts of acid sites on the catalyst surface. Desorption peaks in the range of 180–250 °C, 260–330 °C, and 340–500 °C in the NH_3-TPD profiles are commonly attributed to NH_3 that has been chemisorbed on weak, moderate and strong acid sites, respectively [30]. The NH_3-TPD spectra Fig. 7 of the synthesized γ-alumina nanosphere catalysts contain an intense peak in 175°C (I)

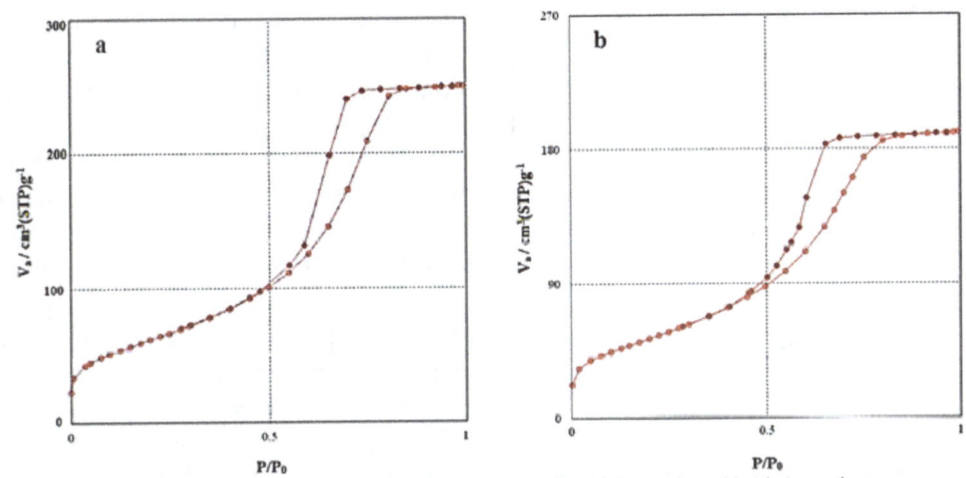

Fig. 4. N_2 adsorption and desorption isotherms, a) Al-B catalyst, b) Al-A catalyst

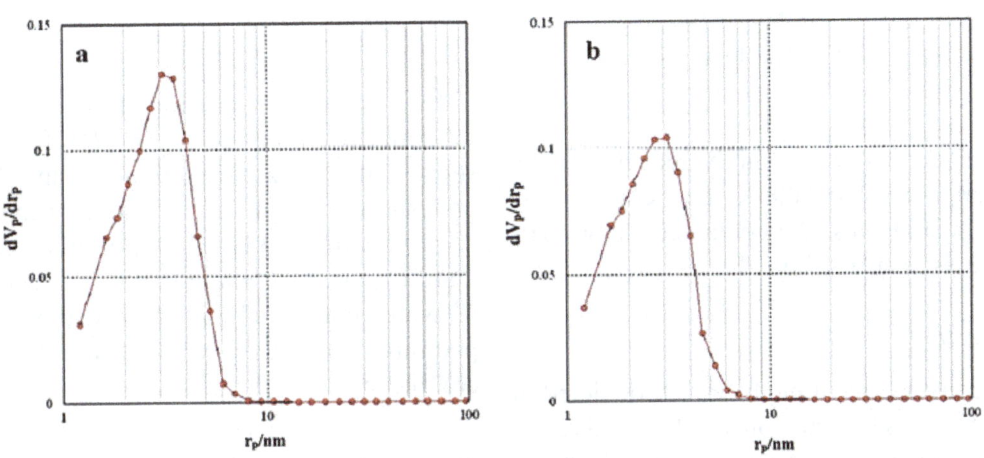

Fig. 5. BJH pore size distribution, a) Al-B catalyst, b) Al-A catalyst

Table.1. BET and NH3-TPD data for catalysts samples

Samples prepared	Specific surface area (m^2/g)	Pore size (nm)	Total Pore volume (cm^3/g)	Totalacidicsite(μmolNH3/g)
(Al-B)	226.25	6.8332	0.3865	741
(Al-A)	196.15	6.184	0.2951	586

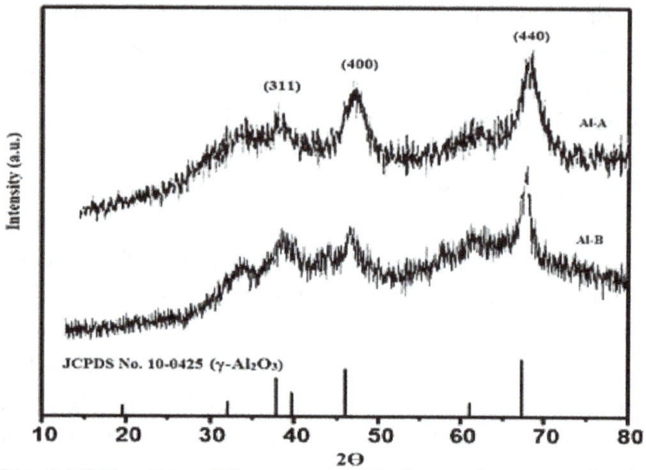

Fig. 6. XRD pattern of the as prepared hollow and porous γ-alumina nanosphere compared with standard γ-alumina

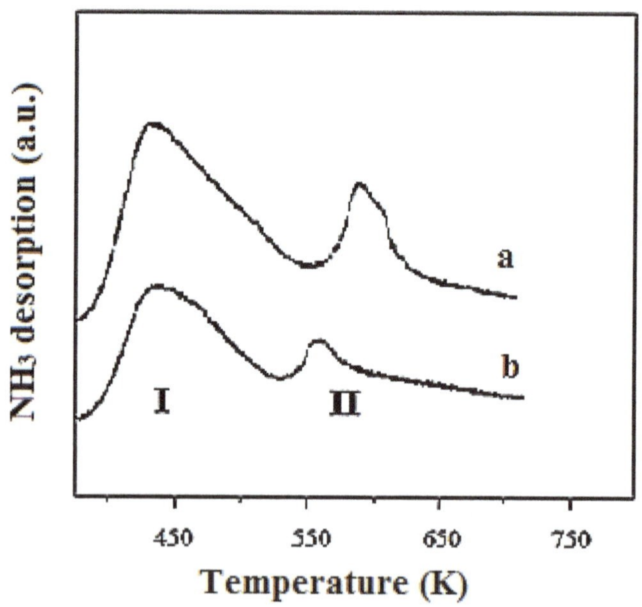

Fig. 7. NH_3-TPD profile a) Al-B catalyst, b) Al-A catalyst

and 325 (II) .This peak was slightly reduced in the temperature range of 200–500 °C that corresponds weak to medium acid sites. Data obtained using NH_3-TPD technique as seen in Table 1.

3.5. Catalytic activity

Experimental conversions of methanol are depicted in table 2, 3 and figure 8,9. Equation 1 bellow was used to measure the conversion of feed methanol (X, %).

$$X_{MeOH}\% = \left[1- \frac{\qquad}{\qquad} \right] \times 100 \qquad (1)$$

Based on the data shown in the above tables, it was found that the A1-B catalyst has a relatively better performance than Al-A catalyst. This means that the greatest A1-B catalyst conversion has 89% occurred on temperature 275 °C, while the maximum Al-A catalyst conversion has 81% occurred on temperature 325 °C.

The performance variation of the two catalysts based on two important parameters of temperature and WHSV are to be discussed later in the paper.

3.5.1. WHSV:

WHSV is an important factor in controlling and optimising the catalyst performance. This is because from one hand the increase in WHSV increases the input feed and decreases feed contact time with the catalyst [31].

Under the circumstances where WHSV decreases the conversion time increases. The most important limitation factor is the catalyst amount. The reaction prior to reaching the equilibrium and at low temperatures is to a great extent depends on WHSV. This dependency tends to decrease on the equilibrium point due to the temperature increment that in turn, compensates the WHSV [32].

The experimental data of this study are clear demonstration of this. The optimum WHSV for both catalysts as the experimental data suggest is 20 h^{-1}.

3.5.2. Temperature

The dehydration reaction from methanol is an exothermic reaction (Equation (2)). Research findings show that temperatures higher than 525 °C have an undesirable effect on reaction, because ΔG^0 takes positive values [33]. The optimum reaction temperature for the Al-B catalyst was observed to be at 275 °C which compared with Al-A catalyst having a temperature of 325 °C has a more appropriate operational condition.

$$2CH_3OH \leftrightarrows CH_3OCH_3 + H_2O \qquad (\Delta H = -24\ kJ/mol) \qquad (2)$$

Table.2. Experimental data for dehydration of methanol over Al-B catalyst

Temperature (°C)	WHSV = 20 h^{-1}	WHSV = 35 h^{-1}	WHSV = 50 h^{-1}
	Conversion %	Conversion %	Conversion %
200	0.36	0.23	0
225	0.56	0.41	0.23
250	0.80	0.68	0.62
275	0.89	0.83	0.73
300	0.86	0.82	0.78
325	0.85	0.82	0.72
350	0.78	0.73	0.67

Table.3. Experimental data for dehydration of methanol over Al-A catalyst

Temperature (°C)	WHSV = 20 h^{-1}	WHSV = 35 h^{-1}	WHSV = 50 h^{-1}
	Conversion %	Conversion %	Conversion %
200	0	0	0
225	0.11	0	0
250	0.34	0.18	0.09
275	0.49	0.41	0.29
300	0.66	0.63	0.48
325	0.81	0.82	0.63
350	0.77	0.76	0.62

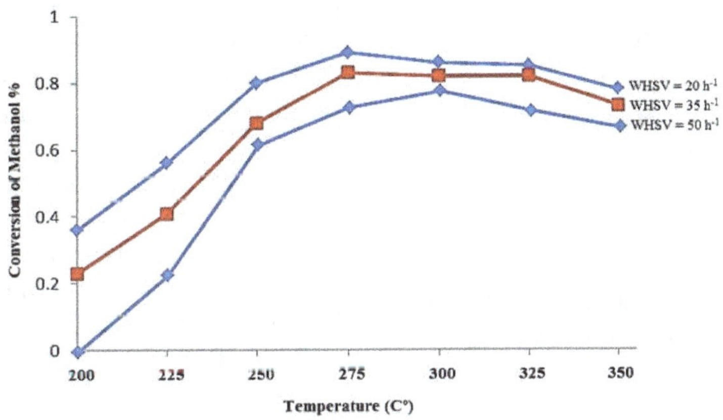

Fig. 8. Conversion of methanol as a function of reaction temperature profile obtained over Al-B catalyst

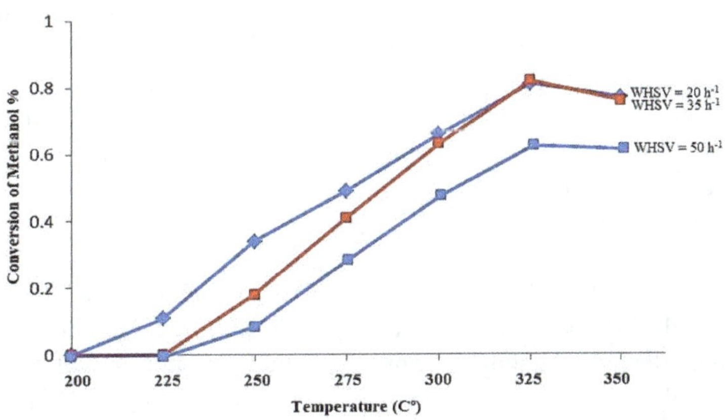

Fig. 9. Conversion of methanol as a function of reaction temperature profile obtained over Al-A catalyst

4. Conclusions

Two synthetic samples, porous and hollow γ-Al_2O_3 nanospheres were applied to successfully synthesize DME from methanol. Quite desirable results were obtained in the process. This was particularly the case for the Al-B which having a higher rate of conversion at the relatively lower temperature compared with the commercial dimethyl ether synthesis samples provided by Engelhand، Sasol and Akzo Nobel as shown in Fig. 10 [34].

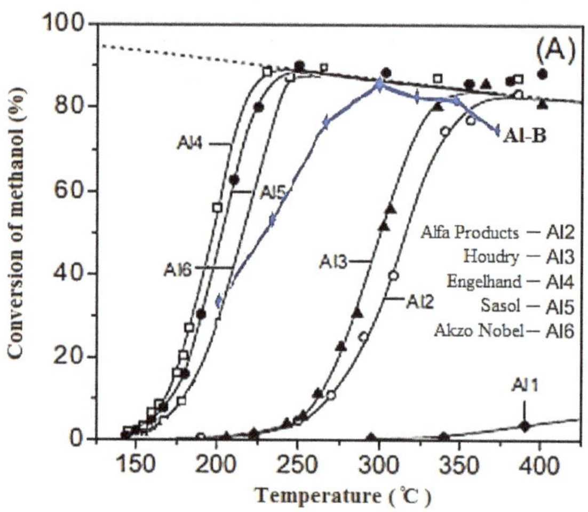

Fig. 10. Optimum conversion of methanol to DME over Al-B catalyst compared with commercial catalysts

Acknowledgments

The authors thanks the Research vice Presidency of Science and Research branch and Dezful branch, Islamic Azad University for their financial and en couragement support.

References

[1] G. Cai, Z. Liu, R. Shi, C. He, L. Yang, C. Sun, Y. Chang, Light alkenes from syngas via dimethyl ether, Appl. Catal. A. 125 (1995) 29-38.

[2] M. Xu, J. M. Lunsford, D.W. Goodman, A. Bhattacharya, Synthesis of dimethyl ether (DME) from methanol over solid-acid catalysts, Appl. Catal. A. 149 (1997) 289-301.

[3] T. Takeguchi, K. Yanagisawa, T. Inui, M. Inoue, Effect of the property of solid acid upon syngas-to-dimethyl ether conversion on the hybrid catalysts com posed of Cu–Zn–Ga and solid acids Appl. Catal. A: Gen. 192 (2000) 201-209.

[4] A.M. Arkharov, S.D. Glukhov, L. V. Grekhov, A. A. Zherdev, N. A. Ivashchenko, D. N Kalinin, Use of Dimethyl Ether as a Motor Fuel and a Refrigerant, Chem. Petrol. Eng. 39 (2003) 330-336.

[5]. Q. Ge, Y. Huang, F. Qiu, S. Li, Bifunctional catalysts for conversion of synthesis gas to dimethyl ether, Appl. Catal. A. 167 (1998) 23-30.

[6] F. Yaripour, F. Baghaei, I. Schmidt, J. Perregaard. Synthesis of dimethyl ether from methanol over aluminium phosphate and silica–titania catalysts, Catal. Commun. 6 (2005) 542-549.

[7] J. Fei, Z. Hou, B. Zhu, H. Lou, X. Zheng, Synthesis of dimethyl ether (DME) on modified HY zeolite and modified HY zeolite-supported Cu–Mn–Zn catalysts, Appl. Catal. A. 304 (2006) 49-54.

[8] J. Xia, D. Mao, B. Zhang, Q. Chen, Y. Zhang, Y. Tang, Catalytic properties of fluorinated alumina for the production of dimethyl ether, Catal. Commun. 7 (2006) 362-366.

[9] J. Khom-in, P. Praserthdam, J. Panpranot, O. Mekasuwandumrong, Dehydration of methanol to dimethyl ether over nanocrystalline Al_2O_3 with mixed γ- and χ-crystalline phases, Catal. Commun.9 (2008) 1955-1958.

[10] R. Vakili, E. Pourazadi, P. Setoodeh, R. Eslamloueyan, M. R. Rahimpour, Direct dimethyl ether (DME) synthesis through a thermally coupled heat exchanger reactor, Appl. Energ. 88 (2011) 1211–1223.

[11] A. García-Trenco, A. Martínez, Direct synthesis of DME from syngas on hybrid CuZnAl/ZSM-5 catalysts: New insights into the role of zeolite acidity, Appl. Catal. A. 411-412 (2012) 170– 179.

[12] M. H. Zhang, Z. M. Liu, G. D. Lin, H. B. Zhang, Pd/CNT-promoted Cu ZrO_2/HZSM-5 hybrid catalysts for direct synthesis of DME from CO_2/H_2 , Appl. Catal. A. 451 (2013) 28– 35.

[13] Y. Zhang, D. Li, Y. Zhang, Y. Cao, S. Zhang, K. Wang, F. Ding, V-modified CuO–ZnO–ZrO_2/HZSM-5 catalyst for efficient direct synthesis of DME from CO_2 hydrogenation, Catal. Commun. 55 (2014) 49–52.

[14] F. Frusteri, G. Bonuraa, C. Cannilla, G. D. Ferrantea, A. Aloise, E. Catizzone, M. Migliori, G. Giordano, Stepwise tuning of metal-oxide and acid sites of CuZnZr-MFI hybrid catalysts for the direct DME synthesis by CO_2 hydrogenation, Appl. Catal. B 176–177 (2015) 522–531.

[15] Y. J. Lee, J. M. Kim, J. W. Bae, C. H. Shin, K. W. Jun, Phosphorus induced hydrothermal stability and enhanced catalytic activity of ZSM-5 in methanol to DME conversion, Fuel 88 (2009) 1915–1921.

[16]L. Liu, W. Huang, Z. h. Gao, L. h. Yin, Synthesis of AlOOH slurry catalyst and catalytic activity for methanol dehydration to dimethyl ether, J. Ind. Eng. Chem. 18 (2012) 123–127.

[17] Y. Sang, H. Liu, S. He, H. Li, Q. Jiao, Q. Wu, K. Sun, Catalytic performance of hierarchical H-ZSM-5/MCM-41 for methanol dehydration to dimethyl ether, J. Energ. Chem. 22(2013)769–777.

[18] Z. Zuo, L. Wang, P. Han, W. Huang, Effect of surface hydroxyls on dimethyl ether synthesis over the γ-Al_2O_3 in liquid paraffin: a computational study, J. Mol. Model 19 (2013) 4959–4967.

[19] S. M. Solyman, M. A. Betiha, The performance of chemically and physically modified local kaolinite in methanol dehydration to dimethyl ether, Egyp. J. Petroleum. 23 (2014) 247-254.

[20] F. Yaripour, Z. Shariatinia, S. Sahebdelfar, A. Irandoukht, The effects of synthesis operation conditions on the properties of modified c-alumina nanocatalysts in methanol dehydration to dimethyl ether using factorial experimental design, Fuel. 139 (2015) 40–50.

[21] F. Caruso, X. Shi, R. A. Caruso, A. Susha, Hollow titania spheres from layered precursor deposition on sacrificial colloidal core particles, Adv. Mater. 13(10) (2001) 740-744.

[22] Y. Sun, B. Mayer, Y. Xia, Metal nanostructures with hollow interiors, Adv. Mater. 15 (2003) 641-646.

[23] U. Jeong, Y. Wang, M. Ibisate, Y. Xia, Some new developments in the synthesis, functionalization, and utilization of monodisperse colloidal spheres, Adv. Funct. Mater. 15 (2005) 1907-1921.

[24] Y. Wang, W. J. Tseng, A novel technique for synthesizing nanoshell hollow alumina particles, J. Am. Ceram. Soc. 92 (2009) S32–S37.

[25] R. G. Chaudhuri, S. Paria, Core/Shell nanoparticles: Classes, properties, synthesis mechanisms, characterization and Applications, Chem. Rev. 112 (2012) 2373–2433.

[26] M. S. Sadjadi, N. Rostamizadeh, A new strategy in the synthesis of hollow γ-Al_2O_3 nanosphere using alginate gel casting process, Res J Biotechnol. 11(4) (2016) 30-35.

[27] M. Firoozi, M. Baghalha, M. Asadi, The effect of micro and nano particle sizes of H ZSM-5 on the selectivity of MTP reaction, Catal. Commun. 10 (2009) 1582–1585.

[28] Z. Zeng, J. Yu, Z. Guo, Preparation of functionalized core-shell alumina/polystyrene composite nanoparticles, Macromol. Chem. Phys. 206 (2005) 1558–1567.

[29] T. Shirai, H. Watanabe, M. Fuji, M. Takahashi, Structural properties and surface characteristics on aluminum oxide powders, Annual Report of the Ceramics Research Laboratory Nagoya Institute of Technology. 9 (2009) 23-31.

[30] F. Arena, R. Dario, A. Parmaliana, A characterization study of the surface acidity of solid catalysts by temperature programmed methods, Appl. Catal. A. 170 (1998) 127-137.

[31] G. R. Moradi, R. Ghanei, F.Yaripour, Determination of the optimum operating conditions for direct synthesis of dimethyl ether from Syngas, Int. J. Chem. Reactor Eng. 5 (2007) A14-19.

[32] M. Fazlollahnejad, M. Taghizadeh, A. Eliassi, G. Bakeri, Experimental study and modeling of an adiabatic fixed-bed reactor for methanol dehydration to dimethyl ether, Chin. J. Chem. Eng. 17 (2009) 630-634.

[33] B.T. Diep, M.S. Wainwright, Thermodynamic equilibrium constants for the methanol-dimethyl ether-water system, J. Chem. Eng. Data. 32 (1987)330–333.

[34] S. S. Akarmazyan, P. Panagiotopoulou, A. Kambolis, C. Papadopoulou, D. I. Kondarides, Methanol dehydration to dimethylether over Al_2O_3 catalysts, Appl. Catal. B. 145 (2014) 136– 148.

Permissions

All chapters in this book were first published in JPST, by Iranian Research Organization for Science and Technology; hereby published with permission under the Creative Commons Attribution License or equivalent. Every chapter published in this book has been scrutinized by our experts. Their significance has been extensively debated. The topics covered herein carry significant findings which will fuel the growth of the discipline. They may even be implemented as practical applications or may be referred to as a beginning point for another development.

The contributors of this book come from diverse backgrounds, making this book a truly international effort. This book will bring forth new frontiers with its revolutionizing research information and detailed analysis of the nascent developments around the world.

We would like to thank all the contributing authors for lending their expertise to make the book truly unique. They have played a crucial role in the development of this book. Without their invaluable contributions this book wouldn't have been possible. They have made vital efforts to compile up to date information on the varied aspects of this subject to make this book a valuable addition to the collection of many professionals and students.

This book was conceptualized with the vision of imparting up-to-date information and advanced data in this field. To ensure the same, a matchless editorial board was set up. Every individual on the board went through rigorous rounds of assessment to prove their worth. After which they invested a large part of their time researching and compiling the most relevant data for our readers.

The editorial board has been involved in producing this book since its inception. They have spent rigorous hours researching and exploring the diverse topics which have resulted in the successful publishing of this book. They have passed on their knowledge of decades through this book. To expedite this challenging task, the publisher supported the team at every step. A small team of assistant editors was also appointed to further simplify the editing procedure and attain best results for the readers.

Apart from the editorial board, the designing team has also invested a significant amount of their time in understanding the subject and creating the most relevant covers. They scrutinized every image to scout for the most suitable representation of the subject and create an appropriate cover for the book.

The publishing team has been an ardent support to the editorial, designing and production team. Their endless efforts to recruit the best for this project, has resulted in the accomplishment of this book. They are a veteran in the field of academics and their pool of knowledge is as vast as their experience in printing. Their expertise and guidance has proved useful at every step. Their uncompromising quality standards have made this book an exceptional effort. Their encouragement from time to time has been an inspiration for everyone.

The publisher and the editorial board hope that this book will prove to be a valuable piece of knowledge for researchers, students, practitioners and scholars across the globe.

List of Contributors

Yasser Vasseghian and Mojtaba Ahmadi
Chemical Engineering Department, Faculty of Engineering, Razi University, Kermanshah, Iran

Mohammad Joshaghani
Faculty of Chemistry, Razi University, Kermanshah 67149, Iran

Maryam Omidvar
Department of Chemical Engineering, Quchan Branch, Islamic Azad University, Quchan, Iran

Esmaeil Koohestanian
Department of Chemical Engineering, University of Sistan and Baluchestan, Zahedan, Iran

Omid Ramezani Azghandi
Department of Mechanical Engineering, Ferdowsi University of Mashhad, Mashhad, Iran

Mostafa Milani and Seyed Mohammad Zahraee
Advanced Materials and Renewable Energies Department, Iranian Research Organization for Science and Technology, Tehran, Iran

Seyed Mohammad Mirkazemi
School of Metallurgy and Materials Engineering, Iran University of Science and Technology, Tehran, Iran

Maryam Ranjbar and S. Heydar Mahmoudi Najafi
Department of Chemical Technologies, Iranian Research Organization for Science and Technology (IROST), Tehran, Iran

Mostafa Yousefi
Department of Chemical Technologies, Iranian Research Organization for Science and Technology (IROST), Tehran, Iran
National Iranian Oil Products Distribution Company (NIOPDC), Zahedan Region, Zahedan, Iran

Mahboobe Lahooti and Azim Malekzadeh
Department of Chemistry, Faculty of Sciences, Damghan University, Damghan, Iran

Saeed Vakilpour and Ahmad Ghaderi Hamidi
Department of Metallurgy and Materials Engineering, Hamedan University of Technology, Hamedan, Iran

Amin Eftekhari and Mohammad Bagher Rahmani
Department of Physics, Shahrood University of Technology, Shahrood, Iran

Fatemeh Masdarolomoor
Department of Chemistry, Shahrood University of Technology, Shahrood, Iran

Fahimeh Ebrahim Asghari, Hossein Ghadamian and Mohammad Aminy
Department of Energy, Materials and Energy Research Center (MERC), Tehran, Iran

Saharnaz Tajik and Omid Moini Jazani
Department of Chemical Engineering, Faculty of Engineering, University of Isfahan, Isfahan, Iran

Soheila Shokrollahzadeh and Seyed Mahdi Latifi
Department of Chemical Technologies, Iranian Research Organization for Science and Technology (IROST), Tehran, Iran

Hamed Baniamerian and Soheila Shokrollahzadeh
Department of Chemical Technologies, Iranian Research Organization for Science and Technology (IROST), Tehran, Iran

Seyed Mohammad Zahraee
Department of Advanced Materials and Renewable Energy, Iranian Research Organization for Science and Technology (IROST), Tehran, Iran

Mohammad Eftekhari Yazdi, Alireza Kalani Nejad and Saeed Dinarvand
Mechanical Engineering Department, Islamic Azad University, Central Tehran Branch, Tehran, Iran

Hossein Tamim
Mechanical Engineering Department, Amirkabir University of Technology, Tehran, Iran

Maryam Arabpour, Ahmad Rahbar-Kelishami and Reza Norouzbeigi
Department of Chemical, Petroleum and Gas Engineering, Iran University of Science and Technology, Tehran, Iran

R.Veys Karami and H. Izadi Vasafi
Department of Polymer Engineering, Shahreza Branch, Islamic Azad University, Isfahan, Iran

O. Moini Jazani
Department of Chemical Engineering, Faculty of Engineering, University of Isfahan, Isfahan, Iran

A. Talaei
Polymer Engineering Department, Amir Kabir University of Technology, Tehran, Iran

Leila Khoshrooyan, Ali Eliassi and Maryam Ranjbar
Chemical Technology Department, Iranian Research Organization for Science and Technology (IROST), Tehran, Iran

Arash Amani and Ali Reza Solaimany Nazar
Department of Chemical Engineering, University of Isfahan, Isfahan, Iran

Hasan Sabzyan and Gholamhassan Azimi
Department of Chemistry, University of Isfahan, Isfahan, Iran

Oner Yusuf Toraman
Mining Engineering Department, Faculty of Engineering, Omer Halisdemir University, 51240 Nigde, Turkey

N. Rostamizadeh and M. S. Sadjadi
Department of Chemistry, Science and Research Branch, Islamic Azad University, Tehran, I. R. Iran

S. A. A. Sadjadi
Institute of Water and Energy, Sharif University of Technology, Tehran, I. R. Iran

Index

www.ingramcontent.com/pod-product-compliance
Lightning Source LLC
LaVergne TN
LVHW060610210926
848485LV00005B/1396